# Cybersecurity Architect's Handbook

An end-to-end guide to implementing and maintaining robust security architecture

**Lester Nichols**

# Cybersecurity Architect's Handbook

**Group Product Manager**: Pavan Ramchandani
**Publishing Product Manager**: Prachi Sawant
**Book Project Manager**: Ashwini Gowda
**Senior Editor**: Runcil Rebello
**Technical Editor**: Yash Bhanushali
**Copy Editor**: Safis Editing
**Proofreader**: Safis Editing
**Indexer**: Rekha Nair
**Production Designer**: Shankar Kalbhor
**Senior DevRel Marketing Coordinator**: Marylou De Mello

First published: March 2024
Production reference: 1230224

Published by
Packt Publishing Ltd.
Grosvenor House
11 St Paul's Square
Birmingham
B3 1RB, UK

ISBN 978-1-80323-584-4

www.packtpub.com

*To my wife, Cherie, my love and best friend, who has been my most profound inspiration.*

*– Lester*

# Contributors

## About the author

**Lester Nichols** brings over 25 years of cybersecurity and technology leadership experience to his role as author. Currently serving as director of security architecture and VP of cybersecurity operations at JPMorgan Chase & Co., he has previously held senior infrastructure and security positions in the government, financial services, healthcare, and IT consulting sectors.

Lester takes an architect's approach to security, implementing holistic defenses tailored to each organization's unique risks and objectives. He is known for his expertise in securing highly complex environments encompassing technologies from mainframes to the cloud.

A recognized thought leader, Lester has contributed to leading publications, including the *Computer Security Handbook*. He holds a master's in information assurance and an array of cybersecurity certifications.

With decades of experience spanning strategic planning, risk management, and hands-on technology, Lester provides a seasoned perspective into the real-world practices of cybersecurity architecture. Through this book, he aims to equip fellow professionals with versatile skills to meet escalating threats with strategic, adaptable security fulfilling unique organizational needs.

*First and foremost, I am grateful to Jesus Christ for the blessings in my life. To my wife and family, your unwavering love, support, and patience as I spent countless hours immersed in writing has been a gift. I dedicate this book to you.*

*I also want to acknowledge the many mentors who have shaped my career journey and instilled knowledge in me. And to the students and colleagues I've had the honor of mentoring, thank you for inspiring me to pay forward what others have so graciously shared. My wish is that this book imparts something of value to you all.*

*To the cybersecurity community, it has been a privilege to grow alongside so many talented professionals working to advance our collective mission. My hope is that this book plays some small role in equipping us to meet the challenges ahead with wisdom, compassion, and purpose.*

*Finally, I'm thankful to everyone at Packt for their dedication and for making this project possible.*

*It is my sincere hope that the book proves useful to you, the reader, as we progress on this ever-evolving cybersecurity journey together.*

# About the reviewers

**Justin Bibee** is a Marine veteran and technology enthusiast with over 20 years of experience in network operations, cybersecurity, and system engineering. He's spent years conducting and leading network and cyber operations within the intelligence community and now serves as a cybersecurity technical PM with Zermount, Inc., leading a spectrum of cybersecurity testing and engineering services for a myriad of government clients. He received a BSc in cybersecurity from UMGC and an MS in cybersecurity from Liberty University, and has taught courses on numerous topics, such as Microsoft security, network security, cyber defense, threat modeling, and incident response, for multiple universities.

*I'd like to thank my family for their patience, understanding, and support throughout the period of time-consuming effort of research and testing that went into the technical review of this publication. I would also like to offer a special thanks to the author for entrusting me with this opportunity—an enriching endeavor I am privileged and honored to be part of!*

**Devender Kumar**, an information security architect, brings 16 years of expertise with certifications such as CISSP, CISM, and ISO Senior Lead Auditor. Having held pivotal architect and manager roles at TCS, HCL Tech, and Canon, he specializes in designing robust security frameworks and implementing advanced technologies. He has led the successful execution of an advanced SOC and ensured compliance with ISO 27001, NIST, CIS, GDPR, and ASD Essential Eight standards. He has worked in multiple geographies and industries, and his diverse skill set spans threat intelligence, risk management, governance, compliance, and proficiency in security technologies.

*I am grateful for the unwavering support of my family and friends who understand the demands of researching ever-changing security landscapes. Thanks to the collaborative security community and trailblazers; your dedication makes this field dynamic and exciting. I appreciate everyone who has contributed to my fulfilling journey in security architecture.*

# Table of Contents

# Part 2: Pathways

4

## Cybersecurity Architecture Principles, Design, and Analysis    103

5

## Threat, Risk, and Governance Considerations as an Architect    133

# 6

# Documentation as a Cybersecurity Architect – Valuable Resources and Guidance for a Cybersecurity Architect Role          171

# 7

## Entry-Level-to-Architect Roadmap                                211

# 8

## The Certification Dilemma                                         245

# Part 3: Advancements

## 9

## 10

## 11

# 12

# Being Adaptable as a Cybersecurity Architect          349

## 13

# Architecture Considerations – Design, Development, and Other Security Strategies – Part 1                                                     375

## 14

# Architecture Considerations – Design, Development, and Other Security Strategies – Part 2                                                     415

# Preface

Cyber threats pose ever-growing risks, yet security measures often lag behind. As organizations increasingly rely on interconnected technologies, the need for robust yet flexible cybersecurity architecture becomes imperative. This book equips you to meet that need. It provides IT and security professionals with a comprehensive guide to becoming proficient cybersecurity architects capable of designing and evolving strategic defenses tailored to unique environments.

Spanning foundations, career pathways, and advancements, the book explores core tenets of security alongside real-world implementation. Early chapters establish critical baseline knowledge regarding key concepts such as confidentiality, networking, risk management, and compliance. The discussion then progresses to navigating career growth as an architect, highlighting crucial skills such as documentation, vendor management, and team collaboration. Advanced sections detail processes for selecting and implementing controls, aligning security with business objectives, and cultivating personal adaptability amid constant change.

Throughout, the emphasis remains practical and actionable. Theories come alive through concrete examples drawn from diverse organizational settings. Labs, diagrams, and exercises immerse you in applying concepts firsthand. Those new to cybersecurity gain indispensable orientation while current professionals discover fresh perspectives.

## Who this book is for

The book is suited to IT administrators, security analysts, developers, and leaders seeking to pivot into architect roles. However, any technology professional wanting to design comprehensive protections will find value. By equipping architects to implement strategic solutions tailored to unique risk landscapes, it enables both novice and seasoned readers to advance architectures to secure our increasingly digital future.

The three main personas who are the target audience of this content are as follows:

- Those new to cybersecurity or **Information Technology** (**IT**) looking to map a career or enhance their current path toward cybersecurity. For those at the onset of their technology or cybersecurity journey, this book provides critical orientation. Whether transitioning from a non-technical background or just embarking on the career path, the content maps a route to becoming a proficient cybersecurity architect.

- Existing IT professionals, at any level, looking to transition toward cybersecurity and, more specifically, toward cybersecurity architecture. For experienced technology professionals such as systems administrators, network engineers, or software developers seeking to transition into cybersecurity, this book bridges connections between familiar concepts and security-focused architecture.

- Existing cybersecurity professionals or entry-level cybersecurity architects looking to enhance and grow within the field and career. For cybersecurity professionals at the outset of their careers, such as analysts or associate-level architects, this book provides pathways to unlock greater responsibilities and leadership.

## What this book covers

*Chapter 1, Introduction to Cybersecurity*, provides foundational concepts and basics to understanding the concepts of cybersecurity and, ultimately, how that plays into the role of the cybersecurity architect. This will provide a foundational level setting for those new to cybersecurity while also providing a fundamental refresher to those who have been working within cybersecurity or IT for some time.

*Chapter 2, Cybersecurity Foundation*, continues on from the introduction to get a bit more granular from a foundational level to discuss some of the main areas that a cybersecurity architect will need to address and understand as it relates to the business and other operational teams. This will be cursory in nature but provides the foundational aspects to progress into the discussion of the cybersecurity career path and the options available to the potential cybersecurity architect in specializing/focusing in a particular area.

*Chapter 3, What Is a Cybersecurity Architect and What Are Their Responsibilities?*, begins with the principle that you have enough understanding of cybersecurity to discuss the role of the cybersecurity architect and how it builds upon other technology roles. Whether that is in the area of enterprise, application, network, or platform architecture, these areas have differing focuses that span everything to a specific subset. This is also in context with the organization and technology. Once the framework of the architect is defined, the responsibilities become more evident, as it relates to the specific area of focus or organization.

*Chapter 4, Cybersecurity Architecture Principles, Design, and Analysis*, provides foundational concepts for cybersecurity architecture, including principles, design, and analysis. It emphasizes using clear terminology and outlining organizational goals and risk tolerance as critical inputs that shape architecture.

*Chapter 5, Threat, Risk, and Governance Considerations as an Architect*, discusses the areas of architecture principles, design, and analysis that will be part of the day-to-day functions of the cybersecurity architect. This will discuss the various approaches to performing the design and analysis of a particular solution or control with an understanding of the principles around the choice one would take over another depending on the situation.

*Chapter 6, Documentation as a Cybersecurity Architect – Valuable Resources and Guidance for a Cybersecurity Architect Role*, takes somewhat of a break from the more detailed concepts to discuss the importance of proper documentation as it relates to the cybersecurity architect role. This will discuss the need for granularity and a detailed approach to documentation through tools such as Microsoft Visio or DrawIO and other similar tools. There will also be a discussion of how to document and/or create scratchpads for notes through tools such as CherryTree. All of this is meant to help propel the visibility of solutioning and architecture design not only within the organization but also for regulatory and compliance requirements.

*Chapter 7, Entry-Level-to-Architect Roadmap*, discusses the journey to get to the top as a cybersecurity architect. It goes without saying that certain career paths are more direct than others for the cybersecurity architect. Like most things in technology, *"it depends"* can be a common answer. This chapter provides various approaches to gaining the experience or skill set to become a cybersecurity architect. Whether that is starting as an IT technician or transitioning from a developer, there are commonalities or skills that need to be gained or used to help shape the path for this career path.

*Chapter 8, The Certification Dilemma*, discusses a number of certifications for security architecture, as well as others to help differentiate yourself from others who are competing for the same position. It also discusses the good, bad, and ugly of the certification process and how to make the choices that will best match your overall career plan and direction.

*Chapter 9, Decluttering the Toolset – Part 1*, explores strategies for cybersecurity architects to thoughtfully assemble their security toolkit by evaluating solutions to find the optimal fit for their organization's specific threat landscape, business needs, and operational constraints. It provides an overview of major security tool categories such as threat modeling, network monitoring, endpoint protection, identity access management, data encryption, vulnerability management, and more. The chapter emphasizes matching defenses to an organization's unique vulnerabilities and risks rather than a one-size-fits-all approach.

*Chapter 10, Decluttering the Toolset – Part 2*, emphasizes the importance of thoughtfully selecting cybersecurity tools tailored to an organization's unique vulnerabilities, infrastructure, and strategic objectives. It advises taking a methodical approach to identifying specific security gaps and requirements first before assessing tools. Tight alignment with frameworks such as NIST CSF, implementing layered defenses, weighing business factors such as cost and usability, and future-proofing selections are highlighted as critical to building an optimal toolkit.

*Chapter 11, Best Practices*, goes into detail about best practices, as it relates to cybersecurity and why it is best to implement solutions using best practices. This includes the use of standards or technology-specific best practices. The chapter will also discuss when one may supersede another and why you may be faced with that scenario.

*Chapter 12, Being Adaptable as a Cybersecurity Architect*, explores how architects can cultivate personal and professional adaptability to implement pragmatic solutions tailored to unique business environments and goals. It builds on previous core concepts to underscore why rigid adherence to "perfect" security often fails, while customizable approaches succeed. Topics span fostering mindsets and strategies to design protection around workflows, manage risks judiciously, and strike balances enabling productivity and innovation. Architects learn how becoming more holistic and adaptable accelerates professional growth while empowering fearless innovation through security tailored to ever-evolving needs.

*Chapter 13, Architecture Considerations – Design, Development, and Other Security Strategies – Part 1*, focuses on core disciplines enabling cybersecurity architects to securely translate organizational needs into tailored technical solutions. It emphasizes aligning security intrinsically with business goals early during conceptualization and design.

*Chapter 14, Architecture Considerations – Design, Development, and Other Security Strategies – Part 2*, serves as a summarizing synthesis tying together the various cybersecurity architecture concepts covered in the book. It emphasizes that architects must have technical expertise as well as versatility to adopt security frameworks amid constant change.

## To get the most out of this book

| Software/hardware covered in the book | Operating system requirements |
| --- | --- |
| Kali Linux | Windows, macOS, or Linux |
| Snort | Processor: Minimum 4 cores/Best results with 8+ cores |
| OPNsense | Memory: Minimum 16 GB/recommended 32+ GB |
| Ansible | Storage: Minimum 500 GB/recommended 1 TB |
| Graylog | Hypervisor: VMware Workstation/Fusion/Oracle VirtualBox/Qemu/Proxmox |
| Veracrypt | |
| OpenVAS/Greenbone | |
| AWS | |
| StackStorm | |
| SecurityOnion | |
| ClamAV | |
| OWASP ZAP and Threat Dragon | |
| Microsoft Threat Modeling Tool | |

# Download the example code files

You can download the example code files for this book from GitHub at `https://github.com/ PacktPublishing/Cybersecurity-Architects-Handbook`. If there's an update to the code, it will be updated in the GitHub repository.

We also have other code bundles from our rich catalog of books and videos available at `https:// github.com/PacktPublishing/`. Check them out!

# Conventions used

There are a number of text conventions used throughout this book.

`Code in text`: Indicates code words in text, database table names, folder names, filenames, file extensions, pathnames, dummy URLs, user input, and Twitter handles. Here is an example: "After running the setup script, run `sudo gvm-check-setup` for validation of the installation and default configuration."

A block of code is set as follows:

```
{
    "v": "1",
    "type": {
      "name": "pipeline_rule",
      "version": "1"
    },
```

When we wish to draw your attention to a particular part of a code block, the relevant lines or items are set in bold:

```
{
    "v": "1",
    "type": {
      "name": "pipeline_rule",
      "version": "1"
    },
```

Any command-line input or output is written as follows:

```
    sudo systemctl enable graylog-server.service
    sudo systemctl start graylog-server.service
 sudo systemctl --type=service --state=active | grep graylog
```

**Bold**: Indicates a new term, an important word, or words that you see onscreen. For instance, words in menus or dialog boxes appear in **bold**. Here is an example: "**Cryptography** is the science of, and some even say the art of, using deception and mathematics to hide data from unwanted access."

> **Tips or important notes**
> Appear like this.

# Get in touch

Feedback from our readers is always welcome.

**General feedback**: If you have questions about any aspect of this book, email us at `customercare@packtpub.com` and mention the book title in the subject of your message.

**Errata**: Although we have taken every care to ensure the accuracy of our content, mistakes do happen. If you have found a mistake in this book, we would be grateful if you would report this to us. Please visit `www.packtpub.com/support/errata` and fill in the form.

**Piracy**: If you come across any illegal copies of our works in any form on the internet, we would be grateful if you would provide us with the location address or website name. Please contact us at `copyright@packt.com` with a link to the material.

**If you are interested in becoming an author**: If there is a topic that you have expertise in and you are interested in either writing or contributing to a book, please visit `authors.packtpub.com`.

# Share Your Thoughts

Once you've read *Cybersecurity Architect's Handbook*, we'd love to hear your thoughts! Scan the QR code below to go straight to the Amazon review page for this book and share your feedback.

`https://packt.link/r/1803235845`

Your review is important to us and the tech community and will help us make sure we're delivering excellent quality content.

# Download a free PDF copy of this book

Thanks for purchasing this book!

Do you like to read on the go but are unable to carry your print books everywhere?  d

Is your eBook purchase not compatible with the device of your choice?

Don't worry, now with every Packt book you get a DRM-free PDF version of that book at no cost.

Read anywhere, any place, on any device. Search, copy, and paste code from your favorite technical books directly into your application.

The perks don't stop there, you can get exclusive access to discounts, newsletters, and great free content in your inbox daily

Follow these simple steps to get the benefits:

1.  Scan the QR code or visit the link below

https://packt.link/free-ebook/9781803235844

2.  Submit your proof of purchase
3.  That's it! We'll send your free PDF and other benefits to your email directly

# Part 1: Foundations

Cybersecurity architecture requires a fusion of strategic perspective and technical detail. Before exploring the specifics of implementation, establishing core foundations proves essential.

This opening part of the book focuses on orienting you with fundamental concepts, principles, and domains underpinning effective cybersecurity architecture. *Chapter 1* provides an accessible overview of key cybersecurity basics, positioning why security matters across increasingly interconnected technology landscapes.

*Chapter 2* delves deeper into foundational areas including access controls, network security, cryptography, and risk management. Practical examples illustrate how each contributes to multilayered protection.

With core building blocks in place, *Chapter 3* delineates what distinguishes the cybersecurity architect role and its responsibilities. It explores the synergies and trade-offs between security strategies and business objectives that architects must balance.

Together, these chapters equip you with baseline security knowledge and clarify the architect's role. By grounding discussions in principles and context, the foundations prepare you to explore pathways to grow architectures strategically in alignment with organizational needs. Even those already familiar will benefit from the concise refresher this part provides on the essential concepts underpinning the latest frameworks, controls, and best practices.

This part has the following chapters:

- *Chapter 1, Introduction to Cybersecurity*
- *Chapter 2, Cybersecurity Foundation*
- *Chapter 3, What Is a Cybersecurity Architect and What Are Their Responsibilities?*

# 1

# Introduction to Cybersecurity

In today's connected world, it is hard to not hear about or unwittingly do something related to cybersecurity. Whether that is the forced password reset associated with your work user account or the notification associated with a data breach, individuals are forced to deal with cybersecurity concepts at all levels. It is for that reason, and without any surprise, that cybersecurity has become a popular career choice and one with growing demand. According to the US Bureau of Labor Statistics (`https://www.bls.gov/ooh/computer-and-information-technology/information-security-analysts.htm#:~:text=Employment%20of%20information%20security%20analysts,on%20average%2C%20over%20the%20decade`), there was an expected growth of 35% in cybersecurity jobs between 2021 and 2023. That is a potential of 3.5 million cybersecurity positions worldwide according to a report by Cybersecurity Ventures (`https://www.esentire.com/resources/library/2023-official-cybersecurity-jobs-report`). This is in contrast with the nearly 175,000 layoffs associated with the tech industry since the beginning of 2022 (`https://layoffs.fyi/`). What does this mean? It means that the cybersecurity industry is not going away anytime soon and the available job opportunities and competition for those jobs is only going to increase.

This means that people, and more specifically those reading this book, are going to be looking for more than a job, but a career in a field that can provide a great deal of growth opportunities and satisfaction. The pinnacle of a cybersecurity technical career is that of the **cybersecurity architect** (**CSA**). The CSA is a role that helps shape, design, and plan the technical aspects of an organization's approach to security at all levels. This chapter provides foundational concepts and basics to understand the concepts of cybersecurity and ultimately how that plays into the role of the CSA. This will provide a foundational-level setting for those new to cybersecurity while also providing a fundamental refresher to those who have been working within cybersecurity or IT for some time.

In this chapter, we're going to cover the following main topics:

- What is cybersecurity?
- Confidentiality/integrity/availability
- Networking and operating systems
- Applications
- **Governance, regulations, and compliance (GRC)**

The reality is that to really get a full understanding of the basic foundations of cybersecurity, it would be longer and in more detail than what you will find in this chapter. That stated, there are some additional resources (books and online resources) that can provide a deeper dive into concepts that are touched upon in this chapter, which you will find in the *Further reading* section at the end of this chapter.

Moreover, while *Part 1* (*Chapters 1-3*) may be old hat for some, it is important to provide a foundational baseline for any reader, beginner, or well-seasoned professional, to effectively have a discussion about the cybersecurity architect. For that reason, those who are familiar with the foundational material can jump to *Part 2*.

## What is cybersecurity?

It is no secret that there are volumes of books written on the topic of cybersecurity, some of which I have been fortunate enough to provide content for. This section is not meant to be a doctoral thesis on cybersecurity, but rather a survey to provide the baseline of information for the remaining topics of the book. As a result, I will periodically reference other material or books to provide you with the ability to do a deeper dive into certain topics to prevent this handbook from becoming a tome.

Let's face it, depending on who you ask, you will get varying definitions of the term cybersecurity. This can range from protecting systems, networks, and programs from digital attacks, to reducing the risk level of an organization, or even calling cybersecurity by another name such as information assurance, security, or cyber, and the list could go on. The reason for the varied definitions or synonyms is it comes down to the perspective of the individual or organization providing the definition or focus. It is also not to say that all the different definitions are incorrect – because most are not – but it shows the focus and priorities as it relates to cybersecurity.

According to the U.S. **Cybersecurity & Infrastructure Security Agency** (**CISA**), cybersecurity is defined as *"the art of protecting networks, devices, and data from unauthorized access or criminal use and the practice of ensuring confidentiality, integrity, and availability of information"* (`https://www.cisa.gov/news-events/news/what-cybersecurity`).

What does this mean practically? It means that, as an individual or business, you are trying to perform whatever tasks or business as efficiently and securely as possible without breaking the bank when it comes to what you are securing. The growth of computers, web-based applications, and information technology has been explosive. The propagation of information around the globe has never been faster and more present at an individual's fingertips than it is today; it is only going to get faster. Technology has brought immense benefits to every facet of society, but unfortunately, there is a dark side to technology too. This dark side comes in the form of data theft, cyber criminals, extortion, identity theft, and much more. It is the dark side of technology that cybersecurity tries to stop or prevent by securing communications, applications, physical access, and so on.

The reality is that the only truly secure computer system is one that is never turned on or used. The moment we enable our new iPhone, boot up the latest tablet or computer, or connect to the internet, we are on a countdown to increasing our risks and reducing the security of the device or application. We could build an almost impenetrable building or castle with a moat and, in honor of Dr. Evil, sharks with lasers, but that would not make life or business any easier or prevent vulnerabilities or risks as they relate to the applications or systems we use.

Previously, we could take defensive measures to the extreme inside the boundaries of traditional tech. But today, and for the foreseeable future, policies like **work from home (WFH)** and **bring your own device (BOYD)** have blurred the boundaries that were traditionally in the sole control of the organization and provide hackers and other bad actors with a much broader target to penetrate or obtain a foothold. Instead, we need to find a middle ground that provides the most security. This comes down to the security of the data we create or modify as an individual, business, or some combination of the two. Cybersecurity looks to strike an acceptable balance between security and the risks that are faced.

With this in mind, most certification bodies, associations, and government entities, such as **International Information Systems Security Certification Consortium (ISC2)**, **Center for Internet Security (CIS)**, **National Institute of Standards and Technology (NIST)**, **Cybersecurity and Infrastructure Security Agency (CISA)**, and others will divide the various domains or subject groupings of cybersecurity into some combination of the following topics:

- Access control
- Secure software development
- **Business continuity planning/Disaster recovery (BCP/DR)**
- Cryptography
- Information security governance/risk management
- Legal/regulatory/compliance and investigations
- Security operations
- Physical and environmental security
- Security architecture
- Telecommunications/network security

The preceding list is the typical breakdown by ISC2 within its body of knowledge for the **Certified Information System Security Professional** (**CISSP**) certification. We will discuss certifications in further detail in *Chapter 8, The Certification Dilemma.*

Cybersecurity is broken down into the following subject areas because of the vast scope of cybersecurity as a whole. By breaking it down, it is easier to group the content for study and further analysis. In addition, many people entering the field of cybersecurity tend to specialize or focus on one area. So, to understand why a person would focus on one area over another, let's define the domains.

## Access control

Access control involves the procedure of permitting solely authorized individuals, programs, or other computer systems to observe, alter, or gain control over a computer system's resources. Furthermore, it acts as a mechanism to restrict the utilization of certain resources to only those users who have been granted authorization.

## Secure software development

Secure software development encompasses a series of procedures and tasks associated with the strategic planning, coding, and administration of software and systems. Furthermore, it encompasses the implementation of protective measures within those systems to guarantee the confidentiality, integrity, and availability of both the software and the data it processes.

## Business continuity planning/disaster recovery (BCP/DR)

BCP and DR encompass the essential measures, procedures, and strategies required to uphold uninterrupted business operations in the face of significant disruptions. This entails recognizing, choosing, executing, testing, and maintaining processes and specific actions aimed at safeguarding vital business infrastructure and operations from system and network interruptions. The ultimate goal is to promptly restore essential services and business activities to their normal functioning state.

## Cryptography

**Cryptography** is the science of, and some even say the art of, using deception and mathematics to hide data from unwanted access. Cryptography has been used for centuries. It addresses the principles, means, and methods to convert plaintext into ciphertext and back again to ensure the confidentiality, integrity, and authenticity or non-repudiation of data.

## Information security governance/risk management

Information security governance and risk management encompasses the multifaceted strategies organizations employ to safeguard critical information assets and systems. This discipline seeks to establish holistic criteria for protection by integrating frameworks, policies, organizational culture, and standards.

Effective governance requires going beyond technology alone to address human behavior. Cultivating security awareness, adhering to best practices, and fostering a culture of responsibility are equally important.

Leading governance frameworks provide guiding models. ITIL outlines IT service management processes. COBIT focuses on IT governance and control. The ISO 27000 family covers information security management systems. NIST's Cybersecurity Framework defines industry standards for security programs.

By leveraging governance principles, organizations can take a strategic approach to managing cyber risks. This means continuously assessing their people, processes, and technology capabilities against standards and then identifying and prioritizing areas for improvement.

Mature security governance is comprehensive yet adaptive. It synthesizes tested frameworks, executive engagement, user education, nimble policies, and robust controls to holistically safeguard systems and information. Organizations must vigilantly govern to evolve governance and stay resilient.

## Legal/regulatory/compliance and investigations

Legal, regulatory, compliance, and investigations comprise the policies, laws, and processes organizations employ to address computer crime and security incidents. This discipline encompasses the following:

- **Computer crime legislation**: Laws prohibiting unauthorized access, hacking, malware distribution, and other cyber offenses

- **Associated regulations**: Mandates around data privacy, breach disclosure, sector-specific requirements, and cybersecurity standards

- **Investigative measures**: Techniques for detecting security incidents through monitoring, log analysis, and forensics

- **Evidence gathering/management methodologies**: Procedures for securely collecting, analyzing, documenting, and preserving evidence for investigations

- **Reporting protocols**: Guidelines for reporting incidents to authorities and impacted parties

Adhering to legal and regulatory obligations is foundational for security. Violations can lead to fines, lawsuits, and reputation damage.

Proactively planning incident response strategies ensures organizations can act swiftly and methodically if breached. Following defined evidence-handling procedures is crucial for accurate forensic investigations.

By integrating lawful compliance into their governance models and preparing principled investigation protocols, organizations reinforce resilience and accountability. This promotes cybersecurity while respecting rights.

## Security operations

Security operations are the ongoing processes and controls implemented to safeguard an organization's information systems and data. This discipline focuses on consistently executing security best practices across centralized and distributed technology environments.

Key responsibilities include the following:

- **Asset protection**: Ensuring hardware, applications, services, and data remain confidential and integral through access controls, encryption, and resilience measures

- **Monitoring and detection**: Employing tools such as SIEMs and IDSs to continuously monitor systems, networks, and user activity to rapidly detect potential incidents

- **Incident response**: Investigating suspected or confirmed events, containing impacts, eradicating threats, recovering systems, and improving future response capabilities

- **Ongoing maintenance**: Keeping security tools and services such as firewalls, antivirus, and log management operating reliably through patches, upgrades, and redundancy

- **Process integration**: Incorporating security processes into IT operations and business workflows to embed good security hygiene

The ultimate goal is to develop mature capabilities to predict, prevent, detect, and respond to threats through technology, processes, and human expertise. Smooth integration of security operations into daily functions creates a resilient institutional immune system.

## Physical and environmental security

Physical and environmental security involves safeguarding facilities housing critical information systems against unauthorized access and environmental hazards. This discipline encompasses the following:

- **Security surveys**: Regularly evaluating facilities' physical access controls, surveillance systems, and vulnerability to threats such as fires or floods

- **Risk and vulnerability assessments**: Identifying physical infrastructure and procedural weaknesses that may enable data breaches or system damage

- **Site planning and design**: Incorporating security into facility layouts through measures such as access control zones, cameras, alarms, and secure equipment rooms

- **Access control systems**: Managing physical access to facilities and critical system components via methods such as ID badges, biometric validation, and multifactor authentication

- **Environmental controls**: Maintaining ideal temperature, humidity, electrical supply, fire suppression, and other environmental conditions to protect systems

- **Procedural security**: Establishing policies for escorting visitors, reporting incidents, performing equipment maintenance, and responding to environmental events

By holistically addressing physical factors alongside digital defenses, organizations can reduce attack surfaces, rapidly detect threats, and improve incident response. Integrating physical and digital security policies creates layered defenses.

## Security architecture

Security architecture involves translating organizational requirements into comprehensive cybersecurity designs encompassing people, processes, and technology controls. This discipline focuses on the following:

- **Security principles and frameworks**: Applying models such as Zero Trust and CIS controls to guide architecture
- **Control translation**: Mapping security requirements to technical safeguards and policies that balance usability and protection
- **Environment design**: Architecting layered defenses tailored to infrastructure, cloud environments, applications, data flows, and diverse access scenarios
- **Monitoring integration**: Incorporating controls and systems to provide robust logging, visibility, analysis, and response capabilities
- **Compliance alignment**: Structuring architecture to adhere to industry regulations, legal obligations, and cybersecurity standards
- **Continuous adaptation**: Evolving architecture to address new threats, business demands, and technology advancements

The architecture serves as a high-level blueprint codifying how security maps to business objectives. It provides the foundation for implementing integrated people, processes, and technology cyber defenses across the enterprise.

Effective architecture requires synthesizing organizational needs with deep security expertise.

## Telecommunications/network security

Telecommunications and network security involve a range of technologies, transmission methods, frameworks, data formats, and protective measures. Their purpose is to ensure the confidentiality, integrity, and availability of data transmitted over both private and public networks and various media. Network security is often regarded as a fundamental aspect of IT and security, as the network serves as a central, if not the most crucial, asset in many environments. The loss of the network often translates to a loss of business and services in most scenarios.

As can be seen in various domains, telecommunications and network security are not only interconnected but deal with risk exposure and mitigating that risk. I have mentioned risk several times, but what is risk? Put simply, **risk** is the possibility of something bad happening. This could be a natural disaster, a hard drive failure, or an advanced persistent threat. With that in mind, cybersecurity is the mitigation of risk to maintain confidentiality, integrity, and availability.

# Confidentiality/integrity/availability

I happen to prefer CISA's definition of cybersecurity, because it is concise and encompasses most other definitions, including my little nutshell. I also like the fact that it includes the CIA triad as the basis of the definition. No, this is not the United States' spy agency, but rather the fundamental foundation of security. That is **Confidentiality, Integrity, and Availability (CIA)**.

We will get to the CIA triad in more detail shortly, but consider our previous discussion about cybersecurity. How does a company maintain its business? Customers support the business because the company provides services acceptable to the customers. What happens if the business is not able to deliver on promised services or the business openly releases customer data? The business would not last long because the customers would quickly transition to competitors. In this example, the business needs to improve reliability or availability and establish a model of confidentiality and integrity to re-establish the trust of the customer. The CIA triad tries to remediate this from the perspective of cybersecurity:

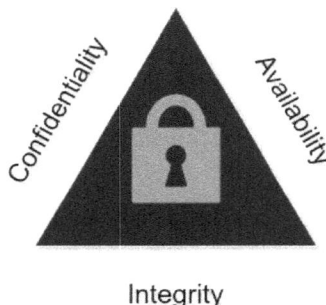

Figure 1.1 – The CIA triad

As previously mentioned, the CIA triad is Confidentiality, Integrity, and Availability. What does this mean? **Confidentiality** refers to protecting information from unauthorized access. **Integrity** refers to the reliability and completeness of data, ensuring that it has not been unintentionally modified or altered by an unauthorized user. Ultimately, integrity ensures that data remains trustworthy, complete, and free from unauthorized changes. **Availability** pertains to the continuous accessibility and optimal functioning of data, systems, and resources as required by authorized users. It guarantees the consistent availability and usability of information and services, ensuring minimal disruptions or downtime. By maintaining reliable operational status, availability enables users to access and utilize resources effectively, thereby supporting business operations and fulfilling organizational requirements. The common thread of any good cybersecurity program or initiative addresses at least one component, and in most cases all three components, of the CIA triad.

In the realm of cybersecurity, maintaining the confidentiality, integrity, and availability of data is paramount. Additionally, non-repudiation ensures that the actions and transactions of individuals cannot be denied. To better understand the different aspects and key concepts, and explain the significance of cybersecurity, the following provides a breakdown of the core components of the CIA triad.

# Confidentiality

Confidentiality involves safeguarding sensitive information from unauthorized access or disclosure, ensuring that only authorized individuals have the ability to access and view such data. It focuses on the protection of sensitive information, preventing it from falling into the wrong hands and maintaining strict control over who can obtain and observe it. Here are key aspects related to confidentiality.

### Data encryption

**Encryption** is the process of converting plaintext data into a coded form (**ciphertext**) that is unreadable without the appropriate decryption key. It prevents unauthorized individuals from understanding the content of the data even if they gain access to it.

### Access controls

**Access controls** involve implementing mechanisms to restrict access to sensitive information based on user roles, permissions, and authentication factors. This prevents unauthorized individuals from accessing confidential data.

### Data classification

**Data classification** involves categorizing data based on its sensitivity level. It allows organizations to prioritize the protection of highly sensitive information and apply appropriate security controls based on the classification.

# Integrity

Integrity ensures that data remains accurate, unaltered, and reliable throughout its life cycle. Maintaining data integrity is crucial to prevent unauthorized modification, corruption, or tampering. Here are key aspects related to integrity.

### Data validation

**Data validation** involves verifying the accuracy and consistency of data. It ensures that data meets specific predefined criteria and is free from errors, omissions, or malicious modifications.

### Hash functions

**Hash functions** are mathematical algorithms that generate a unique string of characters (**hash value**) for a given set of data. By comparing the hash value before and after data transmission or storage, integrity violations can be detected if the hash values do not match.

### Digital signatures

Digital signatures use encryption techniques to provide a mechanism for verifying the authenticity and integrity of electronic documents or messages. They ensure that the sender cannot deny having sent the message and that the content remains unaltered.

## Availability

Availability refers to ensuring that systems, networks, and data are accessible and usable when needed. It involves preventing disruptions, maintaining service continuity, and mitigating the impact of potential incidents. Here are key aspects related to availability.

### Redundancy and fault tolerance

Implementing redundancy and fault-tolerant mechanisms ensures that critical systems and data have backup components or alternate paths, minimizing the impact of hardware failures, natural disasters, or other disruptions.

### Disaster recovery planning

**Disaster recovery planning** involves creating strategies and processes to recover critical systems and data after a disruptive event. It includes regular backups, off-site storage, and documented procedures for system restoration.

### Distributed Denial of Service mitigation

**Distributed Denial of Service** (**DDoS**) attacks aim to overwhelm systems or networks, causing service unavailability. Implementing DDoS mitigation solutions, such as traffic filtering or **content distribution networks** (**CDNs**), helps protect against such attacks and ensures uninterrupted access to services.

## Non-repudiation

**Non-repudiation** ensures that the actions or transactions of individuals cannot be denied or disputed. It provides evidence that a specific action took place and was performed by a specific entity. Here are key aspects related to non-repudiation.

### Digital certificates

**Digital certificates** are electronic documents that validate the identity of individuals or entities in electronic transactions. They are issued by trusted third parties (certificate authorities) and provide assurance of authenticity and non-repudiation.

*Audit trails*

**Audit trails** are records that capture and document the activities and events within a system or network. They serve as evidence of actions performed and can be used to prove the occurrence of specific events or transactions.

*Legal and regulatory compliance*

Non-repudiation has legal and regulatory implications in various industries. Compliance with industry-specific regulations and requirements helps establish accountability and prevents the denial of actions or transactions.

Every cyber-attack or penetration attempts to violate at least one of the CIA triad attributes. The grouping of these three concepts into a triad allows cybersecurity professionals to understand the interconnectedness, overlaps, and conflicts among them. It provides a framework for considering the relationships between confidentiality, integrity, and availability, enabling professionals to analyze how these principles interact with and potentially contradict one another. It is like a three-legged chair. Together, each leg provides a very sturdy platform that is able to stand on its own and under pressure. If one of those legs becomes compromised, the stability and functionality of the platform as a whole becomes untenable. By examining the inherent tension among the components of the triad, security professionals can effectively establish priorities and implement necessary processes. This can be done within a single application or system or across the technology stack collectively. The CIA triad holds significant importance in identifying vulnerabilities and investigating the causes behind network compromises. It serves as a valuable framework for understanding weaknesses and pinpointing areas of improvement after a breach. This information can then be utilized to address vulnerabilities, strengthen security measures, and identify areas of resilience.

Confidentiality, integrity, availability, and non-repudiation are fundamental pillars of cybersecurity. Understanding their significance and implementing appropriate security measures ensures the protection of sensitive information, the reliability of data, uninterrupted access to services, and the establishment of accountability.

# Networking and operating systems

Ultimately, the reason for security is the protection of data at rest or in motion for a business. As such, it requires an objective analysis of the current state of the business or enterprise. The architecture, from a security perspective, is not vendor- or technology-specific but based on best practices. Likewise, it looks at the security requirements by device or technology type to meet the functionality necessary for flexibility in a changing infrastructure while implementing the most appropriate security model for the environment.

In the world of cybersecurity, networking and operating systems play a crucial role in safeguarding digital assets. This aims to provide an accessible overview of networking and operating systems within the context of cybersecurity, explaining their significance, functions, and potential vulnerabilities.

# Networking fundamentals

Networking forms the foundation of modern digital communication and is essential for the functioning of interconnected systems. Understanding networking fundamentals is crucial for comprehending the cybersecurity landscape. Here are the key concepts related to networking.

### Local Area Networks and Wide Area Networks

**Local Area Networks** (LANs) and **Wide Area Networks** (WANs) are two common types of networks. LANs connect devices within a limited geographical area, such as a home or office, while WANs connect geographically dispersed networks. Both types of networks require proper security measures to protect against unauthorized access and data breaches.

### Network devices

Networking devices, such as routers, switches, and firewalls, are responsible for routing, switching, and securing network traffic. Routers direct data packets between different networks, switches connect devices within a network, and firewalls enforce network security policies.

### Network protocols

Network protocols are sets of rules and standards that govern how data is transmitted and received over a network. Common protocols include **Transmission Control Protocol/Internet Protocol** (TCP/IP), which forms the foundation of internet communication, and **Domain Name System** (DNS), which translates domain names into IP addresses.

# Operating systems in cybersecurity

An operating system serves as the software platform that manages computer hardware and software resources. It provides a secure foundation for running applications and plays a crucial role in cybersecurity. Here are the key aspects related to operating systems.

### Types of operating systems

Popular operating systems include Windows, macOS, and Linux. Each operating system has its strengths and vulnerabilities, making it important to understand the specific security considerations for each platform.

### User authentication and access controls

Operating systems employ user authentication mechanisms, such as usernames and passwords, to ensure that only authorized individuals can access the system. Access controls further define permissions and privileges for users, limiting their actions and preventing unauthorized access to sensitive data.

### Patch management and updates

Operating systems regularly release updates and patches to address security vulnerabilities. Timely installation of these updates is critical for protecting against known exploits and ensuring a secure computing environment.

### Antivirus and anti-malware software

Operating systems can be fortified with antivirus and anti-malware software to detect and remove malicious programs that may compromise the system's security. These software solutions help protect against viruses, worms, Trojan horses, and other forms of malware.

## Cybersecurity considerations for networking and operating systems

Securing networks and operating systems is vital to protect against cyber threats. Here are some key considerations.

### Network segmentation

Network segmentation involves dividing a network into smaller, isolated segments to limit the impact of a potential breach. It restricts unauthorized access and contains potential compromises, enhancing overall network security.

**Trust zones**

A **zone** refers to a logical grouping of interfaces or systems that simplifies the management and control of access rules within a network or system. It helps establish and maintain different levels of trust for enhanced security. Each of these zones plays a crucial role in defining and enforcing security policies and controls within a network. By categorizing interfaces and systems into different zones, organizations can streamline their security management processes and ensure appropriate levels of trust and access across their infrastructure. In order to better understand the **trust zone** model, it is necessary to understand the basic concepts of zones. A core principle in modern cybersecurity architecture is **network segmentation** using zones to isolate systems with differing security levels. This recognizes that devices have varying risk profiles and business criticality.

For example, web servers require internet accessibility that exposes attack surfaces. Network zoning isolates vulnerable, public-facing systems from more sensitive assets such as databases or internal services.

Key benefits of network zoning include the following:

- **Tailored security**: Controls and monitoring can be customized per zone, enabling tighter protection for sensitive assets
- **Reduced blast radius**: Threats are confined to one zone rather than propagating across the network

- **Granular access**: Network rules actively limit which zones/systems can communicate
- **Improved visibility**: Traffic flows and anomalies are easier to baseline and monitor within zones
- **Simplified compliance**: Zones help logically group assets aligned to regulations

Effective zoning requires classifying assets by risk, function, and data criticality. Architects can then design zone boundaries leveraging firewalls, switches, VPNs, and tools such as microsegmentation.

By aligning network architecture to security priorities, organizations gain targeted protection and detection, helping fulfill key cybersecurity objectives.

There are four fundamental zones commonly used in network security:

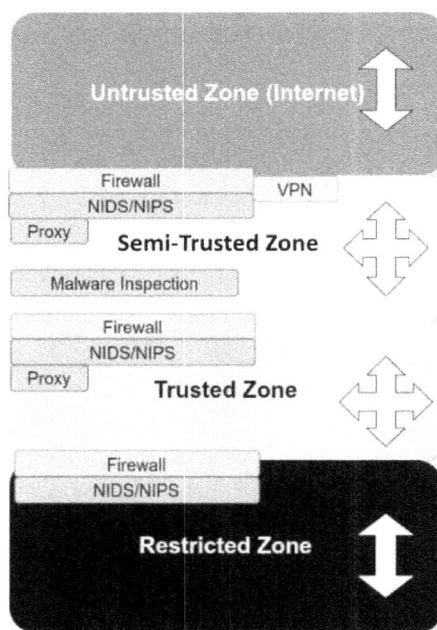

Figure 1.2 – Basic trust zone model

Let us look at these zones in detail:

- **Untrusted Zone** (**UTZ**): The UTZ represents the lowest level of trust within the network. It is typically located on the internet-facing side of a security appliance or network edge. By default, traffic from the UTZ is not allowed to enter other zone types unless explicit rules are defined. However, traffic from the **Trusted Zone** (**TZ**) is usually permitted to communicate with the UTZ through the **Semi-Trusted Zone** (**STZ**), unless specific **access control lists** (**ACLs**) restrict the communication. The UTZ is often associated with the color red, symbolizing caution and potential threats.

- **Semi-Trusted Zone (STZ)**: The STZ offers a higher level of trust compared to the UTZ but is still lower than the TZ. It serves as a secure area between the LAN and the internet. The STZ typically hosts web-tier applications, such as presentation services, reverse-proxy mechanisms, or VPN termination points. It is sometimes referred to as a **Demilitarized Zone (DMZ)**. The STZ is generally represented by the color yellow, indicating a level of caution and limited access.

- **Trusted Zone (TZ)**: The TZ provides the highest level of trust within the network. It is characterized by the least scrutiny and restrictions on traffic. TZs are typically part of the LAN but can extend across an enterprise and WAN connection. This zone encompasses end-user systems such as desktops and laptops. Traffic within the TZ is assumed to be secure and trustworthy. The TZ is commonly associated with the color green, signifying safety and reliability.

- **Restricted Zone (RZ)**: The RZ offers the highest level of security among the four zones. This zone typically contains the most sensitive data/databases and thus only explicit access is allowed to this zone such that direct access to the data within another zone is not allowed except through distinct sources, such as IP addresses and ports. This zone is typically characterized by the color black.

## Beyond the trust zone

The zone model, discussed previously, originally designed to establish trust levels within network environments, can be effectively integrated with the concept of zero trust when adapting to cloud services and a distributed work-from-home model. In a **zero trust framework**, the focus shifts from implicitly trusting certain zones to continuously verifying and authorizing access requests regardless of the user's location or the network they are connected to.

When incorporating cloud services, organizations can leverage the principles of zero trust to redefine the boundaries of each zone. The UTZ expands to include the public cloud, emphasizing the need for strict access controls and authentication mechanisms. By implementing zero trust principles, organizations can enforce granular access policies, employ multi-factor authentication, and conduct continuous monitoring and verification of activities within the cloud environment. The STZ can be re-imagined to encompass the cloud's network perimeter, where zero trust controls are applied to inspect and validate traffic before reaching the protected resources.

In a distributed work-from-home model, zero trust principles are crucial for securing remote employee devices and networks. The TZ evolves to encompass a zero trust architecture, where every user, device, and network connection is treated as un-trusted until explicitly authorized. Organizations can adopt zero trust access solutions, such as **software-defined perimeters** (SDPs) and identity-based access controls, to authenticate and authorize remote users. Continuous monitoring and behavior analysis enable real-time risk assessment, allowing organizations to respond to potential threats promptly. By embracing zero trust, organizations establish a security model that minimizes the risk of lateral movement and unauthorized access, irrespective of the employee's physical location.

Integrating the zone model with zero trust principles enables organizations to adapt to cloud services and a distributed work-from-home model effectively. By redefining zones and implementing zero trust controls, organizations can establish robust security postures that continuously verify and authorize access, ensuring data protection and minimizing the potential for unauthorized activity.

### Perimeter defense

**Perimeter defense** involves implementing security measures at the network's edge to protect against external threats. These measures include firewalls, **intrusion detection systems (IDSs)**, and **intrusion prevention systems (IPSs)** that monitor and filter network traffic.

### Secure protocols and encryption

The use of secure network protocols, such as **Hypertext Transfer Protocol Secure (HTTPS)**, ensures encrypted communication between clients and servers, preventing eavesdropping and data tampering.

### Access controls and least privilege

Implementing access controls and following the principle of least privilege ensures that users have only the necessary privileges to perform their tasks, reducing the risk of unauthorized access and limiting the potential damage from a compromised account.

### Endpoint security

Endpoint security focuses on securing individual devices (endpoints) connected to the network. It involves measures such as antivirus software, host-based firewalls, and regular patching to protect against malware and vulnerabilities.

Networking and operating systems are fundamental components of cybersecurity. Understanding networking concepts, network devices, and protocols enables individuals to comprehend the intricacies of secure communication. Similarly, knowledge of operating systems, authentication mechanisms, and security best practices helps fortify systems against cyber threats. By implementing appropriate security measures, such as network segmentation, perimeter defense, and secure protocols, individuals and organizations can significantly enhance their cybersecurity posture and protect valuable digital assets.

# Applications

Applications play a critical role in today's digital landscape, enabling various tasks and services on computers, smartphones, and other devices. However, they can also pose security risks if not properly designed and secured. This report aims to provide an accessible overview of applications and application security within the context of cybersecurity. The content is tailored for individuals with a high school education level to ensure understanding and comprehension.

# Understanding applications

**Applications**, also known as software programs or apps, are computer programs designed to perform specific tasks or provide specific services. They can range from simple applications such as calculators and word processors to complex applications such as web browsers and online banking platforms. Here are key aspects related to applications.

## Types of applications

Applications can be categorized into various types, including desktop applications, mobile applications, web applications, and enterprise applications. Each type has its unique characteristics and potential security considerations.

## Application development

Applications are developed using programming languages and frameworks. Developers write code to create the desired functionalities and user interfaces. The development process involves multiple stages, including design, coding, testing, and deployment.

## Common application platforms

Popular application platforms include Windows, iOS, Android, and web browsers. Each platform has its own application ecosystem and security considerations. Understanding platform-specific vulnerabilities is crucial for developing and securing applications.

# Importance of application security

Application security is vital in protecting sensitive information, preventing unauthorized access, and ensuring the reliable and secure functioning of applications. Here are key reasons why application security is crucial.

## Data protection

Applications often handle sensitive data, such as personal information, financial details, and intellectual property. Securing applications helps protect this data from unauthorized access, theft, and misuse.

## Prevention of exploits

Vulnerable applications can be exploited by cyber-criminals to gain unauthorized access to systems, execute malicious code, or steal sensitive information. By implementing proper security measures, organizations can mitigate the risk of such exploits.

### Maintaining trust

Secure applications build trust among users, customers, and stakeholders. When users have confidence in the security of an application, they are more likely to use it and share their information, leading to increased adoption and customer satisfaction.

## Common application security challenges

Applications face various security challenges that must be addressed to ensure their resilience against attacks. Here are some common application security challenges.

### Input validation

Applications must properly validate and sanitize user inputs to prevent attacks such as SQL injection, **cross-site scripting** (**XSS**), and command injection. Input validation ensures that the data entered by users does not contain malicious code or unexpected characters.

### Authentication and authorization

Implementing robust authentication mechanisms, such as strong passwords or multi-factor authentication, verifies the identity of users. Authorization ensures that authenticated users have appropriate permissions and access levels within the application.

### Secure coding practices

Developers should follow secure coding practices to minimize vulnerabilities in the application's code. This includes proper handling of user inputs, secure storage of sensitive information, and protection against common code vulnerabilities.

### Secure configuration and patch management

Proper configuration of application components and regular patching of software frameworks and libraries are crucial to address known vulnerabilities and protect against exploits.

## Secure development life cycle

To ensure application security, organizations should adopt a **secure development life cycle** (**SDL**) approach. The SDL encompasses the following key phases.

### Requirements and design

Security considerations should be incorporated into the application's requirements and design phases. This involves identifying potential security risks and defining security requirements.

### Development and testing

Secure coding practices should be followed during the development phase, with regular testing to identify and fix vulnerabilities. This includes unit testing, integration testing, and security testing.

### Deployment and maintenance

Applications should be securely deployed, with proper configuration and hardening of servers and infrastructure. Ongoing maintenance includes applying patches, monitoring for security incidents, and promptly addressing any identified vulnerabilities.

Applications are an integral part of the digital landscape, providing various functionalities and services. Ensuring the security of applications is essential to protect sensitive data, prevent exploits, and maintain user trust. By understanding application types, platforms, and the importance of application security, individuals can comprehend the risks and challenges involved. Implementing secure coding practices, input validation, authentication, and following a secure development life cycle are vital steps in mitigating application security risks. By prioritizing application security, organizations can enhance their overall cybersecurity posture and protect their valuable digital assets.

# Governance, regulations, and compliance (GRC)

In today's complex and interconnected business environment, organizations face numerous challenges related to GRC. GRC refers to the framework and processes that organizations establish to ensure ethical conduct, adhere to laws and regulations, and mitigate risks. This topic aims to provide a comprehensive overview of GRC, explaining its key components, significance, and the role it plays in organizations. The content is presented in a manner that can be easily understood by individuals with a high school education.

## Governance

**Governance** refers to the set of policies, processes, and procedures that guide the overall management and decision-making within an organization. It encompasses the establishment of a clear organizational structure, the definition of roles and responsibilities, and the implementation of effective oversight mechanisms. Good governance ensures that an organization operates in an ethical and transparent manner, aligns its activities with its objectives, and acts in the best interests of stakeholders.

Within the context of GRC, governance focuses on establishing and maintaining appropriate structures and mechanisms to oversee compliance efforts and risk management. This includes the formation of governance bodies, such as boards of directors or steering committees, to provide strategic direction and oversight. Governance also involves defining accountability frameworks, ensuring effective communication channels, and establishing mechanisms for monitoring and reporting on compliance and risk-related matters.

## Regulations

**Regulations** refer to the rules and guidelines established by governmental bodies or industry regulators that organizations must comply with. These regulations are designed to ensure fair business practices, protect consumers, maintain market stability, and address societal concerns. Examples of regulatory bodies include the **Securities and Exchange Commission (SEC)**, the **Federal Communications Commission (FCC)**, and the European Parliament.

Compliance with regulations is crucial for organizations to avoid legal and financial penalties, reputational damage, and loss of customer trust. Regulatory compliance involves understanding the relevant laws and regulations applicable to the organization's industry, implementing processes and controls to adhere to those requirements, and regularly monitoring and reporting on compliance activities. This includes activities such as conducting risk assessments, establishing internal controls, and maintaining proper documentation.

## Compliance

**Compliance**, within the context of GRC, refers to the adherence to laws, regulations, internal policies, and industry standards. It encompasses the processes and activities undertaken by organizations to ensure that their operations are conducted in accordance with applicable requirements. Compliance activities can vary based on the nature of the organization, its industry, and the specific regulations it must adhere to.

Compliance efforts typically involve establishing policies and procedures, conducting regular assessments, monitoring activities for non-compliance, and taking appropriate corrective actions. Compliance programs may also involve training employees on their responsibilities, conducting internal audits, and engaging external auditors for independent assessments. Effective compliance programs not only mitigate legal and regulatory risks but also contribute to maintaining a culture of ethics, integrity, and accountability within organizations.

## The role of GRC in organizations

GRC plays a vital role in organizations by ensuring that they operate within legal and ethical boundaries, manage risks effectively, and maintain the trust of stakeholders. The key benefits of implementing GRC practices include the following:

- **Risk mitigation**: GRC helps organizations identify and assess risks, implement appropriate controls, and monitor risk levels to reduce the likelihood of adverse events and their impact on the organization

- **Regulatory compliance**: By establishing robust compliance programs, organizations can adhere to relevant laws and regulations, avoid penalties, and maintain a positive reputation

- **Enhanced decision-making**: Effective governance structures enable informed decision-making based on accurate and timely information, contributing to the organization's long-term success

- **Stakeholder confidence**: GRC practices foster trust and confidence among stakeholders, including customers, investors, employees, and regulators, as they demonstrate the organization's commitment to ethical conduct and responsible business practices

- **Operational efficiency**: GRC helps streamline processes, eliminate the duplication of efforts, and improve resource allocation, leading to enhanced operational efficiency and cost savings.

GRC is a critical component of organizational management. By implementing effective governance structures, adhering to regulations, and establishing robust compliance programs, organizations can mitigate risks, ensure ethical conduct, and maintain stakeholder confidence. GRC practices contribute to the long-term success and sustainability of organizations in a rapidly changing business landscape. It is essential for organizations to invest in GRC frameworks and allocate resources to ensure ongoing compliance, manage risks, and adapt to evolving regulatory requirements.

## Summary

In this chapter, we discussed some of the foundational concepts around cybersecurity. This included a brief discussion that provided insights into various aspects of cybersecurity. It aimed to ensure that individuals entering into or starting a career can understand the content around the topics covered.

Highlighting the exponential growth expected in the cybersecurity field, making it an attractive career choice with numerous job opportunities, it emphasized the importance of building a strong foundation in cybersecurity and provided additional resources for further learning. The chapter delved into essential areas such as the following:

- **Cybersecurity basics**: It explored the definition of cybersecurity and its significance in today's interconnected world. This highlighted how individuals encounter cybersecurity concepts in their daily lives, emphasizing the growing demand for cybersecurity professionals.

- **CIA Triad**: The CIA triad forms the foundation of cybersecurity. The chapter explained how maintaining the confidentiality of data, ensuring its integrity, and guaranteeing its availability are crucial aspects of protecting sensitive information.

- **Networking and operating systems**: This section provided an overview of how networking and operating systems are intertwined with cybersecurity. It explained the role of networks in facilitating communication and the importance of securing them. It also discussed the vulnerabilities associated with operating systems and the measures to mitigate risks.

- **Applications and application security**: The chapter explored the significance of applications and the potential risks they pose. It discussed the importance of application security in safeguarding against threats such as unauthorized access, data breaches, and malware attacks.

- **GRC**: This section shed light on the role of governance, regulations, and compliance in cybersecurity. It explained how organizations must establish effective governance frameworks, adhere to industry regulations, and ensure compliance with security standards to mitigate risks and protect sensitive data.

Overall, this chapter aimed to equip individuals with the necessary knowledge to understand and engage with cybersecurity concepts. By grasping the fundamentals of cybersecurity, you can pursue a fulfilling career as a cybersecurity architect in this dynamic and evolving field.

In the next chapter, *Cybersecurity Foundation*, we will build on the foundation of cybersecurity started in the introduction and get a bit more granular to discuss some of the main areas that a cybersecurity architect will need to address and understand relating to the business and other operational teams. This will be cursory in nature but will cover foundational aspects to progress into a discussion of the cybersecurity career path and the options available to the potential cybersecurity architect in specializing/focusing on a particular area.

# Further reading

While formal education and training programs provide an excellent foundation, independent reading and research can greatly accelerate learning. Books allow you to immerse yourself in topics at your own pace and as deeply as you desire. They can expose you to new ideas, reinforce concepts, and inspire new directions to explore.

For technologists at any career stage, regularly reading industry books keeps knowledge sharp and perspectives current in a rapidly evolving field. Technical books help build hands-on skills with topics not covered in certifications or coursework. Books on soft skills such as leadership, communication, and career advancement provide crucial complementary knowledge.

Security professionals in particular benefit from reading across a diversity of focus areas to develop well-rounded capabilities. Books on application security, governance frameworks, incident response, secure DevOps, and other domains reveal the interconnectedness and expand expertise.

Immersing yourself in books allows more active rather than passive learning. The ability to underline, take notes, and reference repeatedly aids retention. Don't underestimate the value of building your own professional library. These books and others that I recommend can be found on my GitHub page at `https://github.com/secdoc/Recommended_Reading.git`. Some of these resources are the following:

- *National Institutes of Science and Technology (NIST) Special Publication 800 Series documents* – `https://csrc.nist.rip/publications/PubsSPs.html`

- *Cybersecurity: The Beginner's Guide: A comprehensive guide to getting started in cybersecurity* – `https://www.amazon.com/Cybersecurity-Beginners-comprehensive-getting-cybersecurity/dp/1789616190/ref=pd_bxgy_vft_none_img_sccl_1/140-1711249-4104027?pd_rd_w=jD5Ux&content-id=amzn1.sym.26a5c67f-1a30-486b-bb90-b523ad38d5a0&pf_rd_p=26a5c67f-1a30-486b-bb90-b523ad38d5a0&pf_rd_r=DQZNRTMZ02AB55Q9SV4E&pd_rd_wg=ycpEx&pd_rd_r=adef23f1-13e2-43f1-8be2-20f229fcdf5b&pd_rd_i=1789616190&psc=1`

- *Linux Basics for Hackers: Getting Started with Networking, Scripting, and Security in Kali* – https://www.amazon.com/Linux-Basics-Hackers-Networking-Scripting/dp/1593278551/ref=sr_1_1?crid=3F2F2W0Y9HCP2&keywords=linux+basics+for+hackers&qid=1669167211&sprefix=linux+%2Caps%2C96&sr=8-1

- *Network Basics for Hackers: How Networks Work and How They Break* – https://www.amazon.com/Network-Basics-Hackers-Networks-Break/dp/B0BS3GZ1R9/ref=sr_1_1?crid=1LI7VKVDS6TQA&keywords=network+basics+for+hackers+occupytheweb+2023&qid=1687363423&sprefix=network+basics+for%2Caps%2C102&sr=8-1

- *Cybersecurity Career Master Plan: Proven techniques and effective tips to help you advance in your cybersecurity career* – https://www.amazon.com/gp/product/1801073562/ref=ppx_yo_dt_b_asin_title_o00_s00?ie=UTF8&psc=1

- *Mastering Linux Security and Hardening: Protect your Linux systems from intruders, malware attacks, and other cyber threats, 3rd Edition* – https://www.amazon.com/dp/1837630518?pd_rd_i=1837630518&pf_rd_p=b000e0a0-9e93-480f-bf78-a83c8136dfcb&pf_rd_r=4NYEYTTENEBEADJK3Y5T&pd_rd_wg=dLg1d&pd_rd_w=qBhTt&pd_rd_r=041941a5-8e4b-4b16-8e7e-e857ac0c4f35

- *Mastering Windows Security and Hardening: Secure and protect your Windows environment from cyber threats using zero-trust security principles, 2nd Edition* – https://www.amazon.com/Mastering-Windows-Security-Hardening-environment/dp/180323654X/ref=sr_1_1?crid=RWKRO36CLZ4C&keywords=mastering+windows+security+and+hardening&qid=1687358253&sprefix=mastering+windows%2Caps%2C98&sr=8-1

- *Network Security Principles and Practices* – https://www.amazon.com/Security-Principles-Practices-Professional-Development/dp/1587050250/ref=sr_1_6?crid=3CRERCN9XUYVZ&keywords=network+security+principles+and+practices&qid=1687358450&sprefix=network+security+principles+and+practices%2Caps%2C94&sr=8-6

- *Computer Security Handbook, Set (Volume 1 and 2) 6th Edition* – https://www.amazon.com/Computer-Security-Handbook-Seymour-Bosworth/dp/1118127064

- *CISSP All-in-One Exam Guide, Ninth Edition* – https://www.amazon.com/CISSP-All-One-Guide-Ninth/dp/1260467376/ref=sr_1_1?crid=12E6NH50FG4QA&keywords=shon+harris+cissp&qid=1685654957&sprefix=shon+harris+cissp%2Caps%2C100&sr=8-1&ufe=app_do%3Aamzn1.fos.006c50ae-5d4c-4777-9bc0-4513d670b6bc

# Cybersecurity Foundation

*"If you know the enemy and know yourself, you need not fear the result of a hundred battles. If you know yourself but not the enemy, for every victory gained you will also suffer a defeat. If you know neither the enemy nor yourself, you will succumb in every battle."*

*– Sun Tzu*

Building upon the introduction provided in *Chapter 1*, this chapter delves deeper into the foundational aspects of cybersecurity architecture. It explores key areas that a cybersecurity architect must address and understand concerning the business and operational teams. While the content provided is introductory, it serves as a springboard for future discussions on the cybersecurity career path and the specialization options that are available to aspiring cybersecurity architects.

As quoted from Sun Tzu's *Art of War* at the beginning of this chapter, it is crucial to comprehend your environment and the potential threats posed by both internal and external threat actors. By gaining a comprehensive understanding of your organization's systems, environment, users, and the potential vulnerabilities and threats they may face, you can confidently assess and strengthen your organization's cybersecurity posture. This understanding is vital for effective risk mitigation.

> **Note**
>
> It is important to acknowledge that the following chapters will provide a cursory overview of each topic. To develop a comprehensive understanding and tackle more complex challenges, further exploration and learning beyond this book is encouraged.

This chapter serves as a foundational guide to the main areas of cybersecurity architecture. By exploring access control, network and communication security, cryptography, **business continuity planning (BCP)/disaster recovery planning (DRP)**, and physical security, you will gain insights into key aspects of cybersecurity. The labs included will complement your understanding and provide a stepping stone for future growth and proficiency. Remember, understanding your environment and the potential threats it faces is a fundamental step toward mitigating risks effectively and ensuring the security of your organization.

The following topics will be covered in this chapter:

- Access control
- Network and communication security
- Cryptography
- BCP/DRP
- Physical security

# Access control

Adequate information and system security is a fundamental responsibility of management. Access control plays a vital role in nearly all applications that handle financial, privacy, safety, or defense-related data. It involves determining the permissible actions of authorized users and managing every attempt made by a user to access system resources. While some systems grant complete access after successful authentication, most systems require more sophisticated and complex control mechanisms. In addition to authentication, access control considers how authorizations are structured. This may involve aligning authorizations with the organization's structure or basing them on the sensitivity of documents and the clearance level of users accessing them.

When organizations plan to implement an access control system, they need to consider three crucial abstractions: access control policies, models, and mechanisms.

Access control policies are overarching requirements that define how access to information is managed and who is authorized to access it under specific circumstances. These policies can govern resource usage within or across different organizational units and may be based on factors such as need-to-know, competence, authority, obligation, or conflict of interest.

At a higher level, these access control policies are implemented and enforced through mechanisms. These mechanisms interpret a user's access request, often leveraging predefined structures provided by the system. **Access control lists** (**ACLs**) serve as a familiar example of such mechanisms. Access control models play a crucial role by connecting policy and mechanism, providing a means to describe the security properties of an access control system. These models act as formal representations of the security policy enforced by the system, and they can be valuable for establishing theoretical limitations.

The NIST IR 7316 report titled *Assessment of Access Control Systems* delves into commonly used access control policies, models, and mechanisms in information technology systems, offering a comprehensive understanding of these essential components.

As systems become larger and more complex, access control becomes particularly challenging in distributed systems that span multiple computers. These distributed systems may utilize various access control mechanisms that need to be integrated to align with the organization's policies. For example, in the case of big data processing systems, which handle vast amounts of sensitive information organized in sophisticated clusters, access control requires collaboration among cooperating processing domains to ensure protection. The paper *An Access Control Scheme for Big Data Processing*, written by Vincent C. Hu, Tim Grance, David F. Ferraiolo, and D. Rick Kuhn, presents a general-purpose access control scheme for distributed big data processing clusters, addressing the unique challenges of securing such environments.

Access control is centered around a set of mechanisms that empower systems to regulate behavior, usage, and content. It grants management the ability to define user permissions, resource access privileges, and authorized operations within the system.

Upon acknowledging the significance of information and the necessity to safeguard it from misuse, disclosure, and destruction, organizations employ access controls to uphold the integrity and security of vital business information. Controlling access to computing resources and information can take various forms, whether through technical or administrative means. Regardless of the method used, access controls are essential components of a well-designed and well-managed information security program.

This domain encompasses topics such as user identification and authentication, access control techniques and their administration, and emerging methods of attacking implemented controls. Biometrics, such as voice, handprint, fingerprint, or retinal patterns, are increasingly employed for identification and authentication purposes. Understanding the potential and limitations of biometric technologies is crucial for their appropriate and effective application.

Access controls play a vital role in safeguarding the privacy, confidentiality, and security of patient healthcare information. Outside North America, particularly in European countries, privacy has long been a significant concern. In recent years, American consumers have also become increasingly aware of the need to protect their privacy, especially as their medical information becomes more widespread and potentially vulnerable. Regulations such as the **Health Insurance Portability and Accountability Act (HIPAA)** for medical information and the Gramm-Leach-Bliley Act for financial information demonstrate the US government's recognition of these concerns and the need for protective measures.

Malicious hacking poses a substantial threat to information security by undermining implemented controls. Hackers persistently target organizations, chipping away at their defenses and achieving success far too often. This domain explores advanced attack tools that have led to high-profile incidents, including the defacement of the US Department of Justice's website and **denial-of-service (DoS)** attacks on commercial sites.

Social engineering techniques represent another method used to circumvent controls that have been implemented by exploiting human nature. Unscrupulous individuals employ deceptive tactics to gather information that can be used to bypass security measures. For instance, an unsuspecting user may receive a call from someone posing as a desktop technician, requesting their network password under the guise of diagnosing a technical issue. This password can then be exploited to compromise the system.

It is imperative to stay abreast of the evolving landscape of access controls and security practices to effectively protect sensitive information.

Access control is a critical aspect of cybersecurity architecture that ensures only authorized individuals or systems can access resources, data, and services. As a cybersecurity architect, understanding and addressing access control is crucial for maintaining the confidentiality, integrity, and availability of an organization's information assets. This chapter delves into the foundational aspects of access control and explores its relationship with the business and operational teams.

## Access control fundamentals

Access control is built upon several fundamental principles that govern the enforcement of restrictions on resource access. These principles include the following:

- **Least privilege**: Users and systems should be granted the minimum necessary privileges to perform their assigned tasks, reducing the risk of unauthorized access or misuse

- **Separation of duties**: Critical operations should require the involvement of multiple individuals or systems, preventing any single entity from having complete control or the ability to misuse privileges

- **Need-to-know**: Users and systems should only have access to information necessary for their specific roles and responsibilities

Access control models provide a structured framework for implementing access control mechanisms. Let's look at three commonly used models:

- **Discretionary access control (DAC)**: Access rights are assigned at the discretion of the resource owner, allowing them to control who can access their resources

- **Mandatory access control (MAC)**: Access rights are determined by security classifications and labels assigned to resources and users, ensuring strict enforcement of access policies

- **Role-based access control (RBAC)**: Access rights are granted based on predefined roles, simplifying administration and management by associating permissions with specific job functions

## Aligning access control with the business

To design an effective access control strategy, it is crucial to align it with the specific needs and goals of the business. This involves doing the following:

- **Identifying critical assets**: Determine the organization's most valuable and sensitive resources, such as intellectual property, customer data, or trade secrets

- **Assessing regulatory and compliance requirements**: Understand the industry-specific regulations and legal obligations that govern access control practices

- **Evaluating business processes**: Analyze how different teams and departments collaborate and interact with data and resources to identify access requirements

Based on the business requirements, develop access control policies that define the following aspects:

- **User access levels**: Determine the different access levels required for different roles within the organization, ensuring that privileges are granted based on job responsibilities

- **Data classification and handling**: Establish guidelines for classifying data based on sensitivity, and define how different data classifications should be accessed, stored, and shared

- **Access request and approval processes**: Define the procedures for requesting access permissions, obtaining approvals, and periodically reviewing and revoking access rights

## Collaboration with operational teams

Collaboration with operational teams is vital to ensure the effective implementation of access control measures. Here are some key considerations:

- **Network security**: Work closely with network administrators to implement firewalls, **intrusion detection systems** (**IDSs**), and other network security measures to enforce access control at the network level

- **Identity and access management** (**IAM**): Collaborate with IAM teams to implement robust identity verification processes, **multi-factor authentication** (**MFA**), and centralized user provisioning and deprovisioning

Regular security awareness programs and training initiatives are essential to educate employees about access control best practices, potential risks, and their responsibilities in maintaining security.

Collaborate with the incident response team to develop procedures for managing access control during security incidents. This includes isolating compromised accounts, investigating access logs, and implementing temporary access restrictions.

## Examples of how you can implement access control measures within an enterprise

Implementing access control measures within an enterprise is paramount for safeguarding sensitive information, securing critical resources, and ensuring the integrity of the organization's operations. By defining and enforcing access policies, organizations can grant appropriate privileges to authorized users while restricting access to unauthorized individuals. In this section, we will explore various examples of access control measures that can be employed within an enterprise to strengthen its security posture and protect against potential threats. From ACLs and RBAC to MFA and encryption, these examples showcase the versatility and effectiveness of access control strategies in creating a robust and secure environment for an organization's data and systems. Let's delve into these examples to gain insights into practical implementations of access control measures that align with industry best practices and regulatory requirements.

## Access control systems

Install access control systems at all entry points, including main entrances, sensitive areas, server rooms, and data centers. This can include key card readers, biometric scanners (such as fingerprint or facial recognition), or keypad locks.

Integrate the access control system with the organization's IAM system to centralize user authentication and authorization processes.

Implement an MFA mechanism that requires users to provide multiple forms of identification, such as a key card and a PIN code, a fingerprint scan and a password, or a smart card and a biometric scan.

## User access management

Develop a user access management process that includes user provisioning, deprovisioning, and periodic access reviews.

Establish a clear process for granting and revoking access privileges based on job roles, responsibilities, and the principle of least privilege.

Utilize RBAC or **attribute-based access control** (**ABAC**) to assign access rights and permissions to users based on predefined roles or attributes.

## Physical access controls

Implement physical barriers, such as turnstiles, gates, or security vestibules, at entry points to control the flow of individuals and ensure only authorized personnel gain access.

Utilize access control cards, key fobs, or biometric credentials for employees to gain entry to restricted areas, and enforce strict policies on the issuance and management of these credentials.

Employ visitor management systems that require visitors to register, provide identification, and be escorted by authorized personnel while within the premises.

## Access logging and monitoring

Implement an access logging and monitoring system that captures and logs all access attempts and activities, including successful and failed authentication attempts.

Regularly review access logs to detect any suspicious or unauthorized access attempts, and promptly investigate and respond to any identified anomalies.

Implement real-time monitoring and alerting mechanisms that notify security personnel of any abnormal access patterns or policy violations.

## *Access control policies*

Develop access control policies that clearly define the rules and guidelines for granting and managing access rights within the organization.

Define user access levels and permissions based on job roles, responsibilities, and the principle of least privilege.

Establish procedures for requesting access permissions, obtaining approvals, and periodic reviews of access rights.

## *Secure remote access*

Implement secure remote access solutions, such as **virtual private networks** (**VPNs**) or secure **remote desktop protocols** (**RDPs**), to enable remote workers or authorized individuals to access the organization's resources securely.

Enforce strong authentication measures, such as **two-factor authentication** (**2FA**) or MFA, for remote access.

Implement network segmentation and secure protocols to isolate remote access networks from the rest of the internal network.

## *Physical security integration*

Integrate access control systems with other physical security measures, such as video surveillance, IDSs, or alarm systems.

Configure access control systems to trigger alerts or alarms in the event of unauthorized access attempts or security breaches.

Utilize video surveillance cameras to record entry points and integrate video footage with access control logs for comprehensive monitoring and investigation purposes.

## *Employee education and awareness*

Provide regular training and awareness programs to employees on access control best practices, including password hygiene, secure authentication, and the importance of protecting access credentials.

Educate employees about social engineering techniques, phishing attacks, and the risks associated with sharing access credentials or granting unauthorized access.

Encourage employees to report any suspicious activities, unauthorized access attempts, or potential security incidents through established reporting channels.

### *Access control audits and reviews*

Conduct periodic access control audits to evaluate the effectiveness of implemented measures and identify any vulnerabilities or areas for improvement.

Perform access control reviews to ensure that access rights and permissions are aligned with job roles, responsibilities, and changing business requirements.

Engage external auditors or security professionals to perform penetration testing or vulnerability assessments to identify any weaknesses in the access control system.

These detailed examples provide a starting point for implementing access control measures within an enterprise. It's important to tailor the approach to the organization's specific needs, industry requirements, and risk profile. Regular assessments, audits, and continuous improvement efforts are crucial to ensure that access control measures remain effective in mitigating risks and protecting sensitive information and resources. Access control forms the foundation of a strong cybersecurity architecture, ensuring that only authorized entities can access critical resources and data. By understanding the principles and models of access control, aligning it with business requirements, and collaborating with operational teams, cybersecurity architects can establish effective access control mechanisms that protect the organization's assets and support its objectives. In the next chapter, we will explore the role of the cybersecurity architect and the main areas of focus, including encryption in securing data at rest and in transit.

## Access control lab

Here's a step-by-step lab to help you implement access control in a virtual environment, even if you have little or no experience in cybersecurity. This lab will guide you through the process of setting up access control using a popular open source tool called **pfSense**. pfSense is a firewall and routing platform that provides advanced security features, including access control capabilities.

### *Requirements*

In this hands-on experience, we will explore the intricacies of managing user permissions, controlling resource access, and ensuring data security within a simulated enterprise environment. As we embark on this journey, you will gain practical insights into designing and implementing effective access control measures that align with industry standards and best practices.

Welcome to our pfSense implementation journey! Before we dive into the exciting world of pfSense, let's ensure we have everything we need to create a robust and secure network environment. So, let's gather our requirements and get ready to unlock the potential of pfSense, the open source powerhouse for firewall and routing solutions:

- A computer or virtual machine with at least 4 GB of RAM and 40 GB of disk space

- Virtualization software (for example, VirtualBox, VMware, QEMU/KVM, or Proxmox)

- A pfSense ISO image (available to download from the pfSense website at `https://www.pfsense.org/download/`)

## Step 1 – set up the virtual environment

In this hands-on guide, we will walk you through the process of creating a virtual machine with the optimal specifications to host pfSense, a powerful open source firewall and router platform. By following these steps, you will be well on your way to building a secure and efficient network infrastructure right on your computer:

1. Install your preferred virtualization software on your computer.
2. Create a new virtual machine with the following specifications:

   A. Assign at least 2 GB of RAM to the virtual machine

   B. Create a virtual hard disk with a minimum size of 20 GB

3. Attach the pfSense ISO image to the virtual machine's CD/DVD drive.
4. Start the virtual machine and begin installing pfSense.

## Step 2 – install pfSense

Welcome to *step 2* of our pfSense installation guide:

1. Follow the onscreen instructions to install pfSense on the virtual machine.
2. Configure the network settings during the installation process, ensuring connectivity to your network.

## Step 3 – initial configuration

Now that pfSense has been installed on your virtual machine, it's time for you to set up and fine-tune your network interfaces for optimal performance and security:

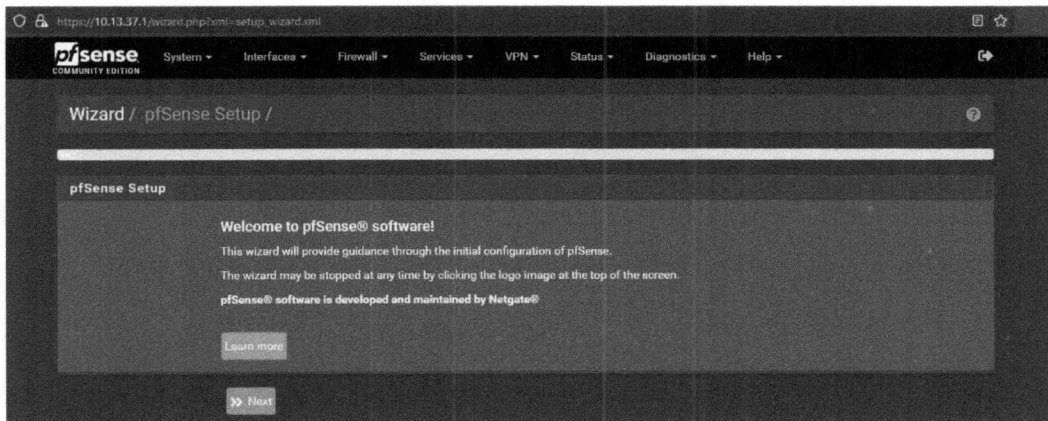

Figure 2.1 – pfSense's initial configuration wizard

Follow these instructions:

1.  After the installation is complete, pfSense will prompt you to configure the WAN and LAN interfaces. Assign appropriate IP addresses to each interface.

2.  Access the pfSense web interface by opening a web browser and entering the LAN IP address you configured in the previous step.

3.  Follow the onscreen wizard to complete the initial configuration of pfSense, including setting the admin password and optional settings.

## Step 4 – create firewall rules

Now that you have successfully configured the initial settings, it's time for you to fine-tune your network's security by creating customized firewall rules:

Figure 2.2 – pfSense – firewall LAN rule example

Follow these instructions:

1.  In the pfSense web interface, navigate to **Firewall** and select **Rules**.

2.  Click on the **LAN** tab and then **Add** to create a new rule.

3.  Define the rule parameters based on your access control requirements. For example, you can create rules to allow or block specific IP addresses, protocols, or ports.

4.  Repeat this process to create additional rules as needed for your access control policy.

## Step 5 – implement network address translation (NAT)

Now, we can delve into advanced network management and put our access control policies to the test:

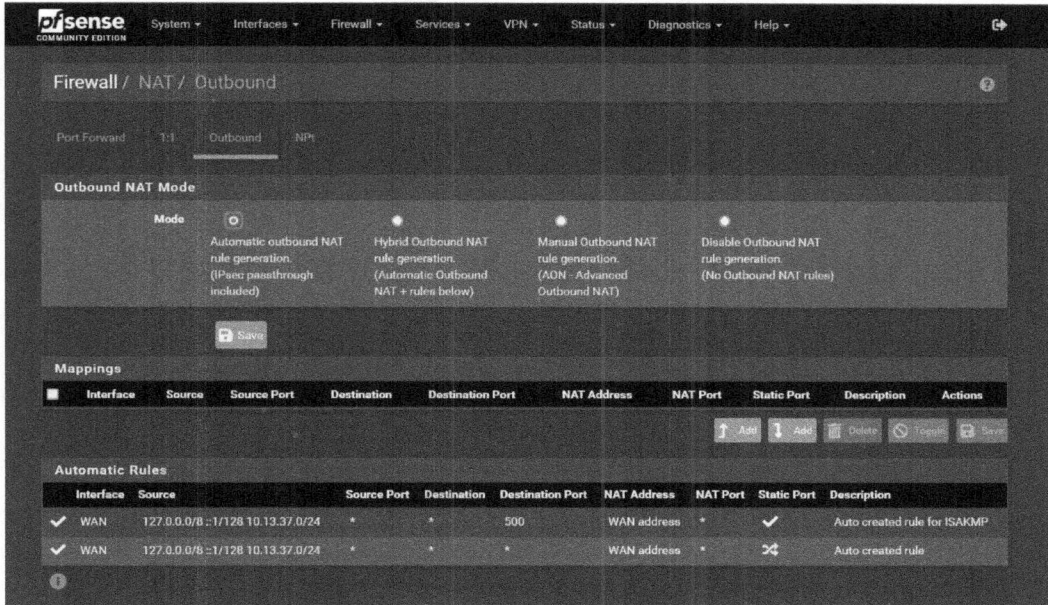

Figure 2.3 – pfSense – firewall NAT configuration example

Follow these instructions:

1. In the pfSense web interface, go to **Firewall** and select **NAT**.

2. Click on the **Outbound** tab and select **Automatic outbound NAT rule generation** or **Manual Outbound NAT rule generation**.

3. Click on **Add** to create a new NAT rule.

4. Configure the NAT rule parameters, including source and destination addresses, ports, and translation settings.

5. Save the rule and apply the changes.

## Step 6 – test access control policies

With your NAT rules in place, it's time to move on to *step 6* – testing your access control policies:

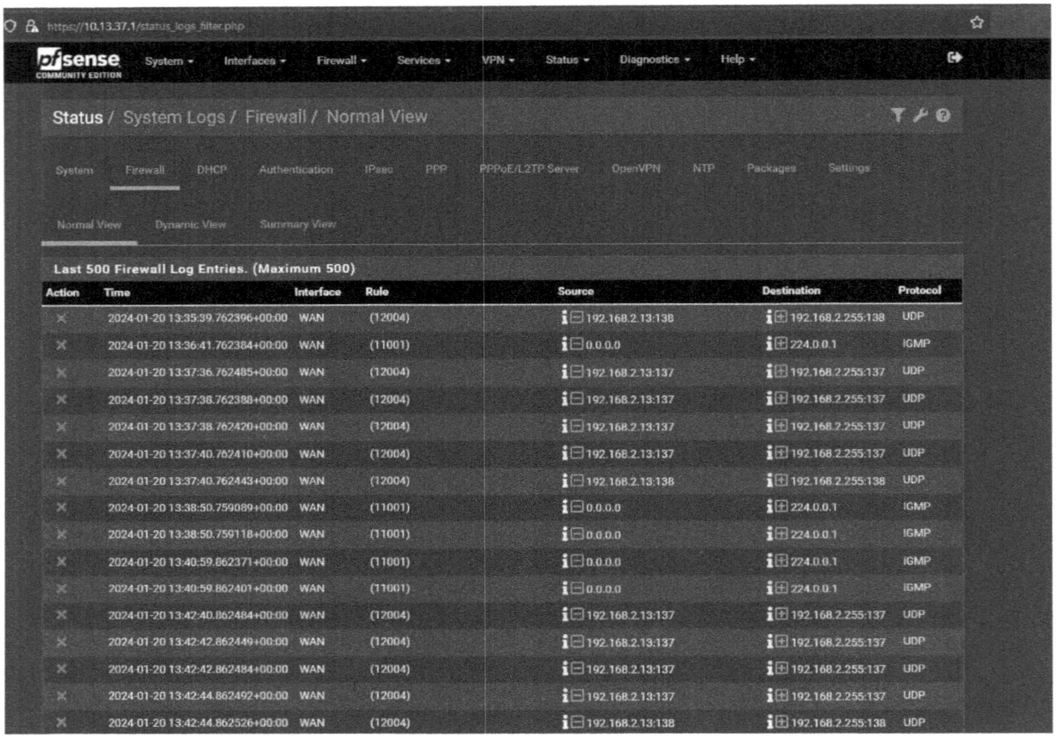

Figure 2.4 – pfSense – firewall logs example

Follow these instructions:

1. Use another computer or virtual machine on the same network to test the access control policies you've set up.

2. Attempt to access resources that should be allowed or blocked based on the rules you've defined in pfSense.

3. Verify that the access control policies are working as intended by observing the response of the network and the behavior of the firewall.

Congratulations! You have successfully implemented access control using pfSense in your virtual environment. This lab provided a basic introduction to access control principles and demonstrated how to configure firewall rules and NAT to enforce access restrictions. Remember to explore this further and familiarize yourself with additional pfSense features and settings to enhance your access control capabilities.

# Network and communication security

Network and communication security is a critical component of a robust cybersecurity architecture. It involves implementing measures to protect the confidentiality, integrity, and availability of data as it traverses networks. As a cybersecurity architect, it is essential to have a deep understanding of network and communication security and its implications for the business and operational teams. This chapter delves into the foundational aspects of network and communication security, providing detailed insights for effective implementation.

## Network security fundamentals

The objectives of network security are centered around safeguarding networks and their resources. The primary goals include the following:

- **Unauthorized access prevention**: Implementing measures to prevent unauthorized individuals or systems from gaining access to the network. This includes securing network perimeters and enforcing strong authentication mechanisms.

- **Data integrity and confidentiality**: Ensuring the integrity and confidentiality of data during transmission. Encryption techniques, secure protocols, and secure channels play a vital role in maintaining data security.

- **Availability assurance**: Protecting network resources from disruptions, ensuring network services are available when needed. This involves implementing redundancy, load balancing, and DoS prevention measures.

## Network security technologies

In this section, we'll explore the diverse and evolving landscape of technologies dedicated to safeguarding networks from cyber threats and unauthorized access. From robust firewalls and IDSs to secure VPNs and advanced encryption mechanisms, we'll dive deep into the tools and techniques used to fortify modern network infrastructures.

Whether you're an aspiring cybersecurity professional, a network administrator, or simply someone curious about the inner workings of network security, this section will provide valuable insights into the arsenal of technologies used to protect data, systems, and users from potential risks. So, let's embark on this enlightening journey through the realm of network security technologies and discover how they contribute to creating resilient and secure networks in an ever-changing digital landscape. Let's get started:

- **Firewalls:** Firewalls act as a first line of defense by monitoring and controlling incoming and outgoing network traffic based on predefined security rules. They can be implemented at both the network and host levels.

- IDSs and **intrusion prevention systems** (**IPSs**): IDS/IPS solutions analyze network traffic in real time to identify and respond to potential security incidents. They detect and prevent unauthorized access, malware, and other malicious activities.

- **VPNs**: VPNs create secure, encrypted tunnels over public networks, such as the internet, to ensure confidential and authenticated communication between remote users or between different office locations.

- **Network segmentation**: Network segmentation is a strategic approach that involves dividing the network into distinct segments or zones, each with security measures and access controls. This practice aims to minimize the impact of a security breach by isolating critical resources from the rest of the network. By doing so, network segmentation effectively restricts unauthorized lateral movement within the network, providing an additional layer of protection against potential cyber threats.

## Securing network communications

To ensure the confidentiality and integrity of data transmitted over networks, it is crucial to implement secure protocols and encryption techniques. Here are some key considerations:

- **Transport layer security** (**TLS**)/**secure sockets layer** (**SSL**): The TLS and SSL protocols play a vital role in ensuring secure communication channels across the internet. By encrypting data and verifying the authenticity of the communicating parties, these protocols establish a robust security framework for various network-based applications. From secure web browsing to encrypted email transmission, TLS and SSL are commonly employed to safeguard sensitive information and protect users from potential cyber threats.

- **VPNs**: VPN technologies, such as **internet protocol security** (**IPsec**) and SSL VPNs, establish encrypted tunnels for secure remote access or site-to-site connections. They prevent unauthorized interception or tampering of data.

- **Secure shell** (**SSH**): SSH is a cryptographic network protocol that's used for secure remote administration, file transfers, and secure access to command-line interfaces. It provides strong encryption and authentication mechanisms.

## Network access control

Network access control mechanisms ensure that only authorized entities can access the network and its resources. Here are some of the key aspects:

- **Network ACLs**: ACLs define rules that filter and control network traffic based on specific criteria, such as IP addresses, protocols, or ports. They enforce access restrictions and allow for granular control over network communication.

- **Network segmentation**: Proper network segmentation is crucial for limiting the potential impact of a security breach. By separating the network into distinct segments, each with its security controls, the spread of threats can be contained, and critical resources can be protected.

- **Network authentication and authorization**: Strong authentication mechanisms, such as MFA, should be implemented to ensure that users are who they claim to be. Additionally, proper authorization processes ensure that users have appropriate permissions and privileges based on their roles and responsibilities.

## Collaboration with operational teams

In the world of cybersecurity, effective collaboration with various operational teams is a key pillar in establishing a strong defense against cyber threats. In this section, we'll explore the significance of working closely with IT, development, DevOps, and other operational teams to integrate security seamlessly into every aspect of the organization's infrastructure and processes.

From fostering a security-conscious culture to aligning security practices with business objectives, we'll delve into the essential strategies that cybersecurity professionals can employ to bridge the gap between security and operations. We'll highlight the benefits of cross-functional collaboration, share best practices for effective communication, and showcase real-world examples of successful collaborations that have bolstered an organization's cybersecurity resilience.

### Network operations and network security

Collaboration with network operations teams is vital for implementing effective network and communication security. Here are some key considerations:

- **Network architecture**: Work closely with network engineers to design and implement a secure network architecture that aligns with security requirements and industry best practices. This includes establishing secure network boundaries, implementing appropriate security controls, and ensuring proper segmentation.

- **Patch management**: Collaborate with operational teams to ensure timely installation of security patches and updates for network devices and infrastructure. Regular patching helps protect against known vulnerabilities and strengthens overall network security.

### Incident response and network security

Collaborate with the incident response team to develop procedures for detecting and responding to network security incidents. Here are some of the key areas of collaboration:

- **Monitoring and analysis**: Establish comprehensive monitoring capabilities, including network traffic analysis, intrusion detection, and log analysis. Timely detection of network security incidents is crucial for effective response.

- **Incident containment and mitigation**: Define procedures to contain and mitigate network security incidents. This includes isolating compromised systems, blocking malicious traffic, and restoring network services.

- **Forensic analysis**: Work together to conduct a thorough forensic analysis of network security incidents. This helps identify the root causes, assess the impact, and implement preventive measures to avoid similar incidents in the future.

## Security monitoring and logging

Collaborate with security operations teams to establish robust monitoring and logging practices. This involves the following aspects:

- **Security information and event management (SIEM) and IDS**: Implement IDS solutions to detect potential threats and anomalies in network traffic. Integrate them with a SIEM system for centralized log collection, correlation, and analysis.

- **Log management**: Ensure that network devices and security systems generate detailed logs that capture relevant information for forensic analysis, incident response, and compliance purposes.

Here are some detailed examples of why you would implement network and communication security measures within the enterprise:

- **Protecting data confidentiality**: Network and communication security measures, such as encryption and secure communication protocols, help ensure the confidentiality of sensitive data transmitted over the network. By implementing these measures, you can prevent unauthorized interception or eavesdropping on sensitive communications, protecting valuable intellectual property, customer data, or trade secrets.

- **Preventing unauthorized access**: Network and communication security measures, such as firewalls, IDSs, and access control mechanisms, help prevent unauthorized individuals or malicious entities from gaining access to the organization's network or systems. By implementing strong network security measures, you reduce the risk of unauthorized access, data breaches, or malicious activities that can disrupt business operations or compromise sensitive information.

- **Mitigating insider threats**: Network and communication security measures help mitigate the risk of insider threats, such as unauthorized access or misuse of network resources by employees or contractors. By implementing access controls, monitoring systems, and user behavior analytics, you can detect and prevent malicious activities or unauthorized data exfiltration by insiders.

- **Protecting against malware and cyber attacks**: Network and communication security measures, such as antivirus software, IPSs, and email filters, help protect against malware, ransomware, phishing attacks, and other cyber threats. By implementing robust security solutions, you can detect and block malicious software, prevent unauthorized access attempts, and safeguard the integrity of the network infrastructure.

- **Ensuring data integrity**: Network and communication security measures, such as data validation, digital signatures, and integrity checks, help ensure the integrity of data transmitted over the network. By implementing these measures, you can verify the authenticity and integrity of data to prevent data tampering, unauthorized modifications, or data corruption during transmission.

- **Compliance with regulatory requirements**: Many industries have specific regulations, such as the **Payment Card Industry Data Security Standard** (**PCI DSS**) or the **General Data Protection Regulation** (**GDPR**), that require organizations to implement network and communication security measures. By adhering to these regulations, you demonstrate compliance, protect sensitive customer information, and avoid legal penalties or reputational damage.

- **Maintaining business continuity**: Network and communication security measures contribute to maintaining business continuity by preventing disruptions caused by cyber attacks, network outages, or unauthorized access attempts. By implementing redundancy, backup systems, and disaster recovery plans, you can ensure the availability of critical network resources and minimize the impact of security incidents on ongoing business operations.

- **Secure remote access**: Network and communication security measures, such as VPNs, secure RDP, or MFA, enable secure remote access to the organization's network and resources. By implementing these measures, you can protect data transmitted between remote locations, ensure secure communication channels, and prevent unauthorized access from external networks.

- **Secure collaboration and communication**: Network and communication security measures enable secure collaboration and communication among employees, partners, and stakeholders. By implementing encrypted email communication, secure messaging platforms, or VPNs, you can protect sensitive information shared within the organization and during external collaborations.

- **Network performance optimization**: Network and communication security measures, such as traffic shaping, **quality-of-service** (**QoS**) mechanisms, or bandwidth management, help optimize network performance and ensure efficient use of network resources. By implementing these measures, you can prioritize critical applications or services, prevent bandwidth abuse, and maintain optimal network performance.

These examples highlight the importance of implementing network and communication security measures within the enterprise to protect data confidentiality, prevent unauthorized access, mitigate cyber threats, comply with regulations, maintain business continuity, and optimize network performance.

Network security measures are essential for establishing a secure and reliable network infrastructure that supports business operations and protects valuable information assets:

- **Firewalls**: Deploy firewalls at network entry and exit points to monitor and control incoming and outgoing network traffic. Configure the firewalls to enforce access policies, filter out malicious traffic, and prevent unauthorized access attempts.

- **IDS/IPS**: Deploying IDS/IPS solutions is a proactive approach to detect and thwart network-based attacks and intrusions effectively. These advanced systems continuously monitor network traffic, analyzing patterns in real time. Upon detecting suspicious activities, they promptly raise alerts or take immediate proactive measures to prevent potential threats from causing harm. By implementing IDS/IPS solutions, organizations can bolster their cybersecurity defenses and ensure a vigilant and responsive network security posture.

- **Secure network architecture**: Design and implement a secure network architecture that segregates different network segments based on security requirements. Use techniques such as network segmentation, **virtual LANs (VLANs)**, and **demilitarized zones (DMZs)** to isolate critical systems and restrict unauthorized access.

- **Encryption**: Use encryption protocols such as SSL/TLS to secure data in transit over the network. Implement secure communication channels, such as VPNs, to ensure secure remote access and protect data transmitted between locations.

- **Access control**: Implement access control mechanisms, such as network authentication protocols (for example, IEEE 802.1X), to verify the identity of devices and users before granting network access. Enforce strong password policies, MFA, or certificate-based authentication for enhanced security.

- **Network monitoring and logging**: Deploy network monitoring tools to capture and analyze network traffic. Monitor for anomalies, suspicious activities, or security incidents. Implement logging mechanisms to record network events, which can aid in incident investigation and forensic analysis.

- **Vulnerability management**: Regularly scan the network infrastructure for vulnerabilities using vulnerability assessment tools. Patch and update network devices, servers, and applications to address known vulnerabilities and reduce the risk of exploitation.

- **Wireless network security**: Secure wireless networks by implementing strong encryption (WPA2 or WPA3), disabling unnecessary services, and using separate guest networks to isolate guest traffic from internal networks. Implement IDSs for wireless networks to detect unauthorized access attempts or rogue access points.

- **Network segmentation**: Implement network segmentation to divide the network into smaller, isolated segments. This reduces the potential impact of a security breach by limiting lateral movement and containing the spread of threats within the network.

- **Employee awareness and training**: Provide ongoing training and awareness programs to educate employees about network security best practices, such as identifying phishing emails, avoiding suspicious websites, and maintaining strong passwords. Promote a culture of security and encourage employees to report any suspicious activities or security incidents.

- **Patch and update management**: Maintain a rigorous patch and update management process for network devices, operating systems, and applications. Regularly apply security patches to address known vulnerabilities and protect against emerging threats.

- **Data loss prevention (DLP)**: Implement DLP solutions to monitor and control the movement of sensitive data within the network. Use techniques such as content filtering, data classification, and data encryption to prevent unauthorized data exfiltration or leakage.

- **Incident response plan**: Develop an incident response plan that outlines the steps to be taken in the event of a security incident or breach. Define roles, responsibilities, and communication channels to ensure a timely and coordinated response.

- **Regular security audits and penetration testing**: Regularly conducting security audits and penetration testing is essential to assess the effectiveness of network security controls. These proactive measures involve identifying potential vulnerabilities, weaknesses, or misconfigurations within the network infrastructure. By doing so, organizations can gain valuable insights into their security posture and take corrective actions promptly. Strengthening network security through these assessments ensures a robust defense against potential cyber threats and provides a proactive approach to maintaining a secure digital environment.

- **Vendor security**: Implement vendor security assessments and due diligence processes to ensure that third-party vendors and service providers adhere to robust network and communication security practices. Review contracts and agreements to ensure security requirements are met.

These detailed examples demonstrate various network and communication security measures that can be implemented within the enterprise to protect against cyber threats, secure data transmission, control access, and maintain the integrity and availability of the network infrastructure. It is crucial to continuously assess, update, and improve network security measures to adapt to evolving threats and protect the organization's sensitive information and assets.

In maintaining network infrastructure, various components, such as servers, workstations, cables, hubs, switches, routers, and firewalls, play vital roles. However, the true value of a network lies not in its equipment but in its data. Data holds paramount importance, often exceeding the cost of replacing network equipment. Therefore, the primary objective of network security is to protect this invaluable data, ensuring its confidentiality, integrity, and availability.

Network security focuses on safeguarding data in various states: storage, transmission, and processing. Vulnerabilities can manifest differently in each state, and any compromise to data's characteristics can pose a threat to the entire network. Threats refer to possible dangers that exploit vulnerabilities.

One approach that's employed by security professionals to enhance network security is penetration testing. Unlike malicious attackers, penetration testers follow a methodology that identifies vulnerabilities without utilizing malicious payloads or unauthorized access. This helps in identifying weaknesses in the system and allows remediation actions. Skilled penetration testers think like attackers and stay updated on new attack techniques, enhancing preparedness for actual attacks.

This section will introduce the tools and techniques that are used in penetration testing and explore various types of malicious code that attackers may use to compromise data's confidentiality, integrity, and availability on a network. By understanding these techniques, organizations can strengthen their network security and protect their valuable data effectively.

## Network security lab

Here's a step-by-step lab to help you implement network security in a virtual environment, even if you have little or no experience in cybersecurity. This lab will guide you through the process of setting up network security using a virtualization platform and basic security measures.

### Requirements

Before diving into our exciting journey, let's ensure we have everything we need to embark on this virtual adventure:

- A computer or virtual machine with at least 4 GB of RAM and 40 GB of disk space
- Virtualization software (e.g., VirtualBox, VMware, QEMU/KVM, Proxmox)

### Step 1 – set up the virtual environment

In this section, we'll walk you through the process of creating a virtual machine that will serve as the gateway to a world of virtual possibilities:

1. Install your preferred virtualization software on your computer.
2. Create a new virtual machine with the following specifications:

   A. Assign at least 2 GB of RAM to the virtual machine

   B. Create a virtual hard disk with a minimum size of 20 GB

### Step 2 – install the operating system

Here, we'll guide you through the process of selecting and installing an operating system on your virtual machine, setting the stage for endless possibilities in the virtual realm:

1. Download and install an operating system of your choice, such as Ubuntu or CentOS, on the virtual machine.
2. Follow the onscreen instructions to complete the installation.

### Step 3 – update the operating system

After successfully installing your preferred operating system, it's essential to keep it up to date to ensure optimal performance, security, and access to the latest features:

1. Once the installation is complete, open a terminal or command prompt on the virtual machine.
2. Update the operating system by running the appropriate update command for your chosen operating system (for example, `sudo apt update` for Ubuntu).
3. Install any available updates by running the appropriate command (for example, `sudo apt upgrade`).

The following screenshot shows the Ubuntu 20.04 APT update process:

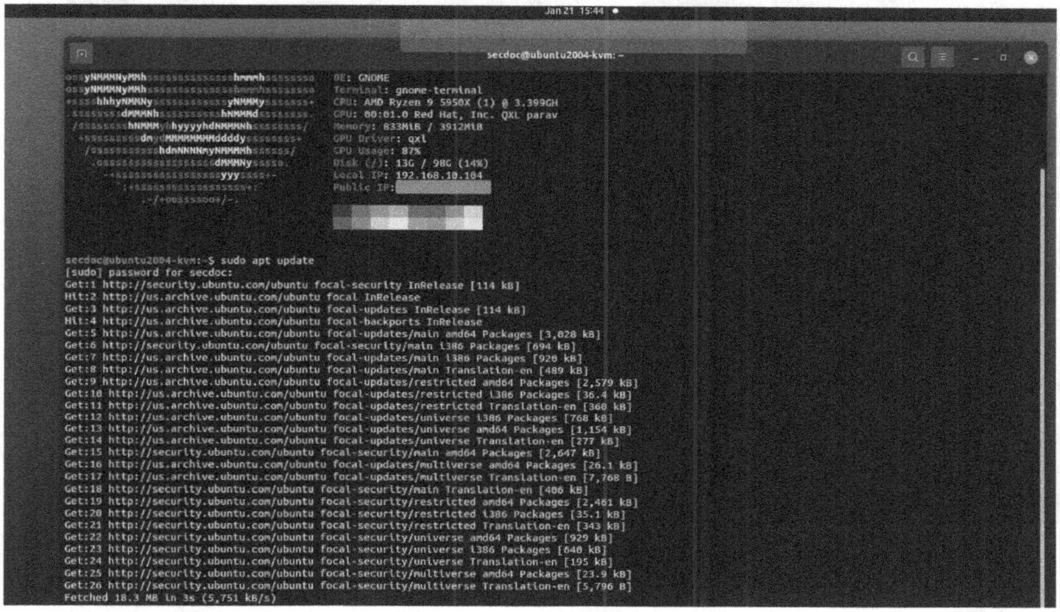

Figure 2.5 – Ubuntu 20.04 APT update

> **Note**
>
> There are several great books you can read to dive deeper into the configuration and hardening of Ubuntu and Linux in general from Packt. These include *Mastering Ubuntu Server*, by Ja LaCroix, *Mastering Linux Security Hardening*, by Donald A. Tevault, and the recently released *The Software Developer's Guide to Linux*, by David Cohen and Christian Sturm.

## Step 4 – enable the firewall

Securing your virtual environment is of utmost importance, and enabling the built-in firewall is a crucial step to enhance its defenses. So, let's dive into the process and fortify your virtual machine against cyber risks:

1.  Still in the terminal or command prompt, enable the built-in firewall for the operating system.

2.  For Ubuntu, run the `sudo ufw enable` command to enable the Uncomplicated Firewall:

Figure 2.6 – Enabling Ubuntu 20.04 UFW

3.  For CentOS, run `sudo systemctl enable firewalld` to enable the `firewalld` service.

4.  Configure the firewall rules to allow only necessary incoming and outgoing connections, such as SSH (port 22) for remote access:

    -  `sudo ufw allow 22/tcp`

    -  `sudo ufw show added`

    -  `sudo ufw reload`

    -  `sudo ufw status`

The following screenshot shows an ACL that's been established to allow SSH TCP port 22 through the firewall and show the status of the ACLs with UFW:

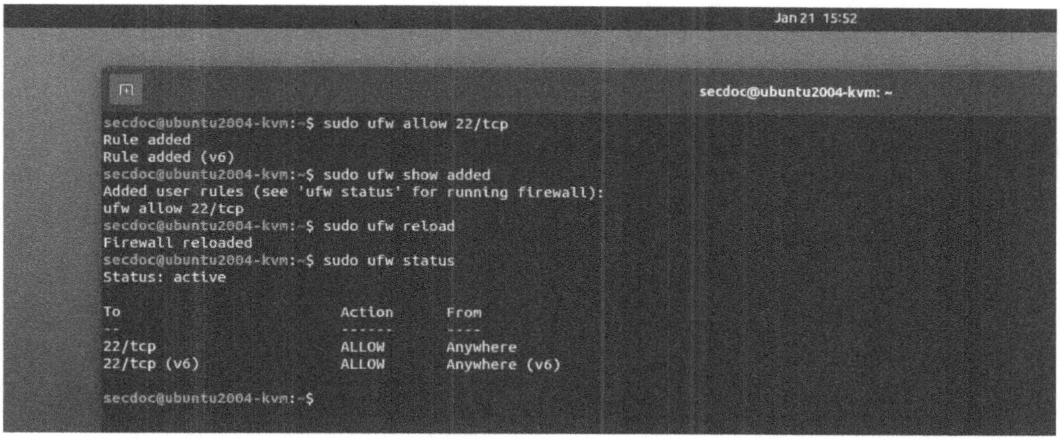

Figure 2.7 – Ubuntu 20.04 UFW ACL

The following screenshot shows a Windows 10 system that can connect to the Ubuntu 20.04 system. You can tell that there is connectivity and that UFW is no longer blocking this connectivity because the system can now obtain a fingerprint and key from the Ubuntu system:

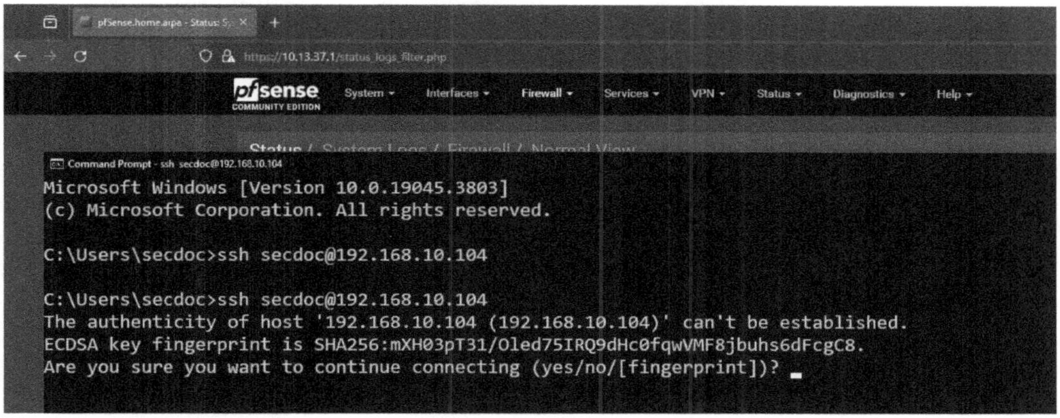

Figure 2.8 – Windows – successful SSH connectivity

## *Step 5 – install and configure antivirus software*

As we continue to prioritize the security of your virtual environment, implementing robust antivirus protection is a key step in safeguarding your system from potential malware threats. So, let's proceed with this crucial step to protect your virtual environment from malware risks:

1.  Install an antivirus software program suitable for your operating system, such as ClamAV for Ubuntu or ClamTk for CentOS.

2.  Follow the installation instructions for the chosen antivirus software.

3.  Configure the antivirus software so that it performs regular scans and updates virus definitions automatically.

## *Step 6 – enable automatic updates*

Keeping your operating system up to date is a crucial aspect of maintaining a secure and stable virtual environment. In this step, we'll guide you through the process of configuring your operating system so that you receive automatic updates effortlessly:

1.  Ensure that the operating system is set to receive automatic updates.

2.  For CentOS, open a terminal, run `sudo yum install yum-cron` to install the `yum-cron` package, and follow the onscreen instructions to configure automatic updates.

3.  For Ubuntu, open the **Software & Updates** application, navigate to the **Updates** tab, and select **Install security updates automatically**.

    To install and configure ClamAV, an open source antivirus software for scanning files and emails for malware, on Ubuntu 20.04, follow these steps:

    A.  Update your system. First, make sure your system is up to date by running the following command:

        ```
        sudo apt update
        sudo apt upgrade
        ```

    B.  Install ClamAV. You can install ClamAV and its command-line tools using the following command:

        ```
        sudo apt install clamav
        sudo apt install clamav-daemon
        ```

The following screenshot shows how to install ClamAV on Ubuntu 20.04:

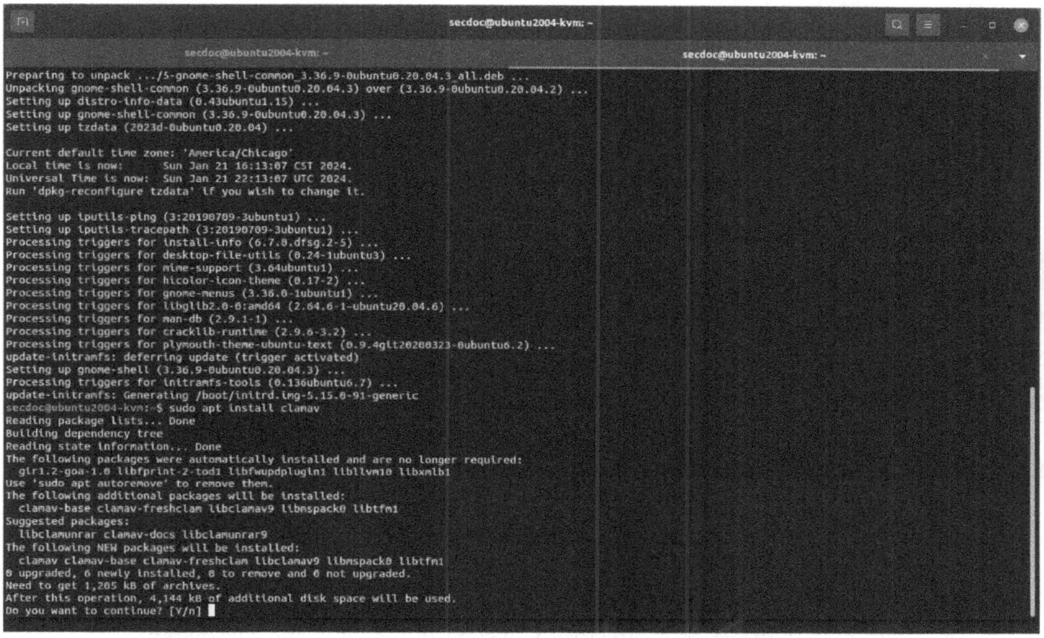

Figure 2.9 – Installing ClamAV on Ubuntu 20.04

C.   Configure ClamAV using the following command:

```
sudo dpkg-reconfigure clamav-daemon
```

The following screenshot shows the configuration window after running the dpkg-reconfigure clamav-daemon command:

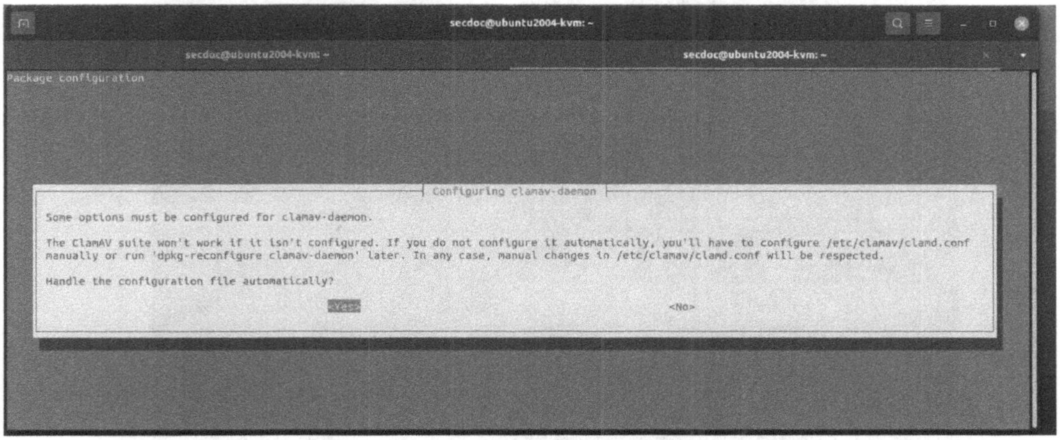

Figure 2.10 – Running the dpkg-reconfigure clamav-daemon command

D.  Update the ClamAV signatures. After its installation, it's essential to update the ClamAV virus signature database regularly. Installing the ClamAV daemon will automatically update the signatures and database. This can be checked and validated by running the following command:

```
systemctl status clamav-freshclam.service
```

The following screenshot shows that the service is running and available for ClamAV:

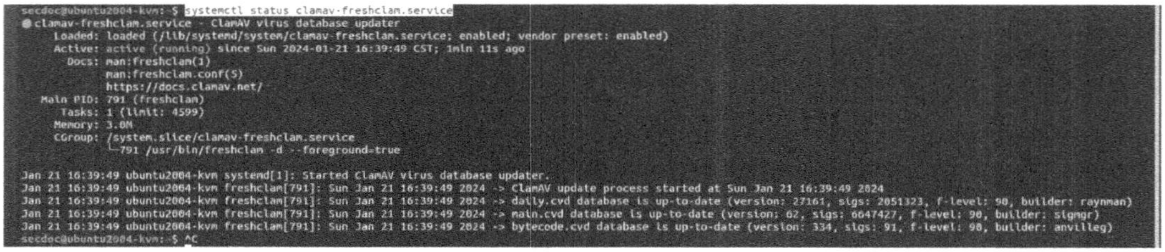

Figure 2.11 – Active ClamAV service

E.  Test ClamAV. You can test ClamAV by running a scan on a specific file or directory. For example, to scan the /home directory, you can run the following command:

```
sudo clamscan -r /home
```

This command will scan the /home directory and report any findings.

That's it! ClamAV is now installed and configured on your Ubuntu 20.04 system:

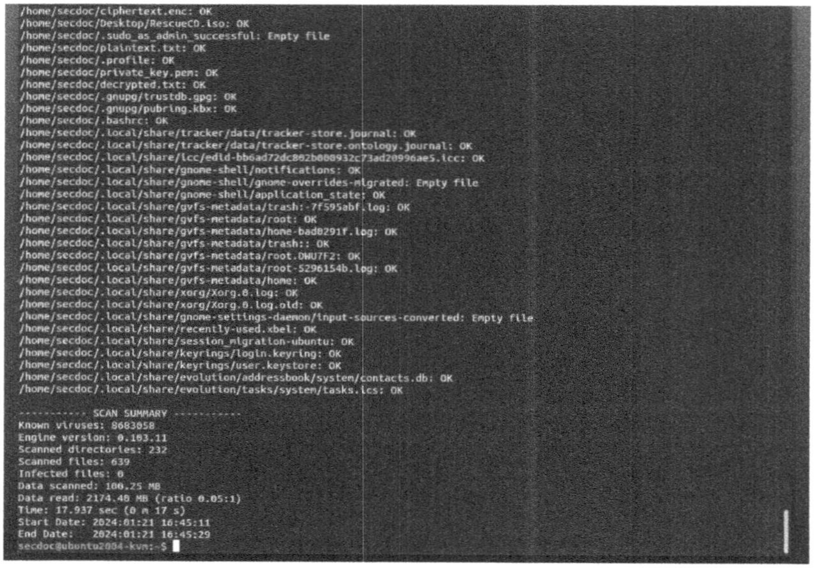

Figure 2.12 – Successful ClamAV scan

You can use it to scan files, directories, or emails for malware.

## Step 7 – implement network segmentation

In this critical step, we'll delve into the powerful concept of network segmentation, which plays a vital role in enhancing the security and efficiency of your virtual environment. So, let's proceed with this vital step and create a robust network architecture that aligns perfectly with your organization's security requirements:

1.  Configure your virtualization platform to create multiple virtual networks or subnets. By dividing your virtual environment into distinct segments, you'll be able to tailor the network settings to the specific needs and security requirements of each virtual machine.

2.  Assign virtual machines to different networks based on their intended purposes or security requirements. This strategic placement ensures that sensitive data or critical systems are isolated from the rest of the environment, reducing the potential impact of a security breach.

3.  Implement routing or firewall rules to control communication between different networks. By carefully managing network traffic, you'll have greater control over data flow and be able to bolster the overall security of your virtual environment.

## Step 8 – secure remote access

As remote access becomes an essential aspect of modern virtual environments, ensuring its proper configuration and security is of utmost importance. So, let's proceed with this crucial step to ensure that your remote access is both convenient and secure:

1.  If you require remote access to the virtual machines, ensure that remote access protocols (for example, SSH) are properly configured and secured.

2.  Disable remote access for unnecessary services or protocols to reduce the attack surface. You'll minimize potential vulnerabilities and mitigate the risk of unauthorized access to your virtual environment.

## Step 9 – regularly back up your virtual machines

As we continue to prioritize the security and reliability of your virtual environment, implementing regular backups is an essential practice to safeguard against potential data loss and system failures. In this step, we'll guide you through the process of setting up regular backups of your virtual machines:

1.  Set up regular backups of your virtual machines to protect against data loss or system failures.

2.  Utilize backup software or built-in virtualization features to create periodic backups of your virtual machines.

Congratulations! You have successfully implemented basic network security measures in your virtual environment. This lab provided a starting point for securing your network by enabling firewalls, installing antivirus software, implementing network segmentation, and securing remote access. Remember to continue exploring and learning more about network security to enhance the protection of your virtual environment.

# Cryptography

In the realm of cybersecurity architecture, cryptography plays a pivotal role in upholding the utmost confidentiality, integrity, and authenticity of sensitive information. As a cybersecurity architect, understanding and effectively utilizing cryptographic techniques is essential for protecting sensitive data and maintaining secure communication channels. This chapter delves into the foundational aspects of cryptography, exploring its significance in the context of the business and operational teams.

## Cryptography fundamentals

In the ever-evolving world of cybersecurity, cryptography stands as a formidable shield, safeguarding sensitive information from prying eyes and malicious threats. In this section, we'll delve deep into the core principles of cryptography, exploring its vital role in ensuring confidentiality, integrity, and authenticity of data.

### Key concepts

Cryptography encompasses various key concepts that form the basis of secure communication:

- **Encryption**: The process of converting plaintext into ciphertext using cryptographic algorithms and keys to ensure confidentiality.

- **Decryption**: The reverse process of encryption, this involves converting ciphertext back into plaintext using the corresponding cryptographic algorithms and keys.

- **Symmetric cryptography**: This involves using a single key for both encryption and decryption. It is efficient but requires a secure method to exchange the key.

- **Asymmetric cryptography**: This involves utilizing a pair of keys – a public key for encryption and a private key – for decryption. It enables secure communication without the need for key exchange.

- **Hashing**: A one-way process that generates a fixed-length cryptographic hash value from input data. It ensures data integrity and can verify data authenticity.

## *Cryptographic algorithms*

Cryptographic algorithms provide the mathematical foundations for securing data. Different types of algorithms serve different purposes:

- **Symmetric key algorithms**: Examples include **advanced encryption standard** (**AES**) and **data encryption standard** (**DES**). They use the same key for both encryption and decryption and are well-suited for fast, efficient encryption of large amounts of data.

- **Asymmetric key algorithms**: Examples include **Rivest-Shamir-Adleman** (**RSA**), Diffie-Hellman, and **elliptic curve cryptography** (**ECC**). These algorithms employ different keys for encryption and decryption, offering secure key exchange and digital signatures.

- **Hashing algorithms**: Common hashing algorithms include MD5, SHA-1, and SHA-256. They generate fixed-length hash values from input data, ensuring data integrity and providing a unique identifier for the input.

# Cryptography in practice

In the realm of cryptography, the security of encrypted data hinges upon effective key management. In this section, we'll discuss critical aspects of key management, exploring key considerations that are essential for safeguarding sensitive information.

By understanding and implementing secure key management practices, you'll be able to enhance the overall security of your encrypted data and fortify your cryptographic defenses. So, let's delve into the world of secure key management and unlock the secrets to safeguarding sensitive information in the digital age.

## *Secure key management*

Effective key management is essential for maintaining the security of encrypted data. Here are some key considerations:

- **Key generation**: Implement secure methods for generating strong cryptographic keys. Randomness and sufficient key length are crucial for resistance against brute-force attacks.

- **Key distribution**: Establish secure mechanisms for distributing encryption keys, especially in symmetric cryptography. This may involve using key exchange protocols or key distribution centers.

- **Key storage**: Safeguard cryptographic keys by employing secure key storage mechanisms, such as **hardware security modules** (**HSMs**) or secure key vaults. Protection against unauthorized access is critical to maintaining the integrity of encrypted data.

## *Secure communication channels*

In today's interconnected world, ensuring the confidentiality, integrity, and authenticity of data exchanged between systems is paramount. In this section, we'll explore two powerful tools – SSL/TLS protocols and VPNs – that play a crucial role in establishing secure communication channels.

By understanding and implementing SSL/TLS protocols and VPNs, you'll be equipped with the tools to establish secure communication channels, safeguard sensitive data, and enhance your organization's cybersecurity posture. So, let's dive into the world of secure communication channels and harness the power of encryption to protect data in transit:

- **SSL/TLS**: SSL/TLS protocols establish secure communication channels over the internet. They ensure the confidentiality, integrity, and authenticity of data exchanged between systems.

- **VPNs**: VPNs create encrypted tunnels to secure communication between remote users or different office locations. They protect data transmitted over public networks.

## *Digital signatures and certificates*

In the rapidly evolving digital landscape, the importance of secure and trustworthy communication cannot be overstated. Enter digital signatures and certificates – the dynamic duo at the forefront of ensuring data integrity and authenticity in the virtual realm:

- **Digital signatures**: Digital signatures provide non-repudiation and data integrity. They are created using the private key of an asymmetric key pair and can be verified using the corresponding public key.

- **Certificates**: Certificates bind a public key to an entity's identity and are issued by trusted third-party entities called **certificate authorities** (**CAs**). They ensure the authenticity and integrity of the public key.

# Collaboration with business and operational teams

In an increasingly regulated and interconnected business landscape, ensuring compliance with industry standards and safeguarding sensitive information has never been more critical. From stringent data protection laws to encryption standards and key management practices, meeting regulatory requirements demands a comprehensive understanding of cryptographic obligations.

To navigate these complexities and fortify communication and collaboration, organizations must adopt a multi-layered approach to secure their digital interactions. Implementing robust email encryption technologies such as **pretty good privacy** (**PGP**) and **secure/multipurpose internet mail extensions** (**S/MIME**) guarantees the confidentiality of sensitive communication, while **secure file transfer protocols** (**SFTPs**) such as SFTP provide a shield against unauthorized access during file transmission.

As legal and compliance teams collaborate to navigate the intricate world of regulatory demands, establishing a secure foundation for communication becomes the key to unlocking success in a compliance-driven world.

## Regulatory and compliance requirements

Collaborate with legal and compliance teams to understand cryptographic requirements mandated by industry regulations. This includes compliance with data protection laws, encryption standards, and key management practices.

## Secure communication and collaboration

By implementing cutting-edge email encryption technologies such as PGP and S/MIME, organizations can shield their sensitive email communication from unauthorized access, preserving the trust and privacy of their digital correspondence.

Furthermore, collaboration with operational teams to deploy SFTPs such as SFTP bolsters the fortress of data protection. This ensures that files traverse the digital realm with an impenetrable layer of security, preventing any unwarranted access or interception.

As we navigate the complexities of a digitally interconnected world, embracing secure communication and collaboration is the cornerstone of establishing a resilient and trustworthy environment for businesses to thrive:

- **Secure email communication**: Implement email encryption technologies, such as PGP or S/MIME, to ensure the confidentiality of sensitive email communication

- **Secure file transfer**: Collaborate with operational teams to implement SFTPs, ensuring the secure transmission of files and preventing unauthorized access

## Cryptography in software and application security

In the digital realm, where software and applications serve as gateways to vast troves of sensitive information, the need for impenetrable security measures has never been greater. With the implementation of secure coding practices and robust cryptographic libraries, development teams can collaborate to weave an unbreachable layer of defense into the very fabric of applications. Operational teams play their part by implementing secure authentication mechanisms, fortified with strong password hashing algorithms and MFA, ensuring that only authorized entities gain access to sensitive resources.

Data confidentiality takes center stage as cryptography steps in to encrypt sensitive information, rendering it unreadable to unauthorized eyes. This proves crucial in protecting customer data, financial records, intellectual property, and trade secrets, thereby safeguarding against data breaches and preserving trust.

Moreover, cryptography's significance extends to compliance with industry-specific regulations such as GDPR or HIPAA, where the use of cryptography becomes mandatory to shield sensitive data and avert legal repercussions.

Data integrity finds its guardian in cryptographic techniques such as digital signatures and hash functions, ensuring the origin and integrity of digital assets and detecting any unauthorized modifications.

From securing remote access and data storage to authentication, identity verification, and protecting financial transactions and e-commerce activities, cryptography is the bedrock of trust and security in the digital landscape.

Being able to collaborate and share sensitive information, both within and beyond the organization, is empowered through encryption, guaranteeing the confidentiality and integrity of shared data.

In an age where data breaches lurk around every corner, cryptography stands firm as the last line of defense. By implementing encryption, the impact of data breaches is mitigated, making stolen information indecipherable and rendering the efforts of attackers futile:

- **Secure coding practices**: Collaborate with development teams to incorporate secure coding practices that utilize strong cryptographic libraries and follow industry best practices.

- **Secure authentication and authorization**: Work with operational teams to implement secure authentication mechanisms, such as strong password hashing algorithms, MFA, and secure session management.

- **Data confidentiality**: Cryptography ensures the confidentiality of sensitive data by encrypting it, making it unreadable to unauthorized individuals. This is crucial for protecting sensitive information such as customer data, financial records, intellectual property, or trade secrets from unauthorized access or data breaches.

- **Secure communication**: Cryptographic protocols, such as SSL/TLS, are used to establish secure communication channels over networks. By encrypting data during transmission, cryptography prevents eavesdropping, tampering, and unauthorized interception of sensitive information exchanged between clients and servers or between remote locations.

- **Compliance with regulations**: Many industries have specific regulations, such as GDPR or HIPAA, which require the use of cryptography to protect sensitive data. Implementing cryptography helps ensure compliance with these regulations, avoiding legal penalties and protecting the organization's reputation.

- **Data integrity**: Cryptographic techniques, such as digital signatures and hash functions, verify the integrity of data. Digital signatures provide a means to authenticate the origin and integrity of digital documents, while hash functions ensure data integrity by generating unique hash values that detect any modifications or tampering with the data.

- **Secure remote access**: Cryptography plays a critical role in securing remote access to the enterprise network. By using VPNs or encrypted RDPs, cryptography enables secure communication and data transfer between remote locations, protecting sensitive information from unauthorized access or interception.

- **Secure storage of data**: Cryptography is used to secure data at rest, such as in databases, filesystems, or backups. By encrypting stored data, cryptography protects sensitive information even if physical storage devices are lost, stolen, or compromised.

- **Authentication and identity verification**: Cryptographic mechanisms, such as digital certificates and **public key infrastructure** (**PKI**), are used for authentication and identity verification. These technologies ensure that communication endpoints, such as clients or servers, can be trusted, and that only authorized entities can access protected resources.

- **Secure transactions**: Cryptography is vital for securing financial transactions and e-commerce activities. Secure protocols such as TLS and SSL ensure the confidentiality, integrity, and authenticity of online transactions, protecting sensitive payment information and preventing unauthorized access.

- **Protection against insider threats**: Cryptography helps protect against insider threats by limiting unauthorized access to sensitive information. By encrypting data, cryptography prevents unauthorized disclosure or misuse of data by employees or contractors who might have legitimate access to the information.

- **Secure collaboration and sharing**: Cryptography enables secure collaboration and sharing of sensitive information within and outside the organization. By encrypting shared documents or using secure file-sharing protocols, cryptography ensures the confidentiality and integrity of shared data, protecting it from unauthorized access or tampering.

- **Protection against data breaches**: Implementing encryption as a data protection measure helps mitigate the impact of data breaches. Even if an attacker gains unauthorized access to encrypted data, encryption makes it extremely difficult to decipher and use the stolen information.

These examples highlight the importance of implementing cryptography within the enterprise to protect data confidentiality, secure communication channels, comply with regulations, maintain data integrity, facilitate secure remote access, and mitigate the impact of data breaches. Cryptography is a crucial component of a comprehensive security strategy, providing robust protection for sensitive information and ensuring the trustworthiness of digital communications and transactions.

Safeguarding sensitive information has become paramount. As cyber threats continue to evolve, data security must remain at the forefront of every organization's priorities. Enter a comprehensive array of encryption and cryptographic techniques, designed to fortify data protection at every level of communication and storage.

From safeguarding data at rest with robust encryption algorithms such as AES and RSA to securing data in transit through cryptographic protocols such as SSL/TLS and IPsec, this multi-layered approach ensures the confidentiality and integrity of information, even if physical storage devices are compromised or data transmissions are intercepted.

Moreover, secure email communication is achieved through techniques such as PGP and S/MIME, allowing only intended recipients to access sensitive email content. File and folder-level encryption further restricts unauthorized access to specific data, providing an additional layer of protection against potential threats.

But it doesn't stop there. Digital signatures, PKI, and secure password storage establish trust and authentication, ensuring the origin and integrity of digital assets, as well as secure user access to systems and resources.

In a world increasingly reliant on mobile, cloud, and IoT technologies, the importance of secure mobile, cloud, and IoT communication cannot be understated. Encryption, data masking, and tokenization techniques extend their protective reach, even in the most complex technological landscapes.

As we venture into an era where data is the lifeblood of every operation, implementing these cutting-edge encryption and cryptographic solutions becomes the cornerstone of maintaining a robust defense against the ever-growing tide of cyber threats:

- **Encryption of data at rest**: Implement encryption algorithms to protect sensitive data stored in databases, filesystems, or backup media. Use strong encryption algorithms such as AES or RSA to ensure the confidentiality and integrity of data even if physical storage devices are compromised.

- **Encryption of data in transit**: Secure data transmission by using cryptographic protocols such as SSL/TLS for secure web communication or IPsec for securing network communication between remote locations. Implement end-to-end encryption for sensitive communication channels, ensuring that data is protected from unauthorized interception or tampering.

- **Secure email communication**: Use email encryption techniques such as PGP or S/MIME to encrypt sensitive email messages. This ensures that only intended recipients can decrypt and access the content of the emails.

- **Secure file and folder encryption**: Implement file and folder-level encryption to protect specific files or directories containing sensitive information. This ensures that even if an unauthorized user gains access to the storage medium, they cannot access the encrypted files without the appropriate decryption key.

- **Digital signatures**: Utilize digital signatures to ensure the authenticity and integrity of digital documents, contracts, or transactions. Digital signatures use asymmetric cryptographic algorithms to verify the origin and integrity of digital assets, assuring that the data has not been tampered with.

- **PKI**: Establish a PKI to manage digital certificates and facilitate secure communication. Use digital certificates to authenticate entities, such as clients or servers, and establish secure connections for activities such as SSL/TLS handshakes or secure VPN connections.

- **Secure password storage**: Apply cryptographic techniques such as hashing and salting to securely store user passwords. Use strong hash functions such as **bcrypt** or **SHA-256** to generate password hashes that are resistant to offline attacks, ensuring that even if the password database is compromised, the actual passwords remain secure.

- **Secure authentication protocols**: Utilize cryptographic authentication protocols such as Kerberos or **secure remote password** (**SRP**) for secure authentication processes. These protocols ensure that authentication credentials are securely transmitted and verified, preventing unauthorized access to systems or resources.

- **Secure key management**: Establish robust key management practices to securely generate, store, distribute, and rotate cryptographic keys. This includes using HSMs or key management systems to protect sensitive cryptographic keys and ensure their proper management and secure use.

- **Data masking and tokenization**: Implement data masking and tokenization techniques to protect sensitive data during testing or development processes. These techniques replace sensitive data with non-sensitive values or tokens, ensuring that the actual data is not exposed and minimizing the risk of data leakage.

- **Secure voice and video communication**: Implement encryption for voice and video communication channels, such as **Voice over IP** (**VoIP**) or video conferencing systems, to protect the confidentiality and integrity of sensitive conversations or meetings.

- **Secure mobile communication**: Utilize encryption techniques for securing mobile communication channels, such as **mobile device management** (**MDM**) solutions or secure messaging apps. Implement secure protocols and encryption algorithms to protect data transmitted between mobile devices and enterprise systems.

- **Secure cloud communication**: Encrypt data before storing it in the cloud to ensure its confidentiality and integrity. Utilize encryption options provided by cloud service providers or implement client-side encryption to have full control over the encryption process.

- **Secure application programming interfaces** (**APIs**): Implement cryptographic protocols such as OAuth or **JSON Web Tokens** (**JWTs**) for secure authentication and authorization between applications and APIs. Use encryption to protect the confidentiality and integrity of data transmitted through API calls.

- **Secure IoT communication**: Implement encryption and secure protocols to protect communication in **Internet of Things** (**IoT**) deployments. Use protocols such as **Message Queuing Telemetry Transport** (**MQTT**) with **transport layer security** (**MQTT-TLS**) or **datagram transport layer security** (**DTLS**) to secure IoT device communication.

These detailed examples demonstrate various ways to implement cryptography within the enterprise to protect data confidentiality, secure communication channels, ensure data integrity, and authenticate entities. Implementing cryptographic techniques requires careful planning, proper key management, and adherence to best practices to ensure the effectiveness and security of the implemented solutions.

Cryptography forms a vital component of a robust cybersecurity architecture, providing the necessary security mechanisms for the confidentiality, integrity, and authenticity of information. By understanding the fundamentals of cryptography, employing secure key management practices, implementing encryption protocols, and collaborating with business and operational teams, cybersecurity architects can ensure the protection of sensitive data and enable secure communication channels. In the next chapter, we will explore the importance of secure network infrastructure and the role of the cybersecurity architect in establishing secure networks.

## Cryptography lab

Here's a step-by-step lab to help you implement cryptography in a virtual environment, even if you have little or no experience in cybersecurity. This lab will guide you through the process of setting up basic encryption and decryption using an open source tool called OpenSSL.

### Requirements

Are you ready to embark on a hands-on journey that will demystify the realm of encryption and decryption? Even if you're new to cybersecurity, fear not – this step-by-step lab has been tailor-made to be your trusted companion.

With these simple requirements, you're all set to embark on a transformative journey in the world of cybersecurity:

- A computer or virtual machine with at least 4 GB of RAM and 40 GB of disk space
- Virtualization software (for example, VirtualBox, VMware, QEMU/KVM, or Proxmox)

### Step 1 – set up the virtual environment

The first step is to install your preferred virtualization software on your computer, opening the gateway to a virtual world of endless possibilities. Once your virtualization software is up and running, we'll embark on creating your virtual machine – a customizable haven where you'll conduct your experiments and explorations.

So, let's dive in and set the stage for your immersive journey in the virtual realm:

1.  Install your preferred virtualization software on your computer.

2.  Create a new virtual machine with the following specifications:

    A.  Assign at least 2 GB of RAM to the virtual machine

    B.  Create a virtual hard disk with a minimum size of 20 GB

### Step 2 – install the operating system

In this step, you'll have the freedom to choose an operating system that perfectly aligns with your goals and preferences:

1.  Download and install an operating system of your choice, such as Ubuntu or CentOS, on the virtual machine.

2.  Follow the onscreen instructions to complete the installation.

### Step 3 – update the operating system

This is a crucial phase that ensures your operating system stays at the forefront of security and performance. Now that your chosen operating system is in place, it's time to take the next important step – updating it to harness the latest advancements and bug fixes:

1.  Once the installation is complete, open a terminal or command prompt on the virtual machine.

2.  Update the operating system by running the appropriate update command for your chosen operating system (for example, `sudo apt update` for Ubuntu).

3.  Install any available updates by running the appropriate command (for example, `sudo apt upgrade`).

### Step 4 – install OpenSSL

As we delve deeper into the realms of security and cryptography, installing OpenSSL will serve as a gateway to a world of encrypted wonders.

In your terminal or command prompt, install OpenSSL by running the appropriate command for your operating system:

*   For Ubuntu, this is `sudo apt install openssl`

*   For CentOS, this is `sudo yum install openssl`

Follow the onscreen instructions to complete the installation:

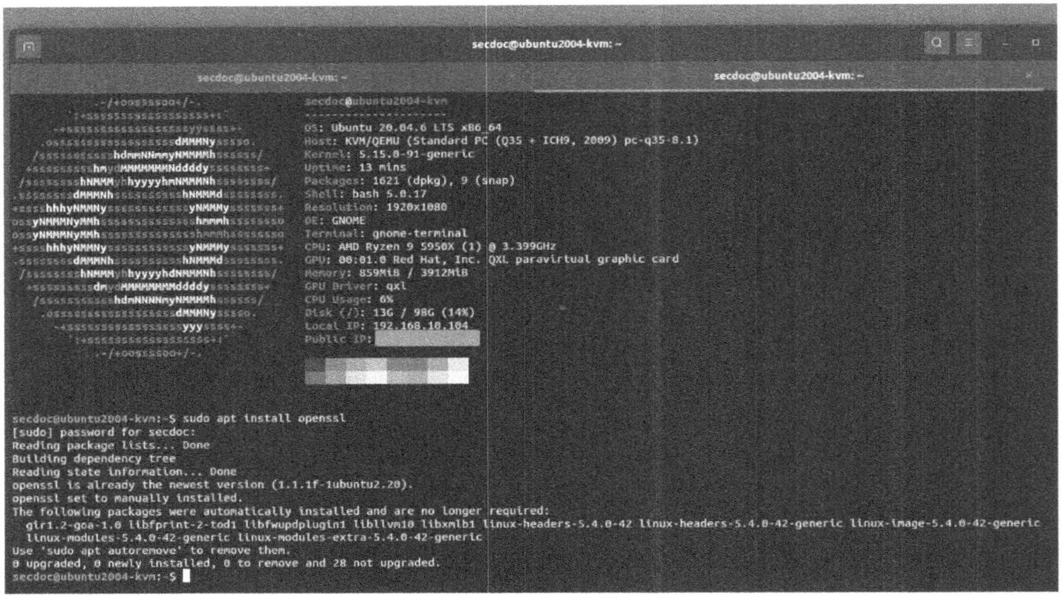

Figure 2.13 – Installing OpenSSL on Ubuntu 20.04

## Step 5 – generate encryption keys

In the heart of the terminal or command prompt, we'll harness the magic of OpenSSL to generate a formidable encryption key pair.

As *step 5* unfolds, encryption keys breathe life into your virtual world, fortifying your data with an impenetrable layer of security:

1.  In the terminal or command prompt, generate an encryption key pair using OpenSSL by running `openssl genpkey -algorithm RSA -out private_key.pem`.

2.  You will be prompted to enter a passphrase to protect the private key. Choose a strong passphrase and remember it.

3.  Next, extract the public key from the generated key pair by running `openssl rsa -pubout -in private_key.pem -out public_key.pem`.

The following screenshot shows the OpenSSL key generation process:

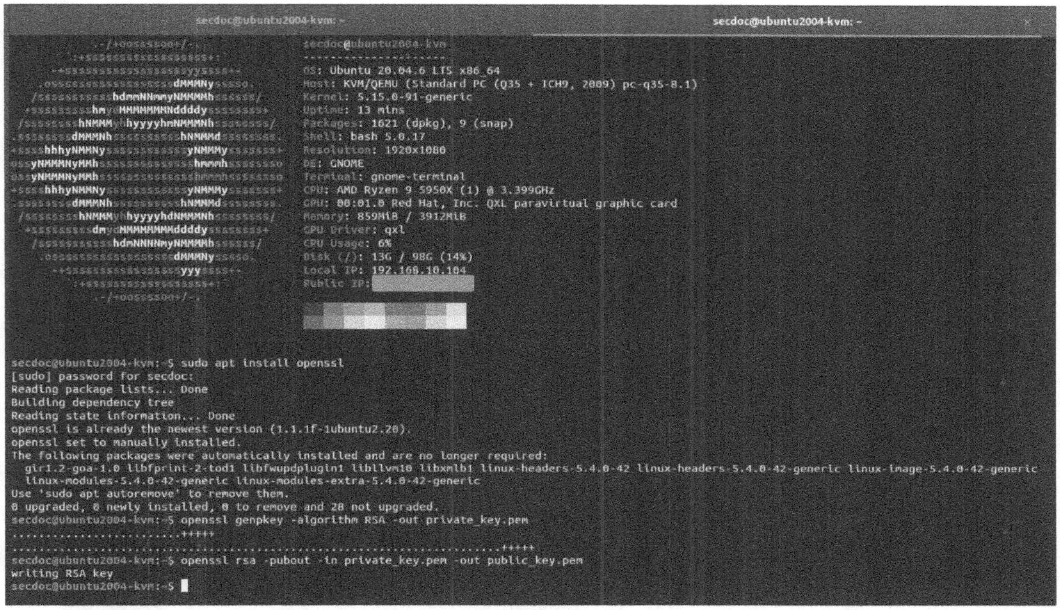

Figure 2.14 – OpenSSL key generation

## Step 6 – encrypt and decrypt data

This step is where data encryption and decryption become your powerful allies in the realm of secure communication. As we delve deeper into the art of cryptography, we'll equip you with the skills to encrypt and decrypt data with ease:

1.   Create a text file containing some sample data that you want to encrypt. For example, create a file named `plaintext.txt` and enter some text:

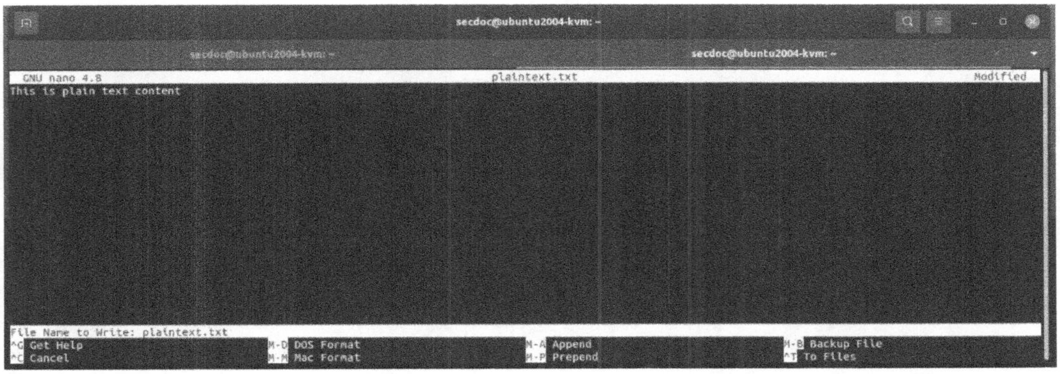

Figure 2.15 – Sample plaintext file

2. Encrypt the contents of the text file using the public key by running the `openssl rsautl -encrypt -pubin -inkey public_key.pem -in plaintext.txt -out ciphertext.enc` command.

3. The encrypted data will be stored in the `ciphertext.enc` file:

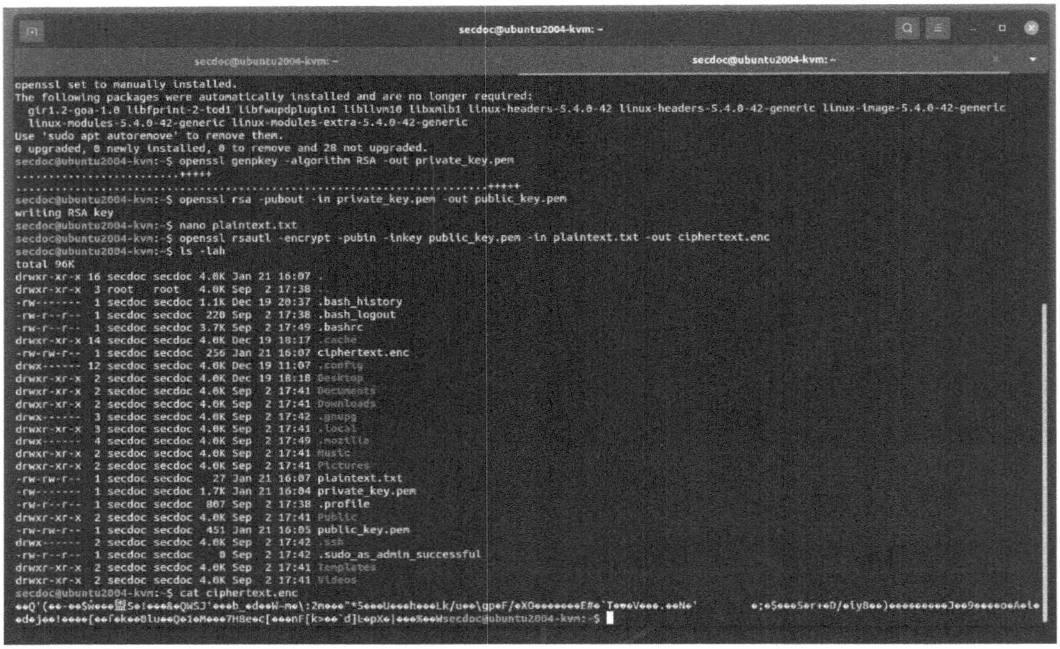

Figure 2.16 – Converting a plaintext file into ciphertext and validating file encryption

4. To decrypt the encrypted data, use the private key and run the `openssl rsautl -decrypt -inkey private_key.pem -in ciphertext.enc -out decrypted.txt` command.

5. The decrypted data will be stored in the `decrypted.txt` file:

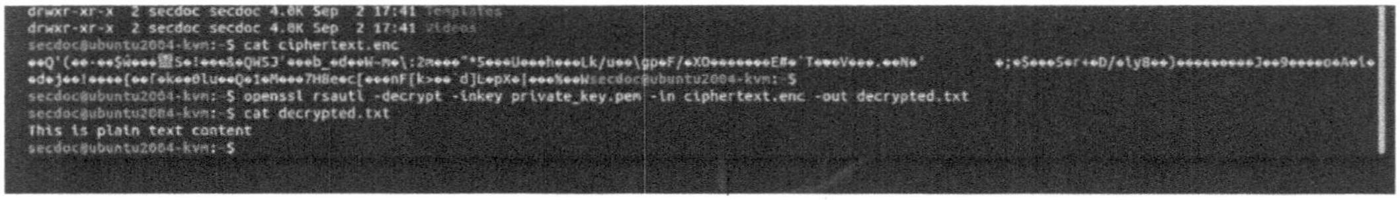

Figure 2.17 – Decrypting ciphertext and successfully validating the plaintext file

### Step 7 – verify encryption and decryption

In this step, we'll put the power of verification in your hands. In this phase, we'll witness the fruits of your labor as you confirm the success of the encryption and decryption processes.

To do so, open the `plaintext.txt` and `decrypted.txt` files to compare their contents. They should be the same, indicating that the encryption and decryption processes were successful.

Congratulations! You have successfully implemented basic cryptography using OpenSSL in your virtual environment. This lab provided a starting point for understanding encryption and decryption processes using public and private keys. Remember to explore and learn more about cryptography to enhance the security of your virtual environment.

# BCP/DRP

BCP/DRP is a critical process that organizations undertake to ensure their ability to continue operations and recover from disruptive incidents or disasters. It involves developing strategies, procedures, and policies to minimize the impact of potential disruptions and maintain business operations in adverse conditions. Let's delve deeper into BCP/DRP.

## BCP

BCP focuses on maintaining essential business functions during and after a disruptive event. The key elements of BCP include the following:

- **Business impact analysis (BIA)**: BIA identifies critical business processes, resources, and dependencies, and assesses the potential impact of disruptions. It helps prioritize recovery efforts and allocate resources effectively.

- **Risk assessment**: Organizations conduct a risk assessment to identify potential threats and vulnerabilities that could impact business operations. This includes natural disasters, cyber attacks, system failures, supply chain disruptions, and human-induced incidents.

- **Recovery strategies**: Recovery strategies involve developing plans and procedures to mitigate the impact of disruptions. This includes identifying alternate facilities, implementing backup systems, establishing redundancy, and arranging for alternative suppliers or service providers.

- **Incident response and communication**: BCP outlines incident response procedures and communication plans to ensure timely and effective response to incidents. It includes establishing emergency response teams, defining escalation protocols, and implementing communication channels for stakeholders, employees, customers, and the media.

- **Training and testing**: Regular training and testing exercises are conducted to validate the effectiveness of BCP measures. These include tabletop exercises, simulations, and full-scale drills to assess the organization's readiness to respond to and recover from various scenarios.

## DRP

DRP focuses on restoring critical IT systems and infrastructure following a disruption. The key elements of DRP include the following:

- **IT systems inventory**: Organizations identify critical IT systems, applications, databases, and infrastructure components that are essential for business operations. This inventory helps prioritize recovery efforts and allocate resources efficiently.

- **Recovery time objective (RTO) and recovery point objective (RPO)**: RTO defines the acceptable downtime for systems and sets the target time for recovery, while RPO defines the maximum acceptable data loss in the event of a disruption. These metrics help you select appropriate recovery strategies and technologies.

- **Backup and recovery solutions**: Organizations implement robust backup and recovery solutions to ensure data integrity and facilitate the restoration of systems. This includes regular backups, offsite storage, replication, snapshots, and cloud-based recovery options.

- **Infrastructure redundancy**: Redundancy measures, such as failover systems, clustering, virtualization, and geographically dispersed data centers, help ensure **high availability (HA)** and minimize downtime during system failures or disasters.

- **Vendor and supplier management**: Organizations establish relationships with technology vendors and service providers to ensure the availability of necessary resources, support, and expertise during recovery efforts. **Service-level agreements (SLAs)** and contracts define the expectations and responsibilities of both parties.

- **Testing and maintenance**: Regularly testing DRPs and infrastructure is crucial to identify and address any vulnerabilities or gaps. Organizations conduct drills, failover tests, and system recovery exercises to validate the effectiveness of their DRP and make necessary improvements.

## Integration with risk management and security

BCP/DRP is closely integrated with risk management and security practices:

- **Risk management**: BCP/DRP is driven by identifying and assessing risks. Risk management processes help identify threats and vulnerabilities, prioritize risks, and inform BCP/DRP decision-making.

- **Information security**: BCP/DRP incorporates information security measures to protect critical systems, data, and resources during a disruption. It includes measures such as access controls, encryption, network segmentation, and incident response procedures.

- **Incident response**: BCP/DRP aligns with incident response processes to ensure a coordinated approach in managing disruptions. Incident response plans often serve as a foundation for BCP/DRP, providing guidance for handling incidents and initiating recovery efforts.

## Compliance and regulatory considerations

Organizations must consider industry-specific compliance requirements and regulations when developing BCP/DRP. This includes data protection laws, privacy regulations, industry standards (for example, ISO 22301), and contractual obligations. Compliance frameworks provide guidance on implementing effective BCP/DRP measures and ensuring regulatory compliance.

BCP/DRP is an ongoing process that requires continuous monitoring, evaluation, and improvement. Organizations should conduct regular reviews, audits, and updates to address changes in the business environment, emerging threats, technological advancements, and lessons learned from incidents.

BCP/DRP is a comprehensive process that enables organizations to proactively prepare for and recover from disruptions. By identifying critical business functions, implementing strategies for resilience, and addressing IT recovery requirements, organizations can mitigate the impact of disruptions and maintain business continuity. Effective BCP/DRP helps protect the organization's reputation, customer trust, and overall business viability in the face of adverse events.

## BCP/DRP lab

Here's a step-by-step lab to help you implement BCP/DRP in a virtual environment, even if you have little or no experience in cybersecurity. This lab will guide you through the process of setting up a basic BCP/DRP using virtualization and backup strategies.

### Requirements

No matter your cybersecurity experience, this step-by-step lab will be your guiding light as we delve into the world of BCP/DRP implementation.

With the power of virtualization and backup strategies, we'll navigate the intricacies of safeguarding your digital assets and fortifying your environment against potential disruptions. We will require the following:

- A computer or virtual machine with at least 8 GB of RAM and 80 GB of disk space

- Virtualization software (for example, VirtualBox, VMware, QEMU/KVM, or Proxmox)

- Virtual machine images for testing purposes (for example, Ubuntu or Windows Server images)

## Step 1 – set up the virtual environment

This phase sets the stage for a resilient and dynamic virtual environment. In this foundational step, we'll walk you through the process of creating your very own virtual machines, each crafted to meet the demanding specifications of cybersecurity:

1.  Install your preferred virtualization software on your computer.

2.  Create a new virtual machine with the following specifications:

    A.  Assign at least 4 GB of RAM to the virtual machine

    B.  Create a virtual hard disk with a minimum size of 40 GB

3.  Repeat this step to create a second virtual machine.

## Step 2 – install operating systems

This is a phase where the heart of your virtual machines comes to life with the installation of operating systems. As we continue our journey toward building a robust environment, the power to choose lies in your hands:

1.  Download and install an operating system of your choice on each virtual machine. For example, you can use Ubuntu for one virtual machine and Windows Server for the other virtual machine.

2.  Follow the onscreen instructions to complete the installation.

## Step 3 – configure network connectivity

Here, the essence of connectivity takes center stage. As we progress through this step, the power of network configuration will unify your virtual machines, fostering seamless communication and cooperation:

1.  In your virtualization software, set up a virtual network to connect the two virtual machines.

2.  Configure the network settings on each virtual machine to ensure they are connected and can communicate with each other.

## Step 4 – create a backup strategy

In this phase, the power of preparedness takes center stage through the creation of a robust backup strategy. In this critical step, we'll fortify your virtual environment against the unexpected with the aid of reliable backup software:

1.  Install backup software on one of the virtual machines. For example, you can use **Veeam Backup & Replication Community Edition**, which is free for virtual environments.

2.  Follow the software's installation instructions to complete the setup.

3.  Configure the backup software to create regular backups of the virtual machines. Set a schedule for automatic backups to occur at desired intervals.

### Step 5 – test disaster recovery

Now, preparedness meets action, and the power of disaster recovery is put to the test. In this critical step, we'll embark on a transformative journey, simulating a disaster scenario to ensure the resilience and efficacy of your backup strategy:

1. Shut down one of the virtual machines to simulate a disaster scenario.
2. Initiate a disaster recovery process by restoring the virtual machine from the backup using the backup software.
3. Verify that the restored virtual machine is functioning properly.

### Step 6 – implement HA

Here, the power of HA takes center stage. In this critical step, we'll fortify your virtual environment against potential disruptions with the implementation of HA settings:

1. Configure HA settings in your virtualization software. This feature ensures that if one virtual machine fails, the other virtual machine will take over automatically.
2. Test the HA configuration by simulating a failure on one virtual machine and verifying that the other virtual machine takes over seamlessly.

### Step 7 – document BCP/DRP procedures

In this final step, the power of documentation takes center stage. We'll embark on creating a detailed document outlining the step-by-step procedures for disaster recovery and business continuity:

1. Create a detailed document outlining the step-by-step procedures for disaster recovery and business continuity.
2. Include information such as contact lists, recovery strategies, backup schedules, and other relevant information.
3. Ensure the document is easily accessible to key personnel responsible for implementing BCP/DRP procedures.

Congratulations! You have successfully implemented a basic BCP/DRP in your virtual environment. This lab provided a starting point for understanding BCP/DRP concepts and testing recovery procedures using virtualization and backup strategies. Remember to explore and learn more about BCP/DRP best practices to enhance the resilience of your virtual environment.

# Physical security

**Physical security** refers to the measures and practices that are implemented to protect physical assets, facilities, and people from unauthorized access, damage, theft, or harm. It encompasses a range of strategies, technologies, and procedures designed to create a secure and safe environment. This section will provide detailed information on physical security.

## Access control

Access control systems ensure that only authorized individuals can enter specific areas or facilities. This includes using techniques such as key cards, biometric authentication (fingerprint or facial recognition), PIN codes, or security personnel to verify and grant access.

Physical access control measures may include gates, turnstiles, locks, and security guards stationed at entrances and sensitive areas.

Access control policies and procedures define who is granted access, when, and under what conditions. It also includes visitor management protocols to track and monitor visitors within the premises.

## Surveillance systems

Surveillance systems help monitor and record activities within and around facilities. They act as deterrents and provide evidence in the event of incidents.

Surveillance systems have evolved with advanced features such as high-definition cameras, **pan-tilt-zoom** (**PTZ**) capabilities, and video analytics.

Video analytics technologies can analyze video footage in real time, automatically detecting suspicious behaviors, abandoned objects, or unauthorized access attempts. This reduces the need for constant manual monitoring and allows security personnel to focus on critical situations.

Cloud-based surveillance systems enable remote monitoring, storage, and access to video feeds from any location, enhancing situational awareness and facilitating investigations.

## Intrusion detection and alarm systems

IDSs detect and alert security personnel about unauthorized attempts to access restricted areas or breaches in physical security.

Alarm systems can include sensors, motion detectors, glass break detectors, or door/window sensors that trigger audible or silent alarms in response to unauthorized access or suspicious activities.

Advanced IDS technologies employ machine learning and behavior analysis to detect anomalous patterns and identify potential threats.

Integration of IDS with SIEM systems allows for centralized log management, correlation of security events, and automated response actions. These systems are often integrated with **security operations centers** (**SOCs**) or monitoring stations for real-time monitoring and response.

## Physical barriers and deterrents

Physical barriers, such as fences, walls, bollards, or vehicle barriers, help restrict access to sensitive areas and prevent unauthorized entry or vehicle intrusion.

Deterrents such as signage, lighting, and landscaping are used to discourage potential intruders or criminals.

## Security personnel and guards

Trained security personnel, including security guards or officers, play a vital role in maintaining physical security. They perform patrols, monitor surveillance systems, and respond to incidents.

Trained security guards may undergo specialized training, including emergency response, conflict resolution, and customer service skills.

Mobile patrols and guard tour systems can be employed to ensure regular checks of critical areas and to maintain visibility across the premises.

## Security policies and procedures

Security policies and procedures establish clear guidelines for employees, contractors, and visitors that outline acceptable behavior, access control protocols, and incident reporting procedures.

Visitor management systems can be utilized to register and track visitors, issue temporary access credentials, and maintain visitor logs for audit purposes.

Secure document disposal procedures, such as shredding or secure bins, should be in place to prevent unauthorized access to confidential or sensitive information.

## Incident response and emergency preparedness

Incident response plans outline specific steps to be taken during security incidents, including communication channels, escalation procedures, and coordination with emergency services.

Emergency preparedness involves regular drills, simulations, or tabletop exercises to test the effectiveness of response plans and ensure personnel are familiar with emergency procedures.

Emergency notification systems, such as mass notification or emergency broadcast systems, can be utilized to quickly disseminate critical information to employees during emergencies.

## Environmental controls

Environmental controls include measures to maintain optimal conditions for equipment and data protection.

Fire detection and suppression systems, including smoke detectors, fire alarms, sprinklers, or clean agent systems, help minimize the risk of fire-related damage.

Temperature and humidity monitoring systems can be employed to ensure that sensitive equipment or storage areas maintain suitable environmental conditions.

## Inventory and asset management

Physical security involves inventory and asset management practices to track and protect valuable assets, such as IT equipment, confidential documents, or high-value items.

Asset tracking systems, asset tagging, and restricted access to storage areas help prevent theft or unauthorized removal.

## Perimeter security

Perimeter security measures should be designed to deter and detect unauthorized access attempts.

Advanced perimeter security technologies include video analytics, thermal imaging cameras, or radar systems to detect and track intrusions along the perimeter.

Integration of perimeter security systems with access control systems and surveillance systems enables a comprehensive security approach.

## Collaboration with law enforcement and first responders

Building relationships and establishing communication channels with local law enforcement agencies, fire departments, and emergency medical services is crucial for effective emergency response.

Conducting joint training exercises or participating in community safety initiatives strengthens collaboration and enhances the response capabilities of both the enterprise and the emergency services.

## Physical security audits and assessments

Physical penetration testing, vulnerability assessments, or security surveys can be conducted by external specialists to identify potential weaknesses and recommend improvements.

## Why implement physical security controls?

Implementing physical security measures within an enterprise is essential for various reasons. Here are some examples of why physical security is implemented:

- **Preventing unauthorized access**: Physical security measures, such as access control systems, surveillance cameras, and security personnel, help prevent unauthorized individuals from gaining access to sensitive areas, facilities, or information. This protects valuable assets, intellectual property, and confidential data from theft, sabotage, or unauthorized disclosure.

- **Protecting employees and visitors**: Physical security measures create a safe and secure environment for employees, clients, and visitors. By implementing measures such as access control, emergency response procedures, and well-trained security personnel, enterprises can mitigate risks, prevent workplace violence, and ensure the physical well-being of individuals within the premises.

- **Safeguarding business continuity**: Physical security measures are crucial for maintaining business continuity. By securing facilities, data centers, or critical infrastructure, organizations can prevent disruptions caused by theft, vandalism, or unauthorized access. This ensures that essential business operations can continue uninterrupted, minimizing downtime and financial losses.

- **Preventing theft and loss**: Physical security measures deter theft of equipment, inventory, or valuable assets. Surveillance cameras, burglar alarms, secure storage facilities, and inventory management systems help prevent internal and external theft, reducing financial losses and protecting the enterprise's profitability.

- **Compliance with regulations**: Many industries have specific regulations and compliance requirements that mandate the implementation of physical security measures. These regulations aim to protect personal information, maintain confidentiality, and safeguard critical infrastructure. By complying with these requirements, enterprises avoid legal penalties, reputational damage, and loss of customer trust.

- **Mitigating insider threats**: Physical security measures also address the risks associated with insider threats, such as unauthorized access, theft, or sabotage by employees or contractors. Access control systems, surveillance, and monitoring tools can detect and deter malicious activities, ensuring that only authorized individuals have access to sensitive areas or information.

- **Ensuring data center security**: Data centers house critical IT infrastructure and store vast amounts of valuable data. Physical security measures such as restricted access, video surveillance, environmental controls, fire suppression systems, and backup power systems protect data centers from unauthorized access, physical damage, natural disasters, or power outages.

- **Enhancing brand reputation and customer trust**: Demonstrating a commitment to physical security enhances an enterprise's reputation and instills trust among customers, partners, and stakeholders. Customers feel confident entrusting their information or conducting business with organizations that prioritize the protection of physical assets and personal data.

- **Protecting intellectual property (IP)**: IP is a valuable asset for many enterprises. Physical security measures safeguard research and development facilities, laboratories, or innovation centers, preventing unauthorized access or theft of proprietary information, trade secrets, or product prototypes.

- **Preventing workplace violence**: Physical security measures, such as access control systems, employee screening procedures, and security awareness programs, contribute to a safer work environment by preventing incidents of workplace violence, unauthorized weapons, or harmful behavior.

- **Ensuring regulatory compliance for safety**: Physical security measures also encompass safety regulations, ensuring compliance with fire safety codes, emergency evacuation plans, or occupational health and safety requirements. This protects employees from physical hazards and minimizes the risk of accidents or injuries.

- **Managing public events or large gatherings**: Physical security measures are crucial during public events or large gatherings organized by enterprises. Implementing access control, crowd management, perimeter security, and emergency response plans helps ensure the safety and well-being of attendees.

By implementing physical security measures, enterprises can mitigate risks, protect assets and individuals, maintain business continuity, comply with regulations, and enhance their reputation. A comprehensive physical security strategy ensures a secure environment that aligns with the overall cybersecurity framework and supports the organization's objectives.

## Physical security lab

Here's a step-by-step lab to help you implement physical security measures, even if you have little or no experience in cybersecurity. This lab will guide you through the process of setting up basic physical security controls within a physical environment. The best way to do this lab is to use your home or apartment as your physical environment.

### Step 1 – conduct a risk assessment

Welcome to the physical security lab, where the power of safeguarding extends beyond the digital realm. In this transformative journey, we'll guide you step-by-step to implement robust physical security measures, even if you have little or no experience in cybersecurity.

The heart of this lab lies in the creation of a fortified physical environment – and what better place to start than your own home or apartment? Together, we'll set up basic physical security controls, ensuring that the essence of protection touches every aspect of your space.

*Step 1* of this immersive lab is all about conducting a comprehensive risk assessment:

1. Identify the assets and areas that require physical security, such as server rooms, storage areas, or sensitive documents.
2. Assess potential risks and threats to these assets, including unauthorized access, theft, or damage.

### Step 2 – implement perimeter security

This phase is where the essence of protection extends to the very boundaries of your facility. In this step, we'll fortify the perimeter of your space with robust security measures, creating a protective shield that guards against potential threats:

- Secure the perimeter of your facility by installing physical barriers, such as fences, gates, or access control systems
- Install security cameras to monitor and record activities around the perimeter
- Consider using motion sensors and alarms to detect unauthorized entry

## Step 3 – control access points

Here, the power of control takes center stage. In this transformative step, we'll focus on fortifying access points to restricted areas, ensuring that only authorized personnel enter with ease:

1. Install access control systems, such as key card readers or biometric scanners, at entry points to restricted areas.

2. Assign unique access credentials to authorized personnel and regularly review and update access privileges.

3. Implement visitor management procedures, including sign-in/out logs or visitor badges, to track and monitor guest access.

## Step 4 – secure server and equipment rooms

At this point, the heart of your digital assets must be fortified with an impenetrable shield. In this step, we'll focus on securing your server and equipment rooms, safeguarding the core of your technological prowess:

- Restrict access to server and equipment rooms by installing access control systems or secure locks

- Use environmental monitoring systems to detect temperature, humidity, or water leaks that could damage equipment

- Implement video surveillance systems to monitor and record activities within these areas

## Step 5 – protect sensitive information

This phase is where the essence of protection extends to safeguarding sensitive information. In this transformative step, we'll focus on fortifying the very heart of your data – the delicate information that demands the utmost security:

- Store sensitive documents or data in locked cabinets or secure rooms

- Implement a document destruction policy to securely dispose of sensitive information when no longer needed

- Use shredders or professional document destruction services to destroy paper documents

## Step 6 – implement security awareness training

At this point, knowledge becomes the foundation of resilience. In this step, we'll focus on the invaluable power of security awareness training, empowering employees to stand as vigilant protectors of your physical environment:

- Educate employees on the importance of physical security and their role in maintaining a secure environment

- Provide training on topics such as recognizing suspicious activities, reporting incidents, and following access control procedures

### Step 7 – establish incident response procedures

In this penultimate step, preparedness meets action through the establishment of incident response procedures. In this transformative step, we'll focus on creating a robust framework to respond to physical security incidents, ensuring that your team is ready to face any challenge that may arise:

- Develop procedures to respond to physical security incidents, such as unauthorized access or theft

- Assign specific roles and responsibilities to staff members to handle different aspects of the incident response

- Regularly test and review the effectiveness of these procedures to ensure they are up to date and well understood

### Step 8 – regularly monitor and maintain physical security

Welcome to the final step of our empowering physical security lab – a phase where vigilance becomes a way of life. We'll focus on the critical task of regularly monitoring and maintaining your physical security measures, ensuring that your protective shield remains strong and steadfast:

- Conduct regular inspections of physical security controls, including locks, access control systems, and surveillance cameras

- Maintain proper lighting in areas where security is essential

- Update security measures based on changes in the environment or identified risks

Congratulations! You have successfully implemented basic physical security measures within your environment. This lab provided a starting point for understanding and implementing physical security controls to protect your assets and maintain a secure environment. Remember to continuously assess and improve your physical security measures to address emerging risks and maintain the integrity of your physical space.

# Summary

Continuing from the introduction, this chapter took a deeper dive into foundational areas that are crucial for cybersecurity architects to understand and address within the context of the business and operational teams. While the coverage remained introductory, it provided the necessary groundwork for discussions on the cybersecurity career path and specialization options.

By engaging in the labs and scenarios provided throughout this chapter, you've developed practical skills and knowledge that can serve as a foundation for further exploration and specialization within the cybersecurity field. This chapter equipped you, as an aspiring cybersecurity architect, with the necessary understanding of these key areas to progress in your career and make informed decisions regarding specialization.

Now that we've discussed and have practical application of foundational concepts, in the next chapter, we'll discuss what the role of a cybersecurity architect is and the associated responsibilities within an organization.

# What Is a Cybersecurity Architect and What Are Their Responsibilities?

*Thus we may know that there are five essentials for victory: (1) He will win who knows when to fight and when not to fight; (2) he will win who knows how to handle both superior and inferior forces; (3) he will win whose army is animated by the same spirit throughout all its ranks; (4) he will win who, prepared himself, waits to take the enemy unprepared; (5) he will win who has military capacity and is not interfered with by the sovereign.*

*–Sun Tzu*

So, in the previous chapters, we covered the foundational and major cybersecurity concepts to help prepare and direct those interested in the field and specifically the **cybersecurity architect** role. In this chapter, the discussion shifts fully to the cybersecurity architect.

A cybersecurity architect is a specialized professional responsible for designing and implementing secure information technology systems and networks within an organization. Their primary role is to create a robust cybersecurity framework that safeguards the organization's digital assets from potential threats, including cyber attacks, data breaches, and other security risks. They work to strike a balance between maintaining the organization's security posture and enabling the efficient flow of information and services.

This chapter will cover the following topics:

- Understanding the role and environment
- What is a cybersecurity architect?
- Areas of focus
- The cybersecurity architect as a part of the bigger team
- Responsibilities
- Scope of vision

## Understanding the role and environment

As quoted from Sun Tzu's *Art of War* at the beginning of this chapter, it is crucial to comprehend your role as the cybersecurity architect and the people or teams you will work with. While the quote is about how to be a successful military leader, the same concepts can be applied to be successful in your career and  role the cybersecurity architect plays a vital role in maintaining a strong security posture for an organization, protecting sensitive information, and reducing the risk of cyber attacks. They are essential in today's digital landscape, where cyber threats continue to evolve and pose significant challenges to businesses and individuals alike. It is also important to understand that the cybersecurity architect interacts with all levels of an organization, and must be able to drive the same passion and commitment in both big projects and little tasks. The cybersecurity architect needs to be the person who can maintain visibility of the big picture while also tackling the in-the-weeds details to ensure that the direction and focus of the organization's security are maintained, regardless of the controls or technologies implemented.

## What is a cybersecurity architect?

While we covered at a high level what a cybersecurity architect is at the start of this chapter, it is important to fully understand what the role entails. I define a cybersecurity architect as follows.

A cybersecurity architect is a specialized professional in the field of cybersecurity who is responsible for designing and implementing comprehensive security solutions to protect an organization's digital assets, information, and systems from potential cyber threats and attacks. Cybersecurity architects are strategic thinkers with expertise in various security domains, and they play a critical role in developing a robust and resilient security infrastructure that aligns with the organization's business goals and risk tolerance. They must also be able to communicate complex technical concepts to non-technical stakeholders and work collaboratively with cross-functional teams to ensure that security is an integral part of the organization's operations.

A successful cybersecurity architect possesses a strong understanding of cybersecurity principles, industry best practices, emerging threats, and the latest security technologies. Overall, a cybersecurity architect plays a vital role in strengthening an organization's security posture and safeguarding its sensitive information and digital assets from potential cyber threats.

> **Note**
>
> As we progress with the discussion on the role and areas of responsibility of a cybersecurity architect, you will notice that there may be the same topic repeated or re-hashed. The reason for this is the fact that those areas, such as a security framework, may have varied focus depending on how or where the framework is applied, and as a result, the approach to other areas may vary slightly to account for the differences in approach.

Let's dive deeper into the role of a cybersecurity architect and further explore the responsibilities required for this critical position from a high level:

- **Threat modeling and analysis**: A cybersecurity architect must understand potential threats and attack vectors that could compromise the organization's systems. They conduct threat modeling exercises to identify vulnerabilities and weaknesses in the infrastructure and applications.

- **Security solution evaluation**: A cybersecurity architect must evaluate and select appropriate security technologies and tools that align with the organization's needs and budget. This includes researching and testing different solutions to ensure they meet security requirements.

- **Secure network design**: A cybersecurity architect must design secure network architectures that segment sensitive data, control access, and monitor traffic to prevent unauthorized access and data exfiltration.

- **Secure application design**: They must work closely with software development teams to integrate security measures into the application development life cycle. They ensure that applications are designed with security in mind and undergo rigorous testing before deployment.

- **Identity and access management (IAM)**: They must implement robust IAM solutions to manage user access and authentication. This includes **multi-factor authentication (MFA)**, **role-based access control (RBAC)**, and **privileged access management (PAM)**.

- **Encryption and data protection**: A cybersecurity architect must use encryption techniques to protect data both at rest and in transit. Cybersecurity architects ensure that sensitive information remains confidential and cannot be easily intercepted or tampered with.

- **Security incident handling**: A cybersecurity architect must develop incident response plans to guide the organization in responding to security incidents promptly and effectively. They lead incident response teams during cyber attacks, working to contain and mitigate the impact of the breach.

- **Cloud security**: If the organization uses cloud services, the cybersecurity architect ensures that cloud environments are appropriately configured and secure. They follow best practices for cloud security and address unique challenges related to cloud-based systems.

- **Mobile security**: A cybersecurity architect must address security concerns related to mobile devices and applications within the organization. This may involve implementing **mobile device management (MDM)** solutions and developing secure mobile application guidelines.

- **Security training and awareness**: They must conduct regular security training sessions for employees to promote a culture of security awareness. This includes educating users about phishing attacks, social engineering, and safe online practices.

- **Vendor risk management**: They must assess the security posture of third-party vendors and service providers to ensure they meet the organization's security standards. This is crucial as vendors may have access to sensitive data or systems.

- **Regulatory compliance**: They must stay informed about relevant data protection laws and industry regulations. Cybersecurity architects ensure that the organization complies with these requirements to avoid legal and financial consequences.

Let us now look at the skills and qualifications required for a cybersecurity architect:

- In-depth knowledge of cybersecurity principles, best practices, and industry standards

- Strong understanding of networking protocols, firewalls, and intrusion detection/prevention systems

- Familiarity with IAM solutions and encryption technologies

- Proficiency in security assessment tools and techniques, including penetration testing

- Experience with security architecture frameworks, such as **The Open Group Architecture Framework (TOGAF)** or **Sherwood Applied Business Security Architecture (SABSA)**

- Excellent communication skills to articulate complex security concepts to technical and non-technical stakeholders

- Ability to work collaboratively with cross-functional teams and senior management

- Analytical mindset to assess security risks and devise appropriate mitigation strategies. This includes being able to deal with what-if concepts and scenarios

- Knowledge of cloud security, mobile security, and emerging technologies in the cybersecurity space

- Relevant certifications, such as **Certified Information Systems Security Professional (CISSP)** and the specialization in architecture, **Certified Information Security Manager (CISM)**, or cloud-based certifications for AWS, Azure, and GCP, are often beneficial

In this section, we looked at the responsibilities and areas of focus at a high level for the cybersecurity architect. As this chapter progresses, we will take a deeper dive into each area to better understand the role of the cybersecurity architect.

## Areas of focus

The role of a cybersecurity architect demands expertise in various areas to design and implement a robust security framework for an organization. While it's essential for the cybersecurity architect to comprehend and operate within these focus areas, it's equally important to recognize that the extent of specialization may vary depending on the organization's size and structure. For instance, some organizations may have designated teams for network security architecture, application security architecture, or enterprise security architecture. Despite being a generalist or specializing in a specific area, the cybersecurity architect must proficiently navigate and address the unique needs of each domain.

In the following subsections are the expanded details on the key areas a cybersecurity architect focuses on.

## Threat landscape analysis and modeling

Understanding the ever-evolving threat landscape is essential for a cybersecurity architect. They continuously monitor and analyze emerging cyber threats, attack vectors, and hacking techniques to proactively adapt security measures and stay ahead of potential risks. This requires thinking about what-if scenarios. What-if scenarios are hypothetical situations or simulations used to explore the potential outcomes of specific events or decisions. They are often employed in various fields, including risk management, business planning, disaster preparedness, and strategic decision-making. By considering different what-if possibilities, individuals or organizations can gain insights into potential consequences, identify vulnerabilities, and develop contingency plans.

Threat landscape analysis is a crucial process in cybersecurity that involves continuously monitoring and assessing the evolving threat landscape to identify potential risks and vulnerabilities. This analysis helps cybersecurity architects and professionals to stay ahead of emerging threats, anticipate potential cyber attacks, and implement proactive security measures. Here's an overview of how threat landscape analysis and modeling is accomplished:

1. **Threat intelligence gathering**: The first step in threat landscape analysis is gathering threat intelligence. This involves collecting information from various sources, such as security vendors, government agencies, cybersecurity forums, dark web monitoring, security blogs, and security incident reports. Threat intelligence includes details about new attack vectors, malware variants, hacking techniques, and vulnerabilities affecting software and hardware.

2. **Cybersecurity news and research**: Cybersecurity professionals continuously monitor news and research from reputable sources to stay informed about the latest cyber threats and trends. This includes following security-focused websites, attending security conferences, and reading research reports and whitepapers.

3. **Security advisory services**: Subscribing to security advisory services provided by organizations such as **computer emergency response teams** (**CERTs**) and security vendors can offer real-time information on emerging threats and vulnerabilities.

4. **Cyber threat hunting**: Organizations may engage in proactive cyber threat hunting exercises to actively search for signs of potential compromise within their networks. This may involve examining logs, network traffic, and system behavior to detect suspicious or anomalous activity.

5. **Collaborative information sharing**: Participation in threat intelligence sharing communities and industry forums allows organizations to share information about recent threats and attacks. Collaborative efforts can help in identifying broader patterns and trends in the threat landscape.

6.  **Vulnerability assessments and penetration testing**: Conducting regular vulnerability assessments and penetration testing exercises provides insights into potential weaknesses in the organization's systems and applications. This helps prioritize security measures and gauge the organization's exposure to specific threats.

7.  **Malware and incident analysis**: Analyzing past security incidents and malware samples provides valuable information about attack methodologies and techniques employed by threat actors. This analysis can assist in preparing for and defending against similar future attacks.

8.  **Monitoring dark web activities**: The dark web is a hotbed for illegal activities, including the buying and selling of stolen data, malware, and hacking tools. Monitoring dark web activities can provide early warnings about potential threats targeting an organization.

9.  **Threat modeling exercises**: Threat modeling involves systematically assessing potential threats and vulnerabilities based on an organization's unique architecture and assets. By identifying possible attack vectors, organizations can design better security measures and prioritize security efforts.

10. **Analyzing security reports and case studies**: Reviewing security reports and case studies related to recent cyber incidents in similar industries can provide insights into the types of threats and attack vectors most relevant to an organization.

11. **Security information and event management (SIEM) solutions**: Implementing SIEM solutions allows organizations to centralize and analyze security event logs from various systems and applications. This helps in detecting abnormal activities and potential security breaches.

Threat landscape analysis is a dynamic and continuous process that requires a proactive approach to stay ahead of cyber threats. Cybersecurity professionals use a combination of threat intelligence sources, collaborative sharing, proactive hunting, vulnerability assessments, incident analysis, and the use of advanced security tools to identify potential risks and vulnerabilities in their organization's environment. By understanding the threat landscape, organizations can implement appropriate security measures to protect their digital assets effectively.

## Security framework development

Cybersecurity architects are responsible for developing and maintaining a comprehensive security framework that aligns with the organization's goals and risk appetite. This framework serves as a guide for implementing security policies, standards, and procedures. Security framework development involves creating a structured and comprehensive set of security policies, standards, procedures, and guidelines to establish a strong security posture within an organization. It provides a strategic approach to addressing cybersecurity challenges and ensures that security measures are aligned with the organization's business goals and risk tolerance. Here's a detailed overview of how security framework development is accomplished:

1.  **Identify security objectives and requirements**: The first step is to define the organization's security objectives and requirements. This involves understanding the organization's business processes, critical assets, data sensitivity, compliance obligations, and the level of acceptable risk.

2. **Conduct risk assessment**: Perform a thorough risk assessment to identify potential threats, vulnerabilities, and potential impacts on the organization's assets and operations. This analysis helps prioritize security efforts and allocate resources effectively.

3. **Define security policies**: Security policies are high-level statements that outline the organization's stance on various security matters. They provide a framework for decision-making and guide employees in adhering to security best practices. Policies cover areas such as data protection, access control, incident response, and acceptable use of resources.

4. **Establish security standards**: Security standards provide specific, detailed guidelines for implementing security controls and best practices. They define how security objectives will be achieved and serve as a baseline for security implementations across the organization. Standards may be industry-specific or derived from security frameworks such as ISO 27001 or NIST Cybersecurity Framework.

5. **Develop security procedures**: Security procedures are step-by-step instructions on how to perform specific security tasks. These procedures help ensure consistent and accurate execution of security measures, such as user access provisioning, password management, and incident response processes.

6. **Incident response planning**: Establish detailed incident response plans and procedures to handle security incidents effectively. Define roles and responsibilities, escalation paths, communication channels, and steps to contain and mitigate the impact of security breaches.

7. **Third-party risk management**: Implement policies and procedures to assess the security posture of third-party vendors and service providers. This ensures that their security practices align with the organization's standards.

8. **Continuous monitoring and improvement**: Security framework development is an ongoing process. Regularly review and update security policies, standards, and procedures to adapt to changing threats and business needs. Implement continuous monitoring to identify security gaps and potential areas for improvement.

9. **Compliance and audit readiness**: Ensure that the security framework aligns with relevant industry regulations and compliance requirements. Prepare for security audits and assessments to demonstrate adherence to security standards.

10. **Employee engagement and communication**: Involve employees across the organization in the security framework development process. Foster a culture of security awareness and encourage employees to be proactive in reporting security incidents and potential risks.

11. **Seek external expertise**: Engage with external cybersecurity experts or consultants to validate the effectiveness and robustness of the security framework and gain valuable insights from experienced professionals.

Security framework development is a comprehensive process that involves understanding the organization's security requirements, conducting risk assessments, defining security policies and standards, implementing access controls and encryption, planning for incident response, and continuously monitoring and improving the security posture. It requires collaboration between various stakeholders, including IT teams, management, and employees, to create a resilient and effective security framework that protects the organization's digital assets from potential threats and risks.

## Network security

Network security is a fundamental aspect of a cybersecurity architect's role. They design and implement secure network architectures that include firewalls, **intrusion detection systems (IDSs)/intrusion prevention systems (IPSs)**, **virtual private networks (VPNs)**, and other security measures to protect against unauthorized access and data breaches. It focuses on protecting the organization's network infrastructure and data from unauthorized access, data breaches, and cyber threats. An aspect of establishing proper security on the network is through the network perimeter. The perimeter is the secured boundary between the internal (trusted) network and the internet (untrusted). Accomplishing effective network security involves implementing various measures and technologies to create a secure and resilient network environment. Here's an overview:

- **Network perimeter defense**:

  - **Firewalls**: Deploying firewalls at network entry and exit points to monitor and control incoming and outgoing traffic based on predefined security rules. Firewalls act as a barrier between the internal network and external entities, filtering and blocking potentially harmful traffic.

  - **IDS/IPS**: Implementing IDS/IPS solutions to detect and prevent suspicious or malicious activities on the network. An IDS identifies potential threats, while an IPS actively blocks or mitigates them.

- **Secure network architecture**:

  - **Network segmentation**: Dividing the network into multiple segments or subnets with restricted access between them. This limits the impact of a security breach and contains potential threats within a specific segment.

  - **Demilitarized zone (DMZ)**: Creating a DMZ, an isolated network zone between the internal network and the internet, where public-facing servers and services are placed. This helps segregate sensitive data from external access.

- **Access control**:

  - **RBAC**: Implementing RBAC to control access rights based on users' roles and responsibilities. This ensures that users can only access resources necessary for their job functions.

  - **Network access control (NAC)**: Enforcing security policies to authenticate and authorize devices before granting access to the network. NAC solutions ensure that only trusted devices can connect to the network.

- **VPN and encryption**: Using VPN technology to create secure, encrypted connections for remote users or branch offices. This ensures that data transmitted over the internet remains confidential and protected from eavesdropping.

- **Network monitoring and logging**:

  - **SIEM**: Deploying SIEM solutions to centralize and analyze logs from various network devices and applications. SIEM helps detect anomalies and potential security incidents.

  - **Network traffic analysis**: Analyzing network traffic patterns and behavior to identify suspicious activities and potential threats. Network traffic analysis tools assist in real-time threat detection.

- **Patch management**: Regularly updating and patching network devices, routers, switches, and other infrastructure components to address known vulnerabilities. Patch management ensures that the network is protected against known exploits.

- **Network security protocols**: Using secure network protocols, such as HTTPS, SSL/TLS, and SSH, to protect data during transmission over the network. Avoiding the use of deprecated and insecure protocols is essential.

- **Network device hardening**: Configuring network devices with secure settings, disabling unnecessary services, and using strong authentication methods. Hardening network devices reduces the attack surface.

- **Network-based anti-virus and malware protection**: Deploying anti-virus and anti-malware solutions at the network level to scan and detect malicious content before it reaches endpoints.

- **Network security policies and employee training**: Defining and enforcing network security policies that guide employees on acceptable network usage and security best practices. Conducting regular security training for employees helps raise awareness of potential network security risks.

- **Incident response planning**: Developing and testing incident response plans to handle network security breaches effectively. This includes defining roles, responsibilities, and communication protocols during a security incident.

- **Continuous monitoring and vulnerability assessment**: Implementing continuous monitoring to detect potential network security breaches and promptly respond to them. Conducting regular vulnerability assessments helps identify weaknesses and potential entry points for attackers.

Accomplishing network security requires a multi-layered approach that involves perimeter defense, secure network architecture, access control, encryption, monitoring, and a proactive stance toward security threats. By employing these measures, organizations can significantly reduce the risk of network-based cyber attacks and safeguard their critical data and infrastructure.

## Application security

Applications are often a prime target for cyber attacks. Cybersecurity architects work closely with software development teams to ensure that security is integrated into the application development life cycle. This involves conducting security code reviews, implementing secure coding practices, and integrating **application security testing** (**AST**) tools. Application security is the practice of identifying and addressing security vulnerabilities and weaknesses in software applications to prevent potential cyber attacks and data breaches. It aims to ensure that applications are designed, developed, and maintained with robust security measures. Accomplishing effective application security involves various techniques, tools, and best practices. Here's a detailed overview of how application security is accomplished:

- **Secure software development life cycle (SDLC):** Implementing security throughout the software development process is critical. This includes integrating security practices in all SDLC phases, such as requirements gathering, design, coding, testing, deployment, and maintenance.

- **Threat modeling:** Conducting threat modeling exercises during the design phase to identify potential threats, attack vectors, and security requirements for the application. This helps in prioritizing security efforts and mitigating vulnerabilities early in the development process.

- **Secure coding practices:** Enforcing secure coding practices and following coding guidelines that prevent common vulnerabilities, such as injection attacks, **cross-site scripting** (**XSS**), and insecure direct object references.

- **Input validation and output encoding:** Implementing input validation to ensure that user-provided data is sanitized and does not lead to code execution vulnerabilities. Output encoding ensures that user-supplied data is correctly rendered and displayed to prevent XSS attacks.

- **Authentication and authorization:** Implementing strong authentication mechanisms to verify user identity. Authorizing users based on their roles and access rights prevents unauthorized access to sensitive functionalities and data.

- **Session management:** Securing session handling to prevent session hijacking and fixation attacks. Ensuring sessions have a timeout, use secure cookies, and are managed securely.

- **Data encryption and protection:** Applying encryption to sensitive data at rest and in transit. Protecting sensitive data against unauthorized access and ensuring compliance with data protection regulations.

- **Error handling and logging:** Implementing proper error handling mechanisms to avoid exposing sensitive information to attackers. Logging security events and error messages assists in monitoring and investigating potential security incidents.

- **Penetration testing and code reviews:** Conducting regular security code reviews and penetration testing to identify vulnerabilities and weaknesses in the application. Fixing discovered issues before deployment is crucial.

- **Web application firewalls (WAFs)**: Deploying WAFs as an additional layer of protection against web application attacks, such as **SQL injection (SQLi)**, **cross-site scripting (XSS)**, and **cross-site request forgery (CSRF)**.

- **Dependency management**: Regularly updating and patching third-party libraries and components to address known vulnerabilities.

- **Cloud-based application security**:

  - **IAM**: Implementing IAM solutions to control access to cloud-based applications and resources.

  - **Security groups and network segmentation**: Configuring security groups and network segmentation to control traffic flow between cloud resources.

  - **WAF services**: Utilizing cloud-based WAF services to protect web applications from common attacks.

  - **Encryption and key management**: Leveraging cloud-based encryption services and key management to secure data at rest and in transit.

  - **Security monitoring and logging**: Using cloud-based monitoring and logging services to track and analyze security events in real time.

- **Popular development tools and languages for application security**: There are multiple development tools and languages used. Some popular **integrated development environments (IDEs)** include Microsoft's VS Code and Visual Studio, JetBrains, Eclipse, Xcode, Atom, and many others. Some IDEs are specific to a particular language or can be used across multiple languages and platforms.

  Programming languages are also varied and numerous. The languages used can be categorized into two types, **compiled languages** or **interpreted languages**.

  Some popular compiled languages include C, C++, C#, Basic, Java, Rust, and Go, to name just a handful. Compiled languages use source code that is then compiled into a program that is machine or byte code. Java is unique in that it gets its source code compiled, but that code then runs in a Java virtual machine, which is an interpreter.

  Interpreted languages are programming languages that can be run from source code through an interpreter. The interpreter is a program that is installed in order for the language's source code to run and then reads the source code line by line to implement the commands of the application.

  Some popular interpreted languages include Java (see previous reference), Perl, Python, PHP, PowerShell, Ruby, and many others.

The selection of the development language comes down to the requirements for the application or the business:

- **Static application security testing (SAST) tools**: Tools such as Fortify, Checkmarx, and SonarQube analyze application source code for security vulnerabilities during development

- **Dynamic application security testing (DAST) tools**: Tools such as Burp Suite, OWASP ZAP, and Acunetix scan running applications to identify security weaknesses from the outside

- **Interactive application security testing (IAST) tools**: Tools such as Contrast Security and Hdiv Security analyze applications during runtime to identify vulnerabilities

- **Secure code review tools**: Manual code review and peer review processes to identify security issues during development

Accomplishing application security requires a proactive and integrated approach throughout the SDLC. By incorporating secure coding practices, threat modeling, regular testing, and the use of appropriate security tools, organizations can build applications that are resilient against various cyber threats and provide a higher level of protection for sensitive data and critical functionalities. Additionally, leveraging cloud-based security components can enhance the security posture of cloud-based applications and services.

## Cloud security

With the increasing adoption of cloud services, ensuring cloud security is crucial. Cybersecurity architects work to secure cloud environments, using encryption, access controls, and monitoring to protect data and applications hosted in the cloud.

Cloud application security is a critical aspect of modern cybersecurity, especially as organizations increasingly adopt cloud computing to host and deliver their applications and services. Securing cloud-based applications involves addressing unique challenges posed by cloud environments, such as shared responsibility models, data residency, multi-tenancy, and remote access. Here's an expanded focus on cloud application security, including the tools and cloud platforms commonly used:

- **Cloud service models and shared responsibility**: Understanding the shared responsibility model provided by different cloud service models (**infrastructure as a service (IaaS)**, **platform as a service (PaaS)**, and **software as a service (SaaS)**). This clarifies the division of security responsibilities between the cloud provider and the customer.

- **Encryption and key management**: Utilizing cloud-based encryption services, such as AWS **Key Management Service** (**KMS**) or Azure Key Vault, to protect sensitive data stored in the cloud. Proper key management ensures that encryption keys are securely stored and managed.

- **Cloud security groups and network segmentation**: Configuring security groups and network segmentation to control traffic flow between cloud resources. This isolates applications and services to reduce the attack surface and limit the potential impact of security breaches.

- **WAF services**: Employing cloud-based WAF services, such as AWS WAF or **Azure Web Application Firewall (WAF)**, to protect web applications from common web-based attacks, such as SQL injection and XSS.

- **Continuous monitoring and logging**: Implementing cloud-based monitoring and logging services, such as AWS CloudTrail and Azure Monitor, to track and analyze security events in real time. This enables quick detection and response to potential security incidents.

- **Serverless application security**: Securing serverless applications by setting appropriate permissions, monitoring function invocations, and implementing serverless-specific security controls.

- **Cloud-based security testing tools**: Cloud-native security testing tools that assess cloud infrastructure and applications for vulnerabilities. For example, AWS Inspector and Azure Security Center offer security assessments and recommendations.

- **Compliance and governance**: Ensuring cloud applications comply with relevant regulations and industry standards, such as the **General Data Protection Regulation (GDPR)**, **Health Insurance Portability and Accountability Act (HIPAA)**, **Payment Card Industry Data Security Standard (PCI DSS)**, and ISO 27001.

- **DevSecOps and automation**: Integrating security practices into DevOps processes to enable rapid and secure application development. Automation tools can help enforce security policies and streamline security assessments.

- **Common cloud platforms and services**:

  - **Amazon Web Services (AWS)**: AWS provides a wide range of security services, including IAM, AWS WAF, AWS Shield for DDoS protection, AWS Inspector for vulnerability assessment, and AWS CloudTrail for logging.

  - **Microsoft Azure**: Azure offers IAM services, Azure WAF, Azure Security Center for monitoring and compliance, and Azure Key Vault for key management.

  - **Google Cloud Platform (GCP)**: GCP provides IAM capabilities, Cloud Armor for DDoS protection, Cloud Security Scanner for web application scanning, and Cloud **Key Management Service (KMS)** for encryption key management.

  - **Cloud-based security providers**: In addition to native cloud services, there are third-party cloud security providers that offer advanced security solutions and services for cloud applications. Examples include Cloudflare, Trend Micro Cloud One, and Palo Alto Networks Prisma Cloud.

Securing cloud-based applications requires a comprehensive approach that leverages the security capabilities provided by cloud platforms, as well as specialized cloud security tools and services. By adopting best practices in IAM, encryption, network segmentation, monitoring, and compliance, organizations can enhance the security of their cloud applications and protect sensitive data and resources from emerging cyber threats.

## Mobile security

In the era of mobile devices, ensuring mobile security is a challenge. Cybersecurity architects implement **mobile device management** (**MDM**) solutions, enforce security policies on mobile devices, and promote secure coding practices for mobile applications. Mobile security refers to the protection of mobile devices, applications, and data from various security threats and risks. As mobile devices become increasingly integral to business operations and personal activities, ensuring mobile security is a crucial aspect of an organization's overall cybersecurity strategy. A cybersecurity architect plays a key role in designing and implementing mobile security measures to protect these devices and the sensitive information they hold. Here's how a cybersecurity architect accomplishes mobile security:

- **MDM**: Implementing an MDM system to centrally manage and monitor mobile devices within the organization. MDM enables security policies, remote wiping, encryption, and authentication mechanisms to be enforced on mobile devices.

- **Mobile application security**: Implementing secure coding practices and conducting security reviews for mobile applications. This includes validating the security of third-party applications and ensuring apps adhere to security standards.

- **Secure mobile app development**: Collaborating with developers to embed security into the mobile app development process. The cybersecurity architect guides the adoption of secure coding practices, vulnerability scanning, and security testing.

- **Secure network connections**: Encouraging the use of secure Wi-Fi networks and VPNs when accessing sensitive data or business applications. This prevents data interception and eavesdropping on public networks.

- **Mobile security policies**: Developing and enforcing mobile security policies that govern the use of mobile devices and applications within the organization. These policies address issues such as device usage, data storage, and **bring-your-own-device** (**BYOD**) policies.

- **Mobile threat defense** (**MTD**): Deploying MTD solutions that actively monitor and protect against mobile-specific threats, such as malware, phishing, and man-in-the-middle attacks.

- **Mobile incident response planning**: Developing incident response plans specific to mobile security incidents, including procedures for reporting and containing mobile-related breaches.

- **Continuous monitoring and risk assessment**: Implementing continuous monitoring of mobile devices and applications to detect anomalies and potential security breaches. Regular risk assessments help identify vulnerabilities and areas for improvement.

By accomplishing mobile security through these measures, a cybersecurity architect helps protect mobile devices, data, and applications from potential threats and vulnerabilities. They play a crucial role in building a secure mobile environment, enabling employees and businesses to use mobile devices effectively without compromising the organization's security.

# Vendor and third-party risk management

Many organizations rely on third-party vendors and service providers. Cybersecurity architects assess the security posture of these external entities to mitigate potential risks associated with their access to sensitive data or systems. A cybersecurity architect plays a vital role in ensuring the security of vendor and third-party relationships within an organization. Their involvement in vendor and third-party risk management can be summarized as follows:

- **Assessment of security risks**: The cybersecurity architect assesses the potential security risks associated with engaging with vendors and third-party partners. They evaluate the security practices, protocols, and infrastructure of these external entities to identify any vulnerabilities or potential threats.

- **Security requirements and due diligence**: They work with procurement and legal teams to define security requirements for vendors and third parties. The cybersecurity architect conducts due diligence on potential partners to ensure that their security practices align with the organization's standards.

- **Contractual security obligations**: The cybersecurity architect helps in drafting contracts and **service-level agreements (SLAs)** that include specific security obligations for vendors and third parties. These contractual clauses outline security responsibilities and set expectations for compliance.

- **Ongoing monitoring and compliance**: The cybersecurity architect establishes a monitoring framework to track the security posture of vendors and third-party partners continuously. Regular assessments and audits are conducted to ensure ongoing compliance with security requirements.

- **Incident response planning**: They collaborate with vendors and third parties to develop incident response plans. The cybersecurity architect ensures that the partners are prepared to handle security incidents effectively and communicate with the organization in case of a breach.

- **Continuous communication**: Building and maintaining open lines of communication with emerging technologies evaluation vendors and third parties is crucial. The cybersecurity architect engages in regular discussions to address security concerns, provide guidance, and foster a collaborative approach to security.

- **Remediation and improvement**: If security gaps are identified in vendor or third-party practices, the cybersecurity architect works with them to implement remediation plans. They provide guidance on enhancing security measures and ensure that necessary improvements are made.

- **Technology integration**: The cybersecurity architect facilitates the integration of security technologies and protocols between the organization and its vendors/third-party partners. This ensures seamless information sharing while maintaining a strong security posture.

- **Incident coordination**: In the event of a security incident involving a vendor or third party, the cybersecurity architect coordinates with the organization's incident response team and the partner's security team to contain the threat and limit its impact.

Overall, the cybersecurity architect's involvement in vendor and third-party risk management is essential for safeguarding the organization's data, systems, and reputation. Their efforts help establish strong security practices throughout the vendor and third-party ecosystem, reducing the organization's exposure to potential cyber risks.

## Emerging technologies evaluation

Keeping up with the latest cybersecurity trends and emerging technologies is vital for a cybersecurity architect. They evaluate the potential impact of new technologies on the organization's security and adapt security measures accordingly. A cybersecurity architect is instrumental in evaluating emerging technologies to determine their potential impact on an organization's security posture. The summary of how they work with emerging technologies evaluation includes the following:

- **Research and analysis**: The cybersecurity architect stays up to date with the latest technological advancements and trends in the cybersecurity landscape. They conduct in-depth research and analysis to understand how emerging technologies could impact the organization's security needs and challenges.

- **Risk assessment**: They assess the potential risks and vulnerabilities associated with adopting new technologies. By evaluating the security features and potential weaknesses, the cybersecurity architect identifies the level of risk that each technology may introduce to the organization.

- **Security requirements**: The cybersecurity architect collaborates with other IT teams and business units to define specific security requirements for adopting emerging technologies. They ensure that security considerations are an integral part of the technology evaluation process.

- **Proof-of-concept (POC) testing**: Before implementation, the cybersecurity architect may conduct POC testing for selected emerging technologies. This helps to evaluate the technology's effectiveness and security capabilities in a controlled environment.

- **Vendor engagement**: If the emerging technology involves third-party vendors, the cybersecurity architect engages with them to assess their security practices and protocols. They ensure that the vendor's security measures align with the organization's standards.

- **Integration with existing security measures**: The cybersecurity architect evaluates how the new technology will integrate with the organization's existing security infrastructure. They ensure that the implementation does not compromise the overall security posture.

- **Scalability and flexibility**: Assessing the scalability and flexibility of emerging technologies is crucial. The cybersecurity architect determines whether the technology can adapt to the organization's changing security needs and growing infrastructure.

- **Compliance considerations**: These considerations judge any regulatory or industry-specific compliance requirements that may apply to the adoption of the new technology. The cybersecurity architect ensures that the organization remains compliant with relevant standards and regulations.

- **Continuous monitoring**: After implementation, the cybersecurity architect monitors the performance and security of the emerging technology regularly. They analyze security data and make adjustments as necessary to maintain a robust security posture.

Overall, the cybersecurity architect's involvement in emerging technologies evaluation is critical for identifying innovative solutions while mitigating potential security risks. By conducting thorough assessments and working collaboratively with various teams, they ensure that the organization adopts emerging technologies securely and effectively.

## Other areas of focus

The cybersecurity architect has to focus on a wide range of subject areas, many of which cross or share common controls or areas. You have probably noticed that many of the items discussed already were discussed in the previous two chapters. That is by design and should underline the importance of understanding and comprehending these subject areas. Some other areas that will cross boundaries are the following:

- **IAM**: Controlling user access to sensitive data and resources is critical. Cybersecurity architects design IAM solutions that include strong authentication mechanisms, **single sign-on** (**SSO**), and RBAC to ensure that users have appropriate levels of access.

- **Data protection and encryption**: Safeguarding sensitive data is a top priority. Cybersecurity architects implement encryption techniques to protect data at rest and in transit. They also establish data classification policies to determine how data should be handled based on its sensitivity.

- **Incident response planning**: Preparing for security incidents is vital. Cybersecurity architects develop detailed incident response plans that outline procedures for detecting, reporting, and responding to security breaches. These plans ensure that the organization can act swiftly and efficiently in the event of an attack.

- **Security training and awareness**: Human error is a significant factor in security breaches. Cybersecurity architects conduct regular security awareness training sessions to educate employees about security best practices, social engineering threats, and the importance of reporting security incidents promptly.

- **Compliance and regulatory adherence**: Cybersecurity architects are responsible for ensuring that the organization complies with relevant data protection laws, industry regulations, and internal security policies. They work closely with legal and compliance teams to meet these requirements.

The cybersecurity architect has many areas of focus. This section provided an overview of these areas and how the cybersecurity architect interacts with or affects these areas. The reality is that the cybersecurity architect is expected to be a subject matter expert when it comes to security and that may be in a specific area of focus or across all areas.

# Cybersecurity architect as a part of the bigger team

A cybersecurity architect is an essential part of a bigger cybersecurity team responsible for safeguarding an organization's digital assets and information. The cybersecurity team typically consists of various roles with different responsibilities, and their inter-relationships are crucial for ensuring comprehensive cybersecurity measures. They work closely with other team members to understand their requirements and provide security guidance during the development and deployment of systems and applications.

Here's how a cybersecurity architect fits into the bigger team and how they interrelate with other roles:

- **Security operations center (SOC) analysts**: SOC analysts are responsible for monitoring and analyzing security alerts and events generated by various security tools, such as SIEM, IDS/IPS, and anti-virus solutions. They investigate potential security incidents and coordinate incident response efforts. The cybersecurity architect collaborates with SOC analysts to fine-tune monitoring rules and ensure that the SOC has access to relevant security data for effective threat detection.

- **Penetration testers (ethical hackers)**: Penetration testers perform authorized simulated cyber attacks to identify vulnerabilities in the organization's systems and applications. The cybersecurity architect works with penetration testers to understand their findings and recommendations, incorporating them into security designs and remediation plans.

- **Information security managers**: Information security managers oversee the overall security program, including policy development, compliance, and risk management. They work closely with the cybersecurity architect to ensure that security practices align with organizational goals and regulatory requirements.

- **Security engineers**: Security engineers implement and manage security technologies, such as firewalls, encryption systems, and IAM solutions. The cybersecurity architect collaborates with security engineers to design and integrate these technologies into the overall security architecture.

- **Compliance and risk management specialists**: Compliance and risk management specialists focus on ensuring that the organization adheres to relevant security standards, laws, and regulations. The cybersecurity architect works with these specialists to identify potential risks and implement appropriate security controls to achieve compliance.

- **Incident response team**: The incident response team is responsible for coordinating and responding to security incidents. The cybersecurity architect plays a crucial role in developing incident response plans, defining roles, and establishing communication channels to ensure an effective response to security breaches.

- **Network and system administrators**: Network and system administrators manage the organization's IT infrastructure. The cybersecurity architect collaborates with these administrators to implement security measures, ensure secure configurations, and conduct regular security updates and patching.

- **Developers and DevOps teams**: Developers and DevOps teams are responsible for building and deploying applications and services. The cybersecurity architect works closely with these teams to embed security into the SDLC and ensure secure coding practices are followed.

- **Business stakeholders and leadership**: The cybersecurity architect communicates security risks, strategies, and requirements to business stakeholders and leadership. They help business leaders understand the importance of cybersecurity and its impact on the organization's overall success.

The effective interrelationships between the cybersecurity architect and other team members facilitate collaboration, knowledge sharing, and a cohesive approach to cybersecurity. By working together, the team can implement robust security measures, detect and respond to security incidents promptly, and continually improve the organization's security posture to defend against ever-evolving cyber threats.

## Responsibilities

Cybersecurity architects carry the weighty responsibility of designing, implementing, and managing robust security measures to protect an organization's digital assets, data, and reputation. Their expertise is crucial in safeguarding against cyber threats and ensuring a proactive and resilient security posture. The responsibilities of a cybersecurity architect typically include the following:

- Security strategy and planning
- Security budgeting
- Security infrastructure design
- Risk assessment and management
- Security policy and standards development
- Security awareness and training
- Security incident response planning
- Collaboration and communication
- Security testing and assessment
- Security monitoring and analysis
- Emerging technologies evaluation
- Compliance and auditing

Cybersecurity architects carry the weighty responsibility of designing, implementing, and managing robust security measures and projects to protect an organization's digital assets, data, and reputation. Their expertise is crucial in safeguarding against cyber threats and ensuring a proactive and resilient security posture.

## Scope of vision

The scope of vision that a cybersecurity architect provides to an enterprise is broad and encompasses various aspects of the organization's security landscape. As a strategic thinker and a subject matter expert in cybersecurity, a cybersecurity architect plays a crucial role in shaping the organization's security posture and ensuring the protection of its digital assets, data, and systems.

The cybersecurity architect's vision is critical for creating a cohesive and effective security environment within the enterprise. They bridge the gap between technical and business aspects of security, ensuring that security measures align with the organization's goals and are integrated into every aspect of its operations. The bridging between the technical and business drivers means that the cybersecurity architect interacts with and assists executive management in driving the goals of the business in a secure way. The cybersecurity architect's scope of vision helps the enterprise build a strong security foundation, proactively manage risks, and respond effectively to security incidents, thereby safeguarding the organization's reputation, customer trust, and competitive advantage.

## Summary

This chapter concludes after offering an in-depth exploration of the cybersecurity architect role and its significance within the realm of cybersecurity. The chapter covered a wide array of essential aspects associated with the role.

You gained a clear understanding of the various areas of focus within the cybersecurity domain, enabling you to identify your specialization or interests effectively. The chapter emphasized the dynamic nature of the cybersecurity architect's role within a larger team context, providing insights into collaborating with other professionals and contributing to the team's overall cybersecurity endeavors.

The core responsibilities of a cybersecurity architect include designing secure systems, implementing robust security measures, and addressing emerging cyber threats. You are now equipped with a comprehensive understanding of the expectations and duties associated with this critical role.

Additionally, the chapter highlighted the importance of the cybersecurity architect's broad vision, which involves aligning their efforts with the broader objectives of the enterprise or business. This strategic perspective ensures that cybersecurity measures play an integral role in supporting the organization's overall goals.

In summary, the chapter imparted valuable knowledge and skills essential to excel as a cybersecurity architect. By grasping the intricacies of this role and its relation to the larger technology team and business context, you will be well prepared to make significant contributions to the field of cybersecurity.

# Part 2: Pathways

With the foundations established, the focus shifts to charting pathways for career progression as a cybersecurity architect. This part explores the multifaceted competencies that architects cultivate across technical, communication, leadership, and strategic domains.

*Chapter 4* dives into core architecture activities – principles, design, and analysis – that guide the implementation of security solutions. *Chapter 5* discusses balancing ideal security with threat models, risk appetite, governance obligations, and business alignment.

*Chapter 6* underscores the vital yet often overlooked skill of meticulous documentation, providing guidance to elevate this strategic capability. *Chapter 7* examines potential roadmaps for progressing into architect roles from various starting points and experience levels.

As certifications are frequently important milestones, *Chapter 8* demystifies the crowded credentialing landscape, providing clarity for navigating decisions tailored to individual growth needs.

Together, these chapters illustrate the extensive yet nuanced expertise that cybersecurity architects integrate to shepherd security programs holistically from concepts to concrete implementation. The pathways illuminate routes to cultivate versatile skills securing technology innovation within complex organizational contexts.

This part has the following chapters:

- *Chapter 4, Cybersecurity Architecture Principles, Design, and Analysis*
- *Chapter 5, Threat, Risk, and Governance Considerations as an Architect*
- *Chapter 6, Documentation as a Cybersecurity Architect – Valuable Resources and Guidance for a Cybersecurity Architect Role*
- *Chapter 7, Entry-Level-to-Architect Roadmap*
- *Chapter 8, The Certification Dilemma*

# Cybersecurity Architecture Principles, Design, and Analysis

*"Let your plans be dark and impenetrable as night, and when you move,
fall like a thunderbolt."*

*– Sun Tzu*

In the previous chapter, we covered the role of the cybersecurity architect and their responsibilities to help prepare and understand the scope of what a cybersecurity architect may do within an organization. In this chapter, the discussion will shift to the principles, design, and analysis of cybersecurity architecture.

**Cybersecurity architecture** is a technical architecture that focuses on achieving specific security goals. Essentially, it focuses on how to systematically, holistically, and repeatedly implement solutions that meet an organization's security and compliance requirements.

You may have noticed I have been using quotes from Sun Tzu's *Art of War* at the beginning of each chapter. This is in part because people are more familiar with Sun Tzu's *Art of War* than they are with Miyamoto Musashi's *The Book of Five Rings*. Both are excellent books on strategy and war. That being said, cybersecurity in itself is a war against malicious actors and threats to an individual and an organization. Cybersecurity architecture can, in turn, associate warfare with the concepts of implementation, design, principles, and analysis. The quote in this chapter could be interpreted to mean that when you plan, make it complete and apply it consistently, like the darkness of night, and when you act on that plan, do so with the power and intensity of lightning. The association with warfare is clear, but the design and implementation of cybersecurity controls can be associated as well. Cybersecurity architecture has the responsibility of providing the vision and approach to the security of an organization, and those plans need to be without compromise – that is, they must be complete but be able to pivot quickly.

This chapter will discuss the areas of architecture principles, design, and analysis that are part of the day-to-day function of any cybersecurity architect. We will discuss the various approaches to designing and analyzing a particular solution or control while understanding the principles around the choice we should take over another, depending on the situation.

This chapter covers the following topics:

- Principles
- Design
- Analysis

# Principles

In today's digital age, cybersecurity has become a critical concern for organizations of all sizes. With cyber threats evolving and becoming more sophisticated, it is imperative to have a robust cybersecurity architecture in place to protect sensitive data and systems. **Cybersecurity architecture** refers to the design and implementation of security measures to safeguard information and technology assets from unauthorized access, use, disclosure, disruption, modification, or destruction.

Before diving into the core concepts of this chapter, it's important to note our intentional choice of terminology. We have opted to use language that is clear and accessible to everyone, even when this deviates from the exact terminology used in some security architecture frameworks and standards.

For example, understanding an organization's goals and linking them to appropriate security outcomes is a crucial concept here. Formally, many frameworks refer to this as *governance* – ensuring resources align with organizational objectives. However, governance means different things to different groups. It could refer to organizational structure, political bodies, technology oversight, or general control.

Moreover, frameworks vary in their terminology. **Open Enterprise Security Architecture (O-ESA)** uses *governance* in the formal sense, as specified previously. **Open Security Architecture (OSA)** uses *governance* too but defines it more broadly as part of the security architecture landscape. **Sherwood Applied Business Security Architecture (SABSA)** prefers *business enablement*.

This variability risks miscommunication if people interpret the same word differently. To avoid ambiguity, we have chosen unambiguous language in this book. We'll directly refer to enabling the organizational mission rather than using the term *governance*.

While this clarity benefits our discussion, it is worth noting our terminology may differ from other sources. As you explore additional architectural materials, keep in mind that the same concepts may be described using different words, depending on the source.

You will find that in technology, in many instances, the answers or responses you will receive will be "*it depends*," depending on the question, without complete context or an explicit example or definition being provided. In architecture, effective communication is critical for both understanding concepts and collaborating with stakeholders. The aim is to directly convey ideas to you using consistent and accessible language. However, as you broaden your knowledge, recognize that other valid resources may use alternative terminology to express similar ideas regarding how security programs align with the overarching organizational goals and risk management needs.

# The importance of cybersecurity architecture

The importance of cybersecurity architecture cannot be overstated. It serves as the foundation for a strong and effective cybersecurity strategy. Without a well-designed architecture, organizations are susceptible to cyber attacks, which can result in financial loss, reputational damage, and legal implications. A comprehensive cybersecurity architecture provides a layered approach to security, mitigating risks and ensuring the confidentiality, integrity, and availability of critical assets.

Cybersecurity architecture has emerged as a business enabler rather than a technical afterthought. By treating cyber risks as core business risks and aligning security with business needs, organizations can maximize their cyber protection while still pursuing their core mission.

# The key principles of cybersecurity architecture

The principles that will be outlined here provide a strategic blueprint for designing, implementing, and operating a robust cybersecurity program. Adhering to these foundational principles is critical because of the following reasons:

- They represent accumulating wisdom from decades of cybersecurity experience in both public and private sectors. Following these tried and tested principles improves the odds of cybersecurity success.
- They align cyber protections with organization-wide risk management rather than just compliance. This enables judicious security investments tailored to the organization.
- They prevent common cybersecurity pitfalls such as lack of visibility, uncontrolled access, and reliance on single defensive layers. Proper architecture avoids these issues by design.
- They balance security with usability through concepts such as standardization, automation, and secure defaults. Usability reduces pushback from business units.
- They emphasize cyber resilience in addition to prevention. No single control is foolproof but layered adaptive defenses minimize impact.
- They drive a holistic and integrated view of cybersecurity. This is key as threats exploit seams and gaps between point capabilities.

In essence, sound cybersecurity architecture principles are timeless guidelines for developing cohesive, comprehensive, and risk-based cyber protections that focus on supporting organizational objectives. Adopting and customizing these principles to meet the business and mission needs of the organization provides a solid foundation for managing cyber risk.

## Defense in depth

Defense in depth is a fundamental principle of cybersecurity architecture. It involves implementing multiple layers of security controls to protect against various types of threats rather than relying on just one. This approach ensures that even if one layer is breached, there are additional layers in place to prevent unauthorized access or damage.

To achieve defense in depth, organizations should implement a combination of preventive, detective, and corrective controls. These include firewalls, intrusion detection systems, antivirus software, strong authentication mechanisms, patch management, and regular security audits. These topics have been referenced and discussed in previous chapters.

## Least privilege

The principle of least privilege is based on the idea that users should only be granted the minimum level of access necessary to perform their tasks. By limiting user privileges, organizations can reduce the potential impact of a security breach or insider threat. This principle involves assigning permissions based on job roles and responsibilities, regularly reviewing access rights, and implementing strong identity and access management controls.

## Separation of duties

The principle of separation of duties aims to prevent conflicts of interest and reduce the risk of fraud or unauthorized activities. It involves dividing responsibilities among multiple individuals to ensure that no one person has complete control over a critical process or system. By implementing this principle, organizations can establish checks and balances, increase accountability, and reduce the likelihood of internal threats.

## Fail-safe defaults

Fail-safe defaults refers to the practice of configuring systems and applications with secure settings as the default option. This principle ensures that even if administrators or users fail to apply specific security measures, the system will still be protected. Organizations should establish secure configurations for operating systems, network devices, and software applications, and regularly review and update them to address new vulnerabilities.

## Secure by design

The principle of secure by design emphasizes integrating security into the design and development of systems, networks, and applications from the ground up. By considering security requirements from the initial stages of the development life cycle, organizations can minimize vulnerabilities and reduce the need for costly security patches and updates. Secure coding practices, secure network architecture, and threat modeling are essential components of secure by design.

## Regular updates and patching

Regular updates and patches are crucial for maintaining the security of systems and applications. Cybercriminals constantly exploit vulnerabilities in software, making it essential for organizations to stay up to date with the latest security patches. Implementing a patch management process, conducting regular vulnerability assessments, and promptly applying updates are essential for protecting against known threats.

### Continuous monitoring

Continuous monitoring involves the real-time collection and analysis of security events to detect and respond to potential security incidents. It involves the use of **security information and event management (SIEM)** solutions, **intrusion detection systems (IDSs)**, and log analysis tools. By continuously monitoring network traffic, system logs, and user activities, organizations can identify and respond to security incidents promptly.

### Incident response planning

Incident response planning is a critical component of cybersecurity architecture. It involves developing a comprehensive plan to address security incidents, including data breaches, malware infections, and other cyber attacks. An effective incident response plan defines roles and responsibilities, outlines the steps to be taken in the event of an incident, and establishes communication channels and escalation procedures. Regularly testing and updating the plan is essential to ensure its effectiveness.

## Implementing the key principles of cybersecurity architecture

Implementing the key principles of cybersecurity architecture requires a systematic and holistic approach. Let's look at the steps that are involved in implementing these principles.

### Identifying and assessing risks

The first step in implementing cybersecurity architecture is to identify and assess risks. This involves conducting a thorough risk assessment to identify potential threats and vulnerabilities that could impact the organization's assets. The risks should be evaluated based on their likelihood and potential impact. Once the risks have been identified, appropriate controls can be implemented to mitigate them.

### Developing a security policy

A security policy serves as a roadmap for implementing and maintaining cybersecurity architecture. It outlines the organization's approach to security, defines roles and responsibilities, and establishes guidelines and procedures for protecting sensitive data and systems. The policy should be aligned with industry best practices and regulatory requirements. Regular reviews and updates are necessary to ensure its effectiveness.

### Designing a secure network infrastructure

Designing a secure network infrastructure is essential for protecting against external threats and unauthorized access. Unauthorized access does not denote just external actors. Internal actors can be a potential source of unauthorized access that a secure network needs to consider. The use of micro-segmentation, ACLs, or other controls, including the concepts around **Zero Trust,** can allow the secure network to treat all zones as potentially hostile, regardless of the internal/external nature of the access. This involves implementing firewalls, **intrusion detection and prevention systems (IDPSs)**, and secure remote access mechanisms. Network segmentation and the use of **virtual private networks (VPNs)** can further enhance network security. Regular security assessments should be conducted to identify and address vulnerabilities.

### Implementing access controls and authentication mechanisms

Access controls and authentication mechanisms play a crucial role in ensuring that only authorized individuals can access sensitive data and systems. This involves implementing strong password policies, multi-factor authentication, and role-based access controls. More complex environments implement context-aware permissions or **just-in-time (JIT)** access management to limit the potential implications associated with password or secrets knowledge in sensitive or high-security environments. Regular user access reviews and account management processes are necessary to maintain the integrity of access controls.

### Implementing encryption and encryption key management

Encryption is an essential component of cybersecurity architecture. It protects data in transit and at rest, making it unreadable to unauthorized individuals. Encryption for data in use is also possible with runtime and RAM encryption mechanisms that leverage **partially homomorphic encryption (PHE)** and **fully homomorphic encryption (FHE)**. Organizations should implement robust encryption algorithms and protocols and ensure that encryption keys are properly managed and protected. Regular encryption audits and key rotation are necessary to maintain the effectiveness of encryption controls.

### Implementing IDPSs

IDPSs help organizations detect and prevent unauthorized access and malicious activities. These systems monitor network traffic, analyze patterns, and alert administrators to potential security incidents. IDPSs should be regularly updated and tuned to detect the latest threats. Integration with incident response mechanisms enables timely responses and the mitigation of security incidents.

### Implementing security monitoring and incident response mechanisms

Security monitoring and incident response mechanisms enable organizations to detect, investigate, and respond to security incidents promptly. This involves implementing SIEM solutions, log analysis tools, and incident response platforms. Regularly testing and updating these mechanisms is necessary to ensure their effectiveness.

## Best practices for maintaining cybersecurity architecture

Sustaining a secure cybersecurity infrastructure demands continuous diligence and compliance with industry standards. Here are some recommended guidelines to keep in mind:

* Consistently update and apply patches to your systems and software to fix known security issues

- Periodically carry out vulnerability evaluations and penetration tests to discover and remediate potential weak points

- Adopt stringent password guidelines and use multi-factor authentication to safeguard against unauthorized entries

- Continually monitor and modify access permissions to guarantee that only approved users can access confidential information and systems

- Utilize network partitioning to minimize the consequences of a security compromise

- Educate staff about cybersecurity best practices and the value of staying vigilant about security

- Create an all-encompassing backup and disaster recovery strategy to maintain ongoing operations in case of a security event

- Routinely examine and modify security protocols and practices to adapt to evolving threats and legal mandates

- Keep abreast of the most recent developments in cybersecurity and emerging risks to proactively mitigate possible hazards

## Challenges and considerations in implementing cybersecurity architecture

Implementing cybersecurity architecture can be challenging due to various factors. Here are some of the key challenges and considerations:

- **Limited resources**: Organizations may face budgetary constraints and a shortage of skilled cybersecurity professionals, making it challenging to implement and maintain a robust cybersecurity architecture.

- **Complexity**: Cybersecurity architecture involves multiple components, technologies, and processes that can be complex to design, implement, and manage.

- **Rapidly evolving threats**: Cyber threats are constantly evolving, requiring organizations to stay updated regarding the latest threats and vulnerabilities and implement timely security measures.

- **Regulatory compliance**: Organizations must comply with industry-specific regulations and data protection laws, which can add complexity to cybersecurity architecture implementation.

- **User awareness and behavior**: Despite having robust security controls in place, user behavior can pose significant risks. Organizations must invest in security awareness training and enforce policies to promote secure computing practices.

# Cybersecurity architecture frameworks

A **cybersecurity architecture framework** is a set of principles, standards, and best practices that organizations can use to design, implement, and manage their cybersecurity architecture. These frameworks can help organizations do the following:

- Understand their cybersecurity risks
- Identify and prioritize security controls
- Implement security controls effectively
- Monitor and improve their security posture

There are many different cybersecurity architecture frameworks available, each with its strengths and weaknesses. Some of the most popular frameworks are as follows:

- The **National Institute of Standards and Technology** (NIST) **Cybersecurity Framework (CSF)**
- **The Open Group Architecture Framework (TOGAF)**
- The **Sherwood Applied Business Security Architecture (SABSA)**
- The ISO/IEC 27001 **Information Security Management System (ISMS)**

The NIST CSF is a voluntary framework that provides organizations with a set of guidelines for managing cybersecurity risk. The CSF is divided into five functions: *Identify*, *Protect*, *Detect*, *Respond*, and *Recover*. Each function has a set of subcategories that organizations can use to assess their cybersecurity risk and implement appropriate security controls.

TOGAF is a framework for enterprise architecture that includes a security architecture component. TOGAF provides a structured approach to designing, implementing, and managing an organization's IT architecture. The security architecture component of TOGAF can be used to help organizations identify and mitigate cybersecurity risks.

SABSA is a framework for information security that provides a comprehensive approach to managing security risks. SABSA is based on the principle of defense in depth, which means that organizations should implement a layered approach to security that includes physical, technical, and administrative controls.

The ISO/IEC 27001 ISMS is an international standard that provides organizations with a framework for managing information security. The standard includes a set of requirements for establishing, implementing, maintaining, and improving an ISMS. The ISO/IEC 27001 ISMS can be used to help organizations protect their information assets from a variety of threats, including cyber attacks.

When choosing a cybersecurity architecture framework, organizations should consider the following factors:

- The size and complexity of the organization

- The organization's industry
- The organization's risk tolerance
- The organization's budget

> **Note**
> No single cybersecurity architecture framework is perfect for every organization. The best framework for an organization will depend on its specific needs and circumstances.

Here are some of the benefits of using a cybersecurity architecture framework:

- **Improved security posture**: By following a framework, organizations can ensure that they are implementing appropriate security controls to protect their information assets.
- **Reduced risk**: Frameworks can help organizations identify and mitigate cybersecurity risks.
- **Increased efficiency**: Frameworks can help organizations save time and money by providing a standardized approach to security.
- **Pseudo-blueprint**: Frameworks provide pseudo-blueprints for approaching security architecture within your organization as opposed to trying to map everything out freehand. This gives you a starting point and somewhat of a roadmap to success regarding the objectives – it just needs to be shaped for your organization.

## Examples of successful cybersecurity architecture implementations

Several organizations have successfully implemented robust cybersecurity architecture to protect their assets. Here are some industries that have implemented successful cybersecurity architecture:

- **The United States Department of Defense (DoD)**: The DoD has implemented a cybersecurity architecture framework called the **Defense Information Systems Agency (DISA) Security Technical Implementation Guide (STIG)**. The DISA STIG is a set of security controls that organizations can use to protect their information systems. The DoD has been very successful in implementing the DISA STIG, and it has helped improve the security of its information systems.
- **The financial services industry**: The financial services industry is one of the most heavily regulated industries in the world, and it is also one of the most targeted by cyber attacks. To protect themselves from cyber attacks, financial services organizations have implemented a variety of cybersecurity architecture frameworks, including the NIST CSF, TOGAF, and ISO/IEC 27001. These frameworks have helped financial services organizations improve their security posture and reduce their risk of being attacked.

- **The healthcare industry**: The healthcare industry is another industry that is heavily regulated and targeted by cyber attacks. Healthcare organizations have implemented a variety of cybersecurity architecture frameworks, including the NIST CSF, TOGAF, and ISO/IEC 27001. These frameworks have helped healthcare organizations improve their security posture and reduce their risk of being attacked.

- **The retail industry**: The retail industry is also a target for cyber attacks since retailers collect and store a lot of personal information about their customers. Retail organizations have implemented a variety of cybersecurity architecture frameworks, including the NIST CSF, TOGAF, and ISO/IEC 27001. These frameworks have helped retail organizations improve their security posture and reduce their risk of being attacked.

## Business considerations for cybersecurity architecture

Implementing effective cybersecurity architecture has become a key business priority rather than just an IT issue. Here are some important business considerations:

- **Cost-benefit analysis**: There needs to be a business case analysis of the cost of implementing security controls versus the losses prevented. Controls should provide a positive ROI but not hinder business performance.

- **Resource allocation**: Cybersecurity competes with other initiatives for limited budget and human resources. Business leaders must allocate resources based on cyber risks versus other business risks.

- **Alignment with objectives**: Ultimately, the cybersecurity program should enable business goals such as improved customer experience or new revenue channels rather than be restrictive.

- **Shared responsibility**: While IT oversees technical controls, business units must own policies around data, asset management, identity management, and application security intrinsic to their operations.

- **Risk acceptance**: Certain risks may need to be accepted to pursue business innovations such as **Internet of Things** (**IoT**) implementations or cloud migrations. Risk acceptance must be an informed business decision.

- **Insurance and transfer**: Some cyber risks can be transferred via cyber insurance policies or outsourcing contracts. The extent of risk transfer should align with business risk appetite.

- **Metrics and reporting**: Cybersecurity spending and effectiveness metrics should be reported to business leadership just as with other operational metrics. Useful metrics include risk reduction, audit findings, and security incidents.

- **Governance**: Cybersecurity architecture decisions should have sign-offs from business stakeholders via mechanisms such as security steering committees. This ensures alignment with business needs.

- **Culture**: Policies mandated by security teams but not embraced by the broader organizational culture will fail. Culture starts with tone at the top from business leaders.

- **Adapting to change**: Flexibility to adapt the cyber program for events such as mergers, new technologies, and market evolutions is critical. Change capability starts with architecture.

### Resources for learning more about cybersecurity architecture

If you are interested in learning more about cybersecurity architecture, here are some valuable resources:

- *Cybersecurity Architecture: An Enterprise Perspective*, by Neil Rerup
- *Practical Cybersecurity Architecture*, by Ed Moyle and Diana Kelly
- *Threat Modeling: Designing for Security*, by Adam Shostack
- NIST Cybersecurity Framework
- The **Cybersecurity and Infrastructure Security Agency (CISA)** website
- The SANS Institute blog

Cybersecurity architecture plays a crucial role in safeguarding organizations' sensitive data and systems from cyber threats. By implementing the key principles discussed in this section, organizations can establish a strong and effective cybersecurity strategy. It is essential to continuously monitor and update cybersecurity architecture to address emerging threats and vulnerabilities. By following best practices and staying informed about the latest trends and technologies, organizations can enhance their cybersecurity posture and protect against potential risks.

## Design

When it comes to securing your cloud, enterprise, application, or network, a well-structured cybersecurity architecture design is of paramount importance. It forms the backbone of any organization's cyber defense strategy and must be meticulously planned and implemented. Cybersecurity architecture design plays a vital role in protecting an organization from potential threats and vulnerabilities. It serves as the blueprint for a robust security strategy, outlining the mechanisms and controls that will be used to secure the organization's digital assets.

Without a well-thought-out cybersecurity architecture design, organizations leave themselves open to various risks, such as data breaches, cyber attacks, and financial losses. Therefore, understanding and implementing an effective cybersecurity architecture design is crucial for any cloud enterprise application network.

# How does cybersecurity architecture design work?

Before developing a cybersecurity architecture, we need to gather foundational information to establish the organizational context for the design work. Specifically, we need a baseline understanding of the organization, including its goals, culture, mission, and unique needs.

Grasping the organizational nuances and specifics is crucial because they will ultimately drive the architecture. The context shapes everything from the design scope to security controls, implementation plans, operating constraints, and functional requirements.

The cybersecurity architecture must align with the practical realities of the organization. All aspects of the architecture should be appropriate and feasible within the organization's operating environment.

In essence, the organization provides the backdrop that informs architectural choices. By thoroughly understanding the organizational landscape first, we can craft targeted designs that support business objectives and account for constraints. The organization context supplies key inputs that allow us to tailor the architecture to the entity's risk profile and resources.

Cybersecurity architecture design is not a one-size-fits-all concept. It varies from one organization to another, based on various factors such as the nature of the business, organizational goals, and specific industry requirements.

## Identifying organizational goals

The first step in designing an effective cybersecurity architecture is to identify the organization's primary business goals. These goals act as the guiding force for the entire design process, ensuring that the security measures that are implemented align with the organization's mission.

## Establishing the context for designs

The context for the designs is established based on the organizational goals that have been identified. This context is crucial in shaping the design and scope of the security measures. It also plays a vital role in determining the implementation, operational constraints, and functional requirements of the cybersecurity architecture.

## Developing security goals

Once the organizational goals and the context for designs have been established, the next step is to develop specific security goals. These goals are derived from the organization's primary goals and the context for designs that have been identified. They serve as the foundation for the entire cybersecurity architecture design, outlining the specific security outcomes that the organization aims to achieve.

## Designing security measures

Based on the established security goals, specific security measures are designed. These measures are the actual components of the cybersecurity architecture, including technologies, processes, and controls that will be used to achieve the security goals.

# The key aspects of cybersecurity architecture design

When planning a cybersecurity architecture design, every decision matters. Some choices influence the outcome more than others, but each contributes to shaping the overall result.

Additionally, logically, certain decisions need to precede others, impacting the sequence of preparation. Other external factors also constrain planning and decisions.

Without accounting for these dynamics, invalid assumptions risk wasted time, money, and a suboptimal experience. The circumstances surrounding the planning process govern which choices make sense and when.

Ignoring decision interdependencies and constraints leads to haphazard preparation. This undermines efficient resource use and fails to maximize the value of the experience for stakeholders.

Therefore, diligent planning requires understanding decision impacts, ordering, and context. Thoughtful sequencing aligned with priorities and objectives allows for an orchestrated effort that makes the most of investments to create an engaging, meaningful result.

Several key aspects should be considered when developing a cybersecurity architecture design. These aspects are crucial in ensuring that the design is effective, robust, and capable of protecting the organization against various cyber threats.

## *Aligning with organizational culture and skills*

The cybersecurity architecture design should align with the organization's culture and the skills of its staff. This means that the design should be developed in a way that it can be easily implemented and managed by the organization's staff while considering their skills and abilities.

Even a technically solid control may not be feasible if it's deployed in a mismatched environment. Consider a movie available only on DVD. Without a DVD player, I cannot watch it, regardless of how much I want to. Similarly, security controls that rely on specialized skills or are intolerable to corporate culture will struggle, even if they are theoretically effective. For example, forensic analysis requires niche expertise to ensure evidence admissibility. Most entities outsource this capability since maintaining it in-house is cost-prohibitive.

In this case, solutions that demand advanced forensics skills would misalign with typical corporate environments. The organization lacks the context required to support the control. Another example is controls infringing on employee privacy. In cultures emphasizing autonomy, these would breed backlash, despite security gains.

Evaluating alignment involves assessing the following aspects:

- The required skill sets and staffing models
- Integration with existing tools and infrastructure
- Impacts on productivity and user experience

- Legal, regulatory, and policy constraints
- Cultural acceptance and change management needs

Misaligned controls risk low adoption, improper use, unsustainable resource demands, and cultural rejection – even if technically sound.

By assessing alignment early, organizations can select solutions that fit their environment. This smooths deployment and maximizes usability.

Alignment evaluates how well a control meshes with organizational realities beyond just technical effectiveness. A control that's only effective in theory provides little real-world security value if it's deployed in a mismatched environment. Evaluating context fit ensures that implemented solutions meet needs sustainably.

When architecting cybersecurity solutions, certain immutable realities govern what designs are viable. These core assumptions and limitations determine which decisions make sense.

Longtime staff may implicitly operate within these ingrained organizational *rules*. However, explicitly identifying constraints creates a shared understanding for aligned planning. Defining these guiding rules allows us to shape architectures feasible for the environment.

Consider the ancient Roman architect Vitruvius' principles of *durability, utility, and beauty*. In particular, the first two rely on context. A structure's durability depends on operating conditions. Its utility stems from user needs.

Likewise, cybersecurity architecture must account for organizational realities. The designs must withstand business environments and enable organizational missions to truly provide utility.

By deliberately enumerating assumptions and limitations, we lay the groundwork for contextually-optimized architectures. Design choices align with realities when constraints are well defined. The resulting architectures implement security principles into practices tailored to the organization.

### *Establishing the organizational context*

When architecting cybersecurity solutions, Vitruvius emphasized durability and utility as key principles. These factors heavily depend on understanding the organizational context where solutions will operate.

This chapter focuses on defining the context that will shape your architectural designs. We will explore three critical areas that provide this contextual backdrop:

- Business/organizational goals
- Existing governance structures
- Risk management needs

Organizational goals offer the purest insight into an entity's priorities. Fully mapping all goals to derive security requirements would provide an ideal design framework. However, limited time often precludes comprehensively eliciting goals directly.

Therefore, we can reference existing governance structures to infer goals more expediently. Items such as policies, procedures, and standards encapsulate previous decisions reflecting organizational objectives.

While not a perfect substitute, governing documents provide readily available context to inform architectural choices. They offer pre-codified guidance steeped in the entity's mission and risk priorities.

This pragmatic approach accounts for real-world time constraints while still leveraging available artifacts to approximate security requirements suited to the organization. The resulting cybersecurity architecture stays anchored to business objectives and practical implementation realities.

## *Ensuring effectiveness*

The effectiveness of the cybersecurity architecture design depends on how well it can achieve the established security goals. The design should be evaluated regularly to ensure that it is performing effectively and is capable of protecting the organization's digital assets. The first key criterion when evaluating security solutions is effectiveness – how well a control achieves its intended outcome.

Just as business strategies are measured by metrics such as profitability and time savings, security controls can be evaluated based on their efficacy in delivering targeted security goals.

Effectiveness is critical because not all security measures are equal, even when aiming for the same result. For example, a six-digit PIN and a passphrase both provide access control to an application, laptop, or phone but differ enormously in their ability to protect confidentiality. A six-digit PIN is weak, allowing a relatively easy bypass in comparison to a passphrase with today's computing systems.

In essence, effectiveness assesses how successfully a control satisfies requirements and security objectives. An effective control delivers the desired security outcome reliably.

When reviewing existing security programs, audits often point out ineffective controls that are failing to provide adequate protection. However, not all ineffective controls are flawed by design.

Implementation choices also impact effectiveness. A control that's deployed incorrectly may not work as intended. Monitoring and testing validate whether implementations match expected effectiveness.

By evaluating current and potential controls based on their effectiveness, organizations can do the following:

- Identify existing gaps that are unable to meet security requirements
- Prioritize the remediation of ineffective controls
- Select new solutions to deliver the required protections
- Validate that the implementations operate as intended

Measuring effectiveness provides crucial insight into how well security solutions meet risk reduction needs. It enables data-driven decisions to maximize security posture through deploying reliably effective controls tailored to organizational goals.

### Considering maturity

The maturity of the cybersecurity architecture design refers to the reproducibility and reliability of the supporting processes. A mature design is consistent, managed, and capable of recovering from interruptions. When evaluating security solutions, the concept of maturity refers not just to how long something has existed or its acceptance. More importantly, it means the reproducibility and reliability of the processes that support the implementation.

In this sense, maturity mirrors frameworks such as **Capability Maturity Model Integration (CMMI)**, which assesses process maturity for software development. You can read more here: `https://cmmiinstitute.com/learning/appraisals/levels`.

Two security controls that fulfill the same goal can have very different maturity profiles. Consider incident response at two firms. One has an ad hoc process without documentation, automation, or metrics. The other employs a robust automated workflow that collects performance data to drive ongoing improvements.

Both meet the function of incident response with similar efficacy. However, the first follows an unstable, unmanaged process, which CMMI would rate at the *Initial* (Level 1) maturity level. The latter exemplifies a mature, consistent, and optimized approach, ranking at Level 4 or 5.

Mature processes offer advantages such as consistency, resilience to personnel changes, and measurability for improvement. However, increasing maturity may require investments in terms of time and budget.

Therefore, evaluating maturity as well as functionality is valuable when assessing security solutions. More mature processes ensure reliability and optimization better but require greater upfront resource commitment. Organizations can weigh these tradeoffs based on their needs and environment.

Regardless of which controls are selected, implementing them via mature, managed processes provides benefits such as reproducibility, resilience, and consistency. As solutions are designed and deployed, architecting the supporting processes using maturity best practices can maximize effectiveness, sustainability, and measurability.

A well-structured cybersecurity architecture design is essential for protecting any cloud, enterprise, application, or network. It involves a thorough understanding of the organization's goals, establishing the context for designs, developing specific security goals, and designing effective and efficient security measures. Furthermore, policies, procedures, and standards play a crucial role in guiding the design process, ensuring consistency, and achieving the desired security outcomes.

## *Maintaining efficiency*

Beyond effectiveness and maturity, efficiency is another key dimension when assessing security solutions. It is also another critical aspect of cybersecurity architecture design. The design should be efficient in terms of cost and time investment. This means that the security measures that have been implemented should be able to achieve the security goals at the least possible cost and within a reasonable time frame. Efficiency considers the time, staffing, and direct costs required for implementation and ongoing operations.

Two controls can be equally effective yet diverge significantly in efficiency. For example, manual code reviews versus automated testing both analyze application source code vulnerabilities. However, manual reviews demand far greater staff, time, and costs

Any security measure carries opportunity costs – that is, what else you could have done with the same resources. With limited budgets and personnel, dedicating resources to one area leaves less for other areas.

Consider web filtering as an example. A team that manually reviews web requests would have high costs and reduce productivity, even if they're effective at identifying malicious sites or websites that are against company policy. Automated filtering provides comparable effectiveness more efficiently.

As a cybersecurity architect, evaluating efficiency implications allows for informed tradeoff decisions to be made when given constrained resources. An expensive but potent control may not be viable if it monopolizes resources, preventing other critical protections.

When assessing controls, weigh up factors such as the following:

- Upfront and ongoing staff time required
- Implementation costs
- Licensing fees
- Maintenance resource needs
- Training demands
- Potential productivity impacts

More efficient solutions maximize benefits while minimizing resource demands. This frees up budget and personnel time to bolster defenses across more areas.

## Cybersecurity architecture design for cloud, enterprise application, and network

Cybersecurity architecture design is crucial for different aspects of an organization's IT landscape, including cloud, enterprise application, and network.

### Cloud cybersecurity architecture design

Cloud cybersecurity architecture design involves designing security controls to protect data and applications in the cloud. It requires understanding the unique security risks associated with cloud computing and designing appropriate security measures.

### Enterprise application cybersecurity architecture design

Enterprise application cybersecurity architecture design focuses on protecting the organization's applications. It involves designing security controls that protect the integrity, availability, and confidentiality of the applications.

*Case study*: Equifax's 2017 breach, attributed to an unpatched software vulnerability, affected 147 million people (`https://www.csoonline.com/article/567833/equifax-data-breach-faq-what-happened-who-was-affected-what-was-the-impact.html`).

### Network cybersecurity architecture design

Network cybersecurity architecture design involves designing security controls that protect the organization's network from threats. It requires understanding the security risks associated with networks and designing appropriate security measures.

No architecture can implement every control. Prioritizing efficient solutions stretches resources further to improve security posture more broadly. Architecting around efficiency also reduces opportunity costs, enabling more comprehensive protections aligned with organizational risk tolerance and resources.

In the digital age, effective cybersecurity architecture design is crucial for protecting an organization's IT landscape. It requires understanding the organization's goals, existing structures, and risk management strategies, as well as establishing a guiding process. By focusing on these elements, organizations can develop robust cybersecurity architecture that aligns with their mission and strategies, enabling them to achieve their goals securely.

Whether it's cloud security, enterprise application security, or network security, a well-thought-out cybersecurity architecture design can significantly enhance an organization's security posture. As the digital landscape continues to evolve, investing in cybersecurity architecture design becomes not just an option, but a necessity for businesses worldwide.

## Analysis

In a world where cyber threats are evolving rapidly, the static defense mechanisms of years past no longer suffice. Cybersecurity architecture analysis emerges as an imperative, continuous process that ensures an organization's digital defenses are calibrated against existing and emerging threats. Cybersecurity architecture analysis is the process of evaluating an organization's cybersecurity architecture to identify potential vulnerabilities and areas for improvement.

The goal of cybersecurity architecture analysis is to ensure that an organization's cybersecurity architecture is effective in protecting its information assets from cyber attacks.

## Business goals

Business goals are the reasons why an organization exists in the first place. They are usually high-level and speak to the organization's mission. For example, a commercial entity might have profitability, shareholder value, or return on investment as business goals.

To identify an organization's business goals, we can use the **seven whys method**. This is a root cause analysis technique that asks the question *Why?* repeatedly until the root cause of something is identified:

- Why?
- Why is this important?
- Why should clients care?
- Why does that confidence matter?
- Why remove these obstacles?
- Why prioritize client retention and acquisition?
- Why aim for profitability?

For example, we might ask "*Why does our organization require multi-factor authentication for employees?*" The answer might be "*So that we know that our employees are who they say they are.*" We can then ask "*Why is that important?*" and the answer might be "*So that our customers trust us.*" By continuing to ask *Why?*, we can eventually identify the organization's business goal of providing a safe and trustworthy environment for its customers. You can gather the answers to the seven whys through the following approaches:

- **Reviewing documentation**: Organizations often have policies, procedures, and standards that document their business goals. By reviewing this documentation, we can get a good understanding of what the organization is trying to achieve.
- **Talking to stakeholders**: Stakeholders are people who have a vested interest in the organization's success. By talking to stakeholders, we can get a firsthand account of what the organization's goals are.

Understanding an organization's core objectives is crucial for a cybersecurity architect in shaping a secure infrastructure. This initial step can be achieved through a multi-pronged approach: reviewing existing documentation such as policies and procedures to grasp business goals and engaging directly with stakeholders for an in-depth perspective. Often, these explorations reveal that most endeavors align with a few foundational goals, usually encapsulated in the organization's mission statement. While a thorough goal-mapping exercise is ideal, it may be time-consuming; starting with established standards can offer a valuable shortcut. A case study of Target's 2013 security breach exemplifies the consequences of not fully integrating cybersecurity goals with the overall enterprise strategy, thereby leaving exploitable vulnerabilities.

Often, when you repeat this exercise across various scenarios, you'll discern that most paths lead back to a handful of foundational objectives. These core objectives are usually embodied in an organization's mission statement, encapsulating the very essence of its existence.

For a cybersecurity architect, the preliminary step is to discern the fundamental philosophies, needs, and goals that will shape their design approach. Ideally, this should be done by meticulously analyzing the company's goals, tracing each goal's trajectory, and consequently crafting the implicit security goals that support their tech utilization. Yet, a comprehensive goal-mapping exercise might be time-intensive. Therefore, starting with already established standards, procedures, and guidelines can be beneficial.

*Case study*: Target's 2013 breach exposed the credit card details of over 40 million customers. Post-breach analysis revealed a lack of integration between cybersecurity and the broader enterprise strategy, resulting in vulnerabilities (`https://jise.org/Volume29/n1/JISEv29n1p11.pdf`).

## Leveraging governance documents to understand organizational goals

Governance documents such as policies, procedures, standards, and guidelines offer valuable shortcuts to inferring organizational goals relevant to security architecture. Though not a full substitute for exhaustive goal analysis, they provide readily available context.

### Policies

Policies codify management expectations on various topics. As articulations of intent, policies directly reflect organizational priorities. Cybersecurity architects can reverse-engineer goals by tracing policy rationale using techniques such as the seven whys.

### Procedures

Procedures outline processes supporting policies but focus on tactical steps rather than intent. While illustrating implementation dynamics, procedures generally provide limited insights into broader goals.

### Standards

Standards specify configurations, tools, and controls to meet policies. However, since standards enable predetermined intent, they offer minimal new revelations of goals beyond what policies state. Standards do not define new objectives. While standards do not define new objectives, they do define the technical requirements that should be met when evaluating technologies and/or changes to the environment.

## *Guidance*

Guidance supplements other governance documents with additional advice or best practices. Like standards, guidance serves policies already in place rather than uncovering new goals.

In summary, while all governance documents inform architecture to some degree, policies offer the most direct window into management priorities. Policies' status as sanctioned, strategic declarations of intent makes them the most useful artifacts for efficiently deducing organizational goals relevant to risk management and security architecture.

At the onset, reviewing the available documentation is crucial. This does not imply combing through every document but prioritizing those pertinent to security. By familiarizing yourself with these documents, you can grasp the organization's approach to security and discern the major objectives your designs must fulfill. Listing these objectives is invaluable.

## Applying documentation to the framework

The first step for a cybersecurity architect is identifying the governing philosophies, needs, and goals that will shape their designs. Ideally, this entails a systematic analysis of all enterprise objectives and deriving corresponding security goals. However, such comprehensive goal mapping requires extensive time.

A more expedient approach is referencing existing governance documentation such as policies, procedures, standards, and guidelines. These codify prior decisions reflecting organizational goals. Reading key security documents provides valuable context for impending designs.

First, governance content offers insights into the organization's general security approach – its flexibility, risk tolerance, and innovation stance. Secondly, it highlights specific security requirements that designs must fulfill, such as encryption mandates. Tracking these key facts informs subsequent planning.

Ultimately, organizational goals explain why the entity exists. Commercial firms may seek shareholder returns or profitability. Non-profits may aim to provide community value. Goals are high-level and relate to the overall mission.

Supplementary business goals such as efficiency, sustainability, or customer experience ladder up to core goals.

In essence, governance documents help cybersecurity architects rapidly discern organizational goals and constraints to anchor designs in the company's realities and priorities. While not replacing exhaustive goal analysis, referenced governance content allows design context to be established quickly.

## Risk tolerance

**Risk tolerance** refers to the level of risk an organization is willing to accept in pursuit of strategic objectives. It represents the degree of uncertainty and potential downside the entity is prepared to withstand.

Risk tolerance is a foundational concept for cybersecurity architecture. Technical controls, budgets, and priorities flow directly from risk appetite. Cybersecurity architects must grasp tolerance to design appropriate protections.

Understanding the organization's risk tolerance is paramount. An organization with a structured risk management process often has a clear risk tolerance statement. Fundamentally, risk management is about optimizing risk, typically reducing it so that it aligns with the established risk tolerance. Steps such as establishing context, risk identification, risk analysis, and risk treatment, as outlined in ISO 31000:2018 (`https://www.iso.org/obp/ui#iso:std:iso:31000:ed-2:v1:en`), are integral to this process:

- **Establish context**: Identify and outline factors to be taken into consideration during the risk management process
- **Risk identification**: Identify potential risk sources
- **Risk analysis**: Analyze the risk, including developing an understanding of consequences, likelihood, and other factors
- **Risk evaluation**: Triage, prioritize, and assign priority to mitigation or other treatment
- **Risk treatment**: Address the risk through mitigation (remediation), acceptance, transference, avoidance, or other measures
- **Monitoring and review**: Monitor the risk over time to ensure that it stays within acceptable parameters

An organization's risk tolerance shapes architecture choices, including the following:

- **Security control selection**: Controls are chosen to reduce risks to acceptable levels based on tolerance. Less tolerance drives more stringent controls.
- **Budgeting**: Funding for cybersecurity is allocated based on the degree of risk the organization will bear and the controls needed to reach target levels.
- **Metrics**: Risk metrics are designed to track progress toward technical risk in line with appetite. Thresholds trigger escalation.
- **Prioritization**: Cybersecurity initiatives are sequenced based on risk urgency relative to tolerance. Quicker action is required for risks exceeding appetite.

In essence, risk tolerance benchmarks guide architecture decisions at both the technical and budgetary levels. This enables appropriate cyber risk management tailored to the organization.

# Assessing risk tolerance

Risk tolerance may be defined quantitatively or expressed qualitatively based on the impacts the organization is willing to absorb. Quantitative tolerance uses specific metrics such as dollar values, outage times, or breach percentages. Qualitative expressions describe risk attitudes such as *moderate* or *aggressive*.

Various methods help gauge organizational risk tolerance:

- **Surveys**: Ask leadership to describe appetite qualitatively or rank hypothetical scenarios

- **History**: Infer tolerance from past decisions and actions in response to realized risks

- **Benchmarking**: Derive relative appetite based on norms for the industry, geography, or size of the organization

- **Loss modeling**: Calculate maximum acceptable losses based on financials such as revenue, margins, and reserves

- **Risk analysis**: Workshop scenarios with estimates of likelihood and impact to find breaking points

Architects should employ multiple techniques to gain a rounded perspective on risk tolerance. Surfacing any gaps between stated and revealed preferences also helps build an accurate picture.

## Setting risk tolerance thresholds

Quantitative thresholds codify risk tolerance into architecture requirements. Here are some examples:

- Maximum annual financial loss from cyber incidents

- Maximum allowable system downtime from attacks

- Minimum required uptime percentage

- Maximum number of record breaches per year

- Maximum impact score for risks accepted versus mitigated

Thresholds provide clear guidance to bound technical decisions and spending. However, care should be taken to avoid arbitrary targets disconnected from actual risk appetite and organizational conditions. Realistic tolerances balance business needs with pragmatic security.

## Cascading tolerance across the organization

Technical architectures represent just one sphere of cyber risk management. Risk tolerance should cascade across other areas, such as the following:

- **Governance**: Risk oversight model, metrics, and reporting

- **Culture**: Degree of risk awareness, accountability, and skepticism

- **Business processes**: Due diligence in risk-bearing activities

- **Investment prioritization**: Focus on managing top risks

- **Insurance**: Coverage limits aligned with appetite

- **Third parties**: Risk-based vendor selection and monitoring

- **Incident response**: Playbooks tailored to expected threats

Extending risk tolerance guidance beyond just technical controls improves holistic resilience. A unified understanding of appetite across the organization also allows for coordinated cyber risk management.

## Optimizing architecture for risk objectives

With risk tolerance established, architects can design and govern technical measures accordingly:

- Control selection condenses to risk mitigation potency relative to cost

- Budgets provide sufficient funding to implement controls, thus reducing risks within tolerance

- Roadmaps sequence initiatives to tackle the biggest tolerance gaps first

- Metrics quantify risk levels compared to targets, triggering an action when exceeded

- Ongoing assessments identify control gaps or efficiency improvements to maintain alignment

- Documentation captures residual risks that are consciously accepted versus mitigated

An architecture that's been optimized for cost-effective organizational risk objectives reduces the most consequential risks to acceptable levels. Adjusting designs based on evolving tolerance and conditions also keeps protections aligned over time.

For cybersecurity architects in such organizations, it's essential to both harness information from the risk management process and ensure that their designs complement and align with it. If the organization lacks a structured risk management process, cybersecurity architects should still strive to comprehend its risk tolerance, potentially through independent analysis. Nonetheless, it's always preferable to have an approved, documented risk tolerance to guide architectural decisions. It's worth noting that some might overestimate their risk tolerance – until adverse outcomes materialize. Hence, even self-assessed risk tolerances should seek endorsement and approval.

Once we have identified the organization's business goals and risks, we can use them to inform our security architecture design. For example, if one of the organization's business goals is to protect its customers' data, we can design security controls that help achieve that goal.

It is also important to understand the organization's risk tolerance. Risk tolerance is the amount of risk that the organization is willing to accept. This will affect the design of our security architecture. For example, if the organization has a low risk tolerance, we will need to design more robust security controls.

By understanding the organization's business goals and risk tolerance, we can design a security architecture that is effective in protecting the organization's assets.

Analyzing cybersecurity architecture is a multi-step process that begins with comprehensively gathering information. This foundational phase taps into various resources, such as the organization's formal security policies, historical logs and reports, network diagrams, asset inventories, and past risk assessments. Such a thorough approach to information gathering forms the bedrock upon which a robust cybersecurity strategy is built. The 2016 Yahoo! security breach serves as a cautionary example, highlighting the importance of regularly reviewing security logs to detect unauthorized activities and prevent vulnerabilities.

Cybersecurity architecture analysis typically involves various steps. Let's take a look.

## Gathering information

The first step in cybersecurity architecture analysis is gathering information about the organization's cybersecurity architecture. This involves meticulous data collection, which then forms the basis for the entire analysis. This information can be gathered from a variety of sources, including the following:

- **Security policies, procedures, and standards**: These help us understand the organization's formal guidelines
- **Security logs and reports**: These provide insight into past security events
- **Network diagrams**: These provide visual representations of the network, showcasing all devices and connections
- **Asset inventories**: These list all software and hardware assets
- **Risk assessments**: These provide historical risk evaluations and offer a glimpse into previous threat landscapes

*Case study*: The 2016 Yahoo! breach could have been minimized had they periodically reviewed their security logs to detect unauthorized access (`https://www.nytimes.com/2016/12/14/technology/yahoo-hack.html`).

## Analyzing the information

Once the information has been gathered, it needs to be analyzed to identify potential vulnerabilities and areas for improvement. This analysis can be done manually or using automated tools:

- **Manual analysis**: This is done by cybersecurity experts, who utilize their experience to spot anomalies
- **Automated tools**: Software such as Nessus, Qualys, Rapid 7, Snort, or NetWitness can scan systems for known vulnerabilities

Once data has been collected from security assessments, structured analysis identifies vulnerabilities, gaps, and areas needing improvement. Thorough analysis involves the following aspects:

- **Organizing data**: Information is compiled, categorized, and filtered to enable useful insights. For example, vulnerability scan results could be grouped by severity, system, or type of finding. This facilitates trend analysis.

- **Correlating data**: Findings across assessments are cross-referenced to uncover intersections. If penetration testing and a gap assessment both reveal poor contractor management, this becomes a priority.

- **Quantifying gaps**: Metrics such as the percentage of systems patched or the number of failed audit controls provide tangible measures of shortfalls. This enables objective benchmarking against standards.

- **Prioritizing**: Factors such as risk levels, compliance impact, and architectural significance help rank findings. This focuses limited resources on addressing the most critical gaps first.

- **Root causing**: Impacted systems and processes are reviewed to determine why gaps exist. This distinguishes between symptoms versus underlying causes for remediation planning.

- **Evaluating compensating controls**: Existing mitigations that reduce exposure from vulnerabilities are documented. For example, an updated IDS may partially offset an unpatched system.

- **Recording progress**: Current findings are compared against past baselines to measure program improvements over time. This demonstrates ROI and helps forecast future needs.

- **Visualizing data**: Dashboards, heat maps, and graphs translate complex data into intuitive formats for stakeholders and leadership.

- **Sharing trends**: Results are summarized into reports, presentations, and meetings to socialize priorities across leadership, technology, and security teams.

Thoughtful analysis distills disparate assessment data into actionable intelligence to help strengthen defenses. Equally as important, it documents positive security advancements over time. Mature programs continually analyze findings to guide strategic roadmaps and communicate progress.

## Prioritizing the findings

Once the vulnerabilities and areas for improvement have been identified, they need to be prioritized. Not all vulnerabilities carry equal risk. Prioritizing them helps allocate resources efficiently. This prioritization can be done based on the following factors:

- The severity of the vulnerability

- The likelihood of the vulnerability being exploited

- The impact of the vulnerability being exploited

After prioritizing findings, cybersecurity architects should provide comprehensive mitigation recommendations, such as the following:

- **Implement new technical controls**: If vulnerability scanning discovers unpatched systems, a recommendation could be to implement automated patch management. For policy gaps, a firewall or IDS could help enforce compliance.

- **Modify policies and processes**: If multiple assessments reveal ineffective access controls, mandating multi-factor authentication for all users could help. Better vetting policies for third parties may mitigate outsourcer risks.

- **Boost employee training**: If social engineering tests succeeded, recommending refreshed awareness training on phishing and pretexting could help strengthen human defenses.

- **Allocate resources**: Demonstrating systemic exposure may require allocating budget and staff for new tools and headcount. Leadership support can pivot the organization toward assessment-driven investment.

- **Assign remediation owners**: Clearly defining owners for fixing findings not only improves accountability but also ensures subject matter experts lead mitigation. A patch management engineer would own new system hardening processes.

- **Track remediation**: Using metrics such as the percentage of findings successfully remediated over time demonstrates progress. This also feeds back into continuous improvement. One way of approaching this is through the **Plan of Action and Milestones (POA&M)**. POA&Ms identify tasks that need to be accomplished and detail the resources that are required to accomplish the elements of the plan, any milestones in meeting the tasks, and scheduled completion dates for the milestones.

- **Review compensating controls**: Existing alternative protections that reduce exposure from an uncovered risk should be documented when complete mitigation is not feasible.

- **Accept residual risk**: Certain risks may be designated as accepted rather than fully mitigated based on measured tolerances. However, this should be an informed decision with leadership approval.

- **Update baselines**: After remediation, repeat assessments validate improvements. The new baseline benchmarks progress for comparison in subsequent assessment cycles.

Effective recommendations exhibit both technical expertise and a grasp of organizational dynamics. This drives credible yet feasible remediation plans that leadership can confidently endorse, fund, and oversee.

## Recommending mitigations

Once the findings have been prioritized, recommendations for mitigations need to be made. Such mitigations can include the following:

- **Implementing new technical controls**: If vulnerability scanning uncovered a significant number of unpatched systems, a recommendation could be made to implement automated patch management software. For policy and compliance gaps, new controls such as a web application firewall or intrusion detection system could help enforce security requirements.

- **Modifying policies and processes**: If multiple assessments reveal ineffective access controls, a recommendation may be made to mandate multi-factor authentication for all users to strengthen identity management. To mitigate third-party risks, enhanced vetting and monitoring policies for outsourcers could be recommended.

- **Boosting security awareness training**: If social engineering testing successfully compromised users through phishing or pretexting, refreshed employee education on secure computing best practices could be recommended to strengthen human defenses. Training could be focused on detected areas of weakness.

- **Allocating resources**: If assessments reveal systemic critical security exposures, increased budget and staff may be recommended to acquire and manage necessary new tools or personnel. Leadership support could be solicited to pivot the organization toward assessment-driven security investments.

- **Assigning remediation ownership**: Clearly defining system, process, and policy owners responsible for implementing remediation promotes accountability. Subject matter experts should be empowered to lead mitigation efforts within their domains. For example, a patch management engineer would own remediation for unpatched system findings.

- **Tracking remediation progress**: Metrics such as the percentage of findings successfully remediated over time provide visibility into progress made. This also feeds continuous improvement initiatives.

- **Reviewing compensating controls**: Existing alternative protections that may help compensate for or reduce risk from an uncovered vulnerability should be documented when complete mitigation is not feasible.

- **Accepting residual risks**: Certain risks may be formally designated as accepted rather than fully mitigated based on measured tolerance thresholds. However, residual risk acceptance should be an informed decision with executive stakeholder approval.

*Illustration*: Think of vulnerabilities as holes in a boat. Mitigations are the efforts to plug these holes.

### Monitoring and improving

Once the mitigations have been implemented, the cybersecurity architecture needs to be monitored and improved on an ongoing basis. Security is a continuous journey. This monitoring can be done by doing the following:

- Reviewing security logs and reports
- Conducting vulnerability assessments
- Testing security controls

Cybersecurity architecture analysis is an important part of maintaining a strong cybersecurity posture. By regularly analyzing the organization's cybersecurity architecture, organizations can identify and address potential vulnerabilities before they are exploited by attackers.

Here are some of the benefits of conducting cybersecurity architecture analysis:

- **Improved security posture**: By identifying and addressing potential vulnerabilities, organizations can improve their security posture and reduce their risk of being attacked
- **Reduced risk**: Cybersecurity architecture analysis can help organizations identify and mitigate cybersecurity risks
- **Increased efficiency**: Cybersecurity architecture analysis can help organizations save time and money by identifying and addressing vulnerabilities before they cause problems
- **Improved compliance**: Cybersecurity architecture analysis can help organizations comply with industry regulations and standards

An accurate understanding of risk appetite provides the foundation for context-specific cybersecurity architecture. Cybersecurity architects should invest the time to properly assess and codify tolerance, at which point they should review all system and compliance documentation related to the framework being implemented and leverage it to inform priorities, controls, budgets, and metrics. This elevates architecture from generic best practices to focused risk management that's tightly aligned with organizational needs.

## Summary

In this chapter, key elements were outlined to help establish the context for cybersecurity architecture design. The aim was to provide a rationale so that the steps that are involved become intuitive based on organizational realities. This allows you to customize your environment since organizational structures vary.

The chapter covered foundational cybersecurity architecture concepts, including principles, design, and analysis. It emphasized using clear, accessible terminology, even when this differs from some frameworks. Understanding organizational goals and risk tolerance is critical for architecture. Design involves steps such as identifying assets, developing security goals, and implementing controls. Analysis evaluates the architecture to uncover gaps, prioritize, and drive improvement. The key principles we outlined included defense in depth, least privilege, and secure defaults.

This chapter stressed the importance of enabling business objectives, managing risk, and tailoring the architecture to the organization's environment and constraints. It noted that communication is vital for architecture, and frameworks may use alternative terminology for similar concepts. Overall, this chapter provided a high-level overview of core architecture elements that focus on effectively meeting organizational security needs within business realities.

In summary, this chapter equipped you with the knowledge you need to establish a solid contextual basis. The remaining chapters build on this by progressing through requirements, logical design, physical design, and implementation planning. The goal is to provide you with an end-to-end methodology while explaining the rationale behind each step so that you can adapt approaches as a cybersecurity architect. A thorough context setting now enables subsequent phases to produce a tailored cybersecurity architecture.

In the next chapter, we'll discuss the threat, risk, and governance considerations based on the context defined in this and the previous chapters and how cybersecurity architects must navigate the various hurdles presented.

# 5

# Threat, Risk, and Governance Considerations as an Architect

*"We cannot enter into alliances until we are acquainted with the designs of our neighbors."*

*– Sun Tzu*

In the previous chapter, we covered areas of architecture principles, design, and analysis that will be part of the day-to-day function of a **cybersecurity architect (CSA)**. The chapter discussed these areas and equipped you to establish a solid contextual basis. The remaining parts build on this by progressing through requirements, logical design, physical design, and implementation planning. The goal is to provide an end-to-end methodology while explaining the rationale behind each step so that you can adapt approaches as a CSA.

With an understanding of the principles, design, and analysis related to architecture, the next step is applying that understanding as regards threats, risks, and governance. As an architect, it is important not to provide designs or implement technologies without an understanding of organizational risk, threats, and governance requirements.

This chapter begins the discussion and journey of understanding, building upon the previous chapter to take into account threats, risks, and governance in the design and architecture scope. Sometimes, the more secure approach is not the best choice from an organizational or business perspective. Alternatively, meeting compliance or regulatory requirements may not make the solution more secure. As a result, this chapter tries to address potential challenges a CSA may face in mitigating or level-setting controls against the threat/risk/governance of the organization.

The chapter covers the following topics:

- Threats
- Risks
- Governance

- How it all relates to the business
- CSAs' balancing act

# Threats

The digital landscape has drastically expanded, making cybersecurity a significant concern for organizations worldwide. The heart of an effective defense against cyber threats lies in comprehensive threat cybersecurity architecture. This architecture is a set of systems and protocols designed to protect and monitor both the physical and digital assets of an organization.

In this section, we delve deep into the concept of threat cybersecurity architecture, exploring its elements, benefits, and how organizations can create a robust framework for enhanced cyber resilience.

## Understanding the threat landscape

Before commencing an examination of an organization's security architecture, a thorough understanding of the threat landscape is imperative. The term **cyber threats** encapsulates a spectrum of possible adversarial actions that imperil the confidentiality, integrity, and availability of an information system. Threat actors range from cyber criminals seeking financial gains to hacktivists propelled by ideological goals to state-sponsored entities engaging in espionage or cyber warfare, as well as insiders who may be motivated by a myriad of personal or professional grievances. Other threats can also include the unintentional insider through accidents, negligence, or complacency.

### Common types of cyber threats

In an era marked by escalating cyber risks, it is essential to delineate the diverse typologies of threats. The subsequent elaboration of common cyber threats aims to offer granularity, with each category distinguished by its modus operandi, associated risk vectors, and resultant impact on the target system.

#### Malware

**Malware**, or **malicious software**, encompasses a set of programs deliberately designed to compromise the operations, data integrity, or user experience of a computing environment.

Its technical characteristics are the following:

- **Viruses**: Self-replicating code segments that attach themselves to legitimate software and execute covertly
- **Worms**: Standalone malware that propagates autonomously across networks
- **Ransomware**: Encrypts files or systems, demanding a ransom for decryption keys

Mitigation strategies are signature-based and behavioral detection, heuristic analysis, endpoint security solutions, and regular software patching.

## Phishing

**Phishing** refers to a class of social engineering attacks that manipulate individuals into divulging confidential information, typically through fraudulent communications that masquerade as trustworthy entities.

Its technical characteristics are the following:

- **Spear phishing**: Targeted phishing aimed at specific individuals or organizations
- **Credential harvesting**: Use of fake login pages to collect user credentials

Mitigation strategies are **multi-factor authentication** (**MFA**), security awareness training, and email filtering algorithms that detect malicious or anomalous patterns.

## Man-in-the-middle attacks

**Man-in-the-middle** (**MitM**) attacks are attacks in which an unauthorized entity intercepts, relays, and potentially modifies data packets traversing between two communication endpoints. While this is the common vernacular, a more modern set of terms for MitM is **adversary-in-the-middle** (**AitM**) or **on-path** attacks.

Their technical characteristics are the following:

- **Address resolution protocol** (**ARP**) **spoofing**: Manipulating the ARP cache to control network traffic
- **Secure socket layer** (**SSL**) **stripping**: Downgrading HTTPS connections to unencrypted HTTP

Mitigation strategies are the implementation of strong encryption protocols such as **Transport Layer Security** (**TLS**), **virtual private networks** (**VPNs**), and authenticated public key exchanges.

## Distributed denial-of-service attacks

**Distributed denial-of-service** (**DDoS**) attacks aim to incapacitate network resources by flooding them with an overwhelming volume of requests or malformed packets.

Their technical characteristics are the following:

- **Amplification attacks**: Exploiting vulnerable protocols to magnify the volume of attack traffic
- **Botnet-driven attacks**: Using compromised machines to generate attack traffic

Mitigation strategies are traffic shaping, rate limiting, deployment of specialized DDoS mitigation appliances, and utilization of cloud-based DDoS protection services.

### Advanced persistent threats

**Advanced persistent threats** (**APTs**) are intricately orchestrated, long-term cyber-espionage campaigns, typically enacted by state-sponsored entities.

Their technical characteristics are the following:

- **Multi-stage exploits**: Sequential exploitation of multiple vulnerabilities

- **Lateral movement**: Internal network traversal to access sensitive data

- **Data exfiltration**: Stealthy transmission of confidential data to external servers

Mitigation strategies are **intrusion detection systems** (**IDS**), proactive threat hunting, **zero-trust architecture** (**ZTA**), and comprehensive logging and monitoring.

Understanding the nuances of these threats is a prerequisite for architecting a resilient cybersecurity framework capable of mitigating risks and minimizing potential damages.

## The imperative for a proactive cybersecurity posture

In light of multifarious and increasingly sophisticated cyber threats, organizations must transition from reactive defense mechanisms to a proactive cybersecurity paradigm. This involves an amalgamation of state-of-the-art **threat intelligence** (**TI**) systems, periodic risk assessments employing techniques such as **Monte Carlo simulations** or **Bayesian network models** for predictive analysis, and the deployment of multi-layered, adaptive security controls.

Monte Carlo simulations utilize computational techniques and statistical methods to analyze complex, unpredictable systems too intricate for purely analytical solutions. By randomly sampling possible outcomes numerous times, the simulations can model overall behavior patterns and key metrics. Bayesian networks are a related technique leveraging probability theory and graphs to map interdependencies and uncertainties between various variables in a system. These probabilistic models help quantify potential scenarios for making data-driven decisions amid complexity. Together, Monte Carlo and Bayesian models offer versatile tools for cybersecurity architects to simulate myriad attack permutations based on system vulnerabilities and threat actor motivations in order to strategically strengthen defenses proactively. Just as battle strategists cannot anticipate every contingency but can forecast key challenges through intelligence gathering, creative modeling and simulation empower security architects to preempt threats by approximating risks and then preparing for them pragmatically.

### Components of a proactive approach

These are the components of a proactive approach:

- **TI platforms** (**TIPs**): Utilize **machine learning** (**ML**) algorithms to aggregate and analyze data from various sources, thereby facilitating anticipatory threat modeling

- **Automated risk assessments**: Leverage real-time analytics tools that scrutinize network traffic, endpoint activities, and application vulnerabilities to forecast potential security lapses

- **Preventive measures**: Implement advanced security protocols such as **endpoint detection and response (EDR)** solutions, **just-in-time (JIT)** access provisioning, and **data loss prevention (DLP)** technologies

## Constructing an integrated threat-security cybersecurity architecture

The erection of an exhaustive, threat-security cybersecurity architecture necessitates a multidimensional planning strategy, incorporating an array of variables that range from business objectives to computational limitations.

### Preliminary considerations

Before embarking on the architectural design phase, it is imperative to comprehensively understand the organizational landscape. This includes elucidating the following:

- **Business objectives**: Identify **key performance indicators (KPIs)** and strategic objectives to ensure that the security framework acts as a facilitator rather than a barrier to achieving these goals

- **Operational workflows**: Analyze workflow diagrams and process flowcharts to grasp the intricacies of business operations

- **User behavior analytics (UBA)**: Employ ML techniques to model normal user behavior, thereby facilitating the detection of anomalous activities

- **Data flow mapping**: Conduct detailed data flow analyses using formal methods such as **Petri Nets** (a mathematical modeling language used to describe distributed systems) or **data flow diagrams (DFDs)** to visualize the movement and transformation of data across the organization

- **System dependencies and constraints**: Employ graph theory algorithms to model dependencies between various software and hardware components and assess potential bottlenecks and failure points

- **Regulatory constraints**: Maintain a catalog of applicable legal and compliance standards that the architecture must adhere to, such as the **General Data Protection Regulation (GDPR)** or the **Health Insurance Portability and Accountability Act (HIPAA)**

By meticulously understanding these variables, security architects can construct a cybersecurity framework that is intricately aligned with organizational needs and constraints, thereby ensuring that security measures augment rather than inhibit business functionality. This alignment not only enables robust security but also promotes operational efficiency and regulatory compliance.

## Elaborating on security objectives

Upon gaining an intricate understanding of the business context, articulating clearly defined security objectives is the subsequent task. These objectives should span a gamut of information security principles, namely **Confidentiality, Integrity, Availability, Accountability, and Non-repudiation (CIAAN)**, or you may remember this from our previous discussion as the more commonly referred to **CIA Triad**. Utilizing multi-criteria decision-making methods such as the **analytic hierarchy process** (**AHP**) or **weighted sum model** (**WSM**) can aid in prioritizing these objectives based on their significance and exigency.

### Key components of security objectives

The following are the key components:

- **Confidentiality**: Implement cryptographic algorithms, such as AES-256 or RSA, to safeguard sensitive information against unauthorized access

- **Integrity**: Employ cryptographic hash functions such as SHA-256 to ensure data is unaltered during storage or transmission

- **Availability**: Utilize redundant systems and load balancers to guarantee uninterrupted service access

- **Accountability**: Incorporate robust logging mechanisms and **user and entity behavior analytics** (**UEBA**) to trace activities back to specific actors

- **Non-repudiation**: Leverage digital signatures and **public key infrastructure** (**PKI**) to ascertain the origin and receipt of data, thereby preventing the dispute of actions performed

## Identification and evaluation of security risks

A meticulous risk assessment is imperative for the development of a resilient security architecture. This entails a rigorous identification of potential threats, vulnerabilities, and consequent impacts. Leveraging structured methodologies such as **Spoofing, Tampering, Repudiation, Information disclosure, Denial of service, Elevation of privilege** (**STRIDE**) for threat modeling and the **National Institute of Standards and Technology's** (**NIST's**) **Risk Management Framework** (**RMF**) for risk evaluation can lend rigor and comprehensiveness to this process.

### Assessment methodologies

The following are the methodologies:

- **Threat modeling**: Employ tools such as Microsoft's **Threat Modeling Tool** or **OWASP Threat Dragon** to systematically identify, quantify, and prioritize risks

- **Risk assessment frameworks**: Utilize established frameworks such as *NIST SP 800-53* or *ISO/ IEC 27005* to methodologically assess and manage risks

## Selection and deployment of security controls

Post-risk identification, the subsequent phase involves the selection of pertinent security controls. These controls act as procedural, technical, or physical safeguards designed to mitigate the identified risks. Decisions regarding control selection should be underpinned by established security benchmarks such as **Center for Internet Security (CIS)** controls, adherence to industry best practices, and a rigorous cost-benefit analysis employing methods such as **net present value (NPV)** or **internal rate of return (IRR)**.

Types of security controls are set out here:

- **Procedural controls**: Governance frameworks, policies, and procedures
- **Technical controls**: Firewalls, IDS, and encryption mechanisms
- **Physical controls**: Biometric authentication systems and secure physical access to facilities

# Continual monitoring and revision

Once the security architecture has been instantiated, perpetual monitoring and revision are quintessential. Employing tools such as **security information and event management (SIEM)** systems and techniques such as statistical anomaly detection can facilitate the real-time tracking of security incidents and KPIs. This data should then be rigorously analyzed using ML algorithms or statistical methods such as **Chi-Square tests** to identify trends or aberrations, thereby informing necessary architectural modifications or policy adjustments.

By embracing this holistic approach, organizations can achieve a security posture that is not only robust and resilient but also intricately aligned with their business objectives and operational requirements.

## The primacy of preventive measures in security architecture

While the detection and remediation of cyber threats remain pivotal, the overarching emphasis should be on proactive prevention mechanisms. Implementing a robust preventive strategy alleviates not only fiscal ramifications but also reputational repercussions concomitant with cyber incidents.

## Efficacy and efficiency gains through preventive measures

Implementing preventive controls such as ZTA or heuristic-based **intrusion prevention systems (IPS)** can dramatically enhance the efficiency of security apparatus. Advanced ML algorithms can automate the mitigation of over 99% of all threats, thereby channeling the security team's cognitive resources exclusively toward APTs that necessitate human analytical capabilities. This dual benefit of workload reduction and enhanced threat mitigation amplifies the overall security efficacy.

This can be seen as an example through the proactive nature of dynamic firewall and IDS rules. If we use the following screenshot as an example, you can see dynamic firewall and IDS blocks within an environment over the past 30 minutes:

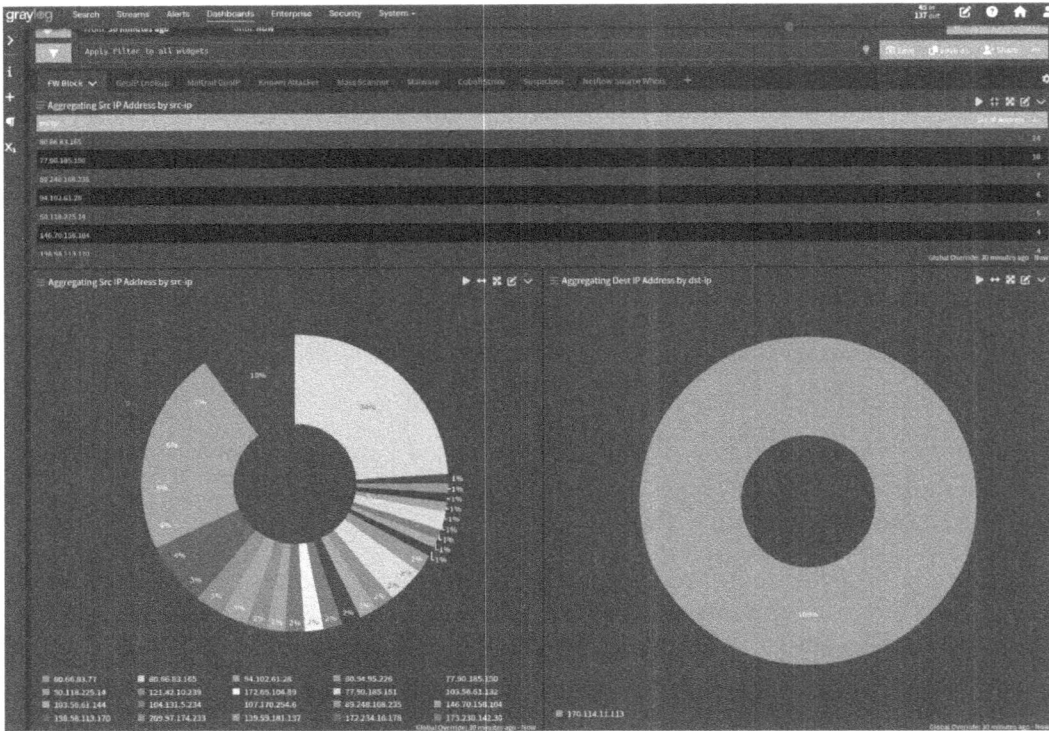

Figure 5.1 – Example Graylog dashboard

The specific query for this is the following within Graylog:

```
action:block AND NOT(dst-ip:255.255.255.255 AND src-ip:0.0.0.0) AND
NOT(src-ip:192.168.* OR src-ip:127.0.0.1)
```

As you can see from *Figure 5.1*, using proactive measures that include correlation with various TI feeds allows dynamic firewall and IDS rules to block potential threat traffic or connection to malicious systems.

## Imperative for architectural agility in contemporary digital environments

Given the velocity of technological change and the dynamism of modern threat landscapes, security architectures must embody a high degree of agility. Adaptable frameworks such as **secure access service edge** (**SASE**) can be deployed expeditiously, often within a matter of hours, and are equipped with an integrated full-stack security suite.

## Leveraging cloud-native constructs for architectural flexibility

The cloud-native features inherent in SASE architectures offer flexibility and scalability. This allows for real-time adjustments to the architecture in response to shifts in the operational landscape or emergent threat vectors, thereby providing a higher degree of business agility.

## Augmenting cyber resilience through structured architectural frameworks

The construct of cyber resilience pertains to an organization's inherent capability to sustain intended operational outcomes, irrespective of adverse cyber occurrences. Well-articulated security architecture, designed using methodologies such as the **Open Group Architecture Framework (TOGAF)** or the **Sherwood Applied Business Security Architecture (SABSA)**, contributes significantly to bolstering an organization's cyber resilience.

## Strategic alignment of security architecture and business objectives

A cardinal principle in augmenting cyber resilience is the meticulous alignment of the security architecture with an organization's predefined business objectives. This requires a comprehensive understanding of the organization's risk tolerance levels and the customization of security controls to meet nuanced requirements and targets of the business.

## Facilitating operational efficiency through automation

Automation technologies, such as **infrastructure as code (IaC)** for automated provisioning or orchestration solutions such as **security orchestration, automation, and response (SOAR)** for automated **incident response (IR)**, drastically reduce the operational overhead for security teams. This enables the human elements of these teams to concentrate on strategic decision-making and threat-hunting exercises.

# Regulatory compliance as an intrinsic outcome

Conformance with regulatory paradigms is another compelling byproduct of a well-executed security architecture. By aligning security controls with industry benchmarks such as the **Payment Card Industry Data Security Standard (PCI DSS)** or GDPR, organizations can substantiate their adherence to best practices and regulatory standards, thereby circumventing potential legal sanctions and reputational impairments.

By synthesizing these diverse but interconnected facets—preventive focus, architectural agility, business alignment, automation, and regulatory compliance—an organization can construct a security architecture that is not only robust but also resilient and agile, fully supporting both operational requirements and strategic objectives.

In an era of escalating cyber threats, building a robust threat security cybersecurity architecture is no longer optional but a necessity for organizations. By taking a proactive approach to cybersecurity and carefully considering various factors, organizations can create a security architecture that not only protects their digital assets but also aligns with their business objectives, enhances their cyber resilience, and ensures regulatory compliance. Remember—the goal of a comprehensive security architecture is not just to prevent cyber attacks but also to enable the organization to swiftly and effectively respond when such incidents occur.

## Threat considerations – examples

To architect effective defenses, cybersecurity professionals must cultivate an intimate understanding of threats facing their organization. This requires identifying high-value assets and data, profiling potential adversary groups and their tactics, continuously monitoring TI, and modeling attack vectors against systems and environments. Practical exercises bring these elements together into an actionable threat model that informs architectural decisions. Hands-on threat modeling exercises guide analysts to create DFDs mapping system interactions, perform STRIDE analysis to find vulnerabilities, integrate TI feeds, and roleplay as actors probing defenses. Applying threat knowledge in simulated scenarios transforms theoretical concepts into operational security wisdom. Just as military units perform wargames, cyber professionals should continually workshop threats and refine strategies. Well-architected defenses derive from informed anticipation of the enemy rather than reaction. The following exercises will equip architects with practical threat modeling experience to enhance architectures.

### *Identification of key assets, data, and systems*

The first step in establishing a robust cybersecurity architecture is to identify critical assets, data, and systems that necessitate protection. These assets may include servers, databases, applications, **intellectual property** (**IP**), employee data, and customer information.

Examples include the following:

- **Financial sector**: Credit card databases, transaction histories, and customer **personal identification information** (**PII**)
- **Healthcare**: **Electronic health records** (**EHRs**), medical imaging data, and patient demographics
- **Manufacturing**: IP such as patents, production processes, and schematics

Let us look at an exercise regarding asset identification:

- Utilize asset management software to catalog all physical and digital assets
- Assign value scores to each asset based on their criticality to the organization
- Use DLP tools to classify data types and their importance

## Understanding threat actors, motivations, tactics, techniques, and procedures

Understanding who potential threat actors are, their motivations, and their modus operandi is crucial for effective threat mitigation.

Examples include the following:

- **Nation-state actors**: Motivated by geopolitical objectives; often employ APTs
- **Cybercriminals**: Primarily motivated by financial gains; employ tactics such as ransomware and phishing
- **Insider threats**: Motivated by grievances or financial gains; employ techniques such as data exfiltration

Let us look at an exercise regarding threat actor profiling.

Conduct a role-playing exercise where team members assume the role of different threat actors to expose potential attack vectors:

- Use TIPs to gather information on known **tactics, techniques, and procedures** (**TTPs**)
- Create actor profiles detailing their common motivations and tactics

## Analyzing TI to stay updated on new and emerging threats

To stay ahead of threat actors, continuous monitoring and analysis of TI feeds are imperative. This includes data on new malware variants, zero-day vulnerabilities, and newly observed TTPs.

Examples include the following:

- **Common Vulnerabilities and Exposures (CVE) databases**: Constantly updated with information on new vulnerabilities
- **TI feeds**: Real-time information on emerging threats and attack indicators
- **Security blogs and forums**: Often the first to report on new types of attacks or vulnerabilities

Let us look at an exercise regarding TI analysis:

- Integrate TI feeds into an SIEM system
- Run periodic reports to identify new threats relevant to your organization
- Conduct tabletop exercises to simulate responses to new threats

### *Conducting threat modeling to identify vulnerabilities and attack vectors*

Threat modeling involves identifying potential vulnerabilities in your systems and understanding ways in which threats could exploit these vulnerabilities. The purpose is to develop a comprehensive understanding of the risk landscape to inform the design of protective measures.

Examples include the following:

- **DFDs**: Use these to map how data moves through your systems and identify potential chokepoints or areas of vulnerability

- **STRIDE methodology**: Identifies threats in six categories—Spoofing, Tampering, Repudiation, Information disclosure, Denial of service, and Elevation of privileges

Let us look at an exercise regarding threat modeling:

- Create DFDs for key systems within the organization

- Conduct a STRIDE analysis on these diagrams to identify potential vulnerabilities

- Use tools such as Microsoft's Threat Modeling Tool or OWASP Threat Dragon to automate the threat modeling process

## Summarizing threats

A multifaceted approach to threat considerations is indispensable for establishing a robust cybersecurity posture. This involves identifying key assets and their associated vulnerabilities, understanding the diverse array of threat actors and their motivations, staying abreast of emerging threats through continuous intelligence gathering, and adopting a proactive stance through threat modeling exercises. These activities collectively form a foundational framework for the development and implementation of effective cybersecurity strategies, thereby enabling organizations to safeguard their assets while minimizing risks and potential damages.

# Risks

The application of **risk cybersecurity architecture** is a pivotal aspect of the digital universe, aimed at safeguarding business operations against potential cyber threats. This comprehensive guide will delve into the nuances of devising a risk cybersecurity architecture, underlining the importance of threat definition and considerations when designing security architecture.

Cyber threats are an inherent part of the digital landscape. As organizations continue to integrate technology into their operations, the need for robust and resilient cybersecurity architecture becomes more critical. Understanding potential risks and designing a security architecture to mitigate them is a fundamental part of an organization's cybersecurity strategy.

# Risk cybersecurity architecture – an overview

Risk cybersecurity architecture serves as the cornerstone of an organization's cybersecurity strategy. It's a holistic approach that embeds security considerations into the design, development, and implementation stages of an organization's IT infrastructure. The ultimate goal of risk cybersecurity architecture is to minimize the likelihood and impact of cyber threats on business operations.

## Understanding risk in cybersecurity

The concept of risk in cybersecurity revolves around the probability of a cyber threat successfully exploiting a vulnerability in an organization's IT infrastructure, leading to business disruption. A risk assessment process helps in identifying these vulnerabilities and devising strategies to mitigate them.

## Importance of risk cybersecurity architecture

Risk cybersecurity architecture plays a critical role in safeguarding an organization's digital assets. By implementing a risk-based approach to cybersecurity, organizations can proactively identify and mitigate potential threats, thereby reducing the likelihood of successful cyber attacks.

## The three pillars of risk cybersecurity architecture

Risk cybersecurity architecture rests on three fundamental pillars:

- **Secure-by-design**: This principle emphasizes integrating security aspects right from the design phase of IT systems, ensuring cybersecurity is a core business objective rather than an afterthought

- **Secure-by-default**: This entails that IT systems are resilient against common exploitation techniques right out of the box without necessitating additional security configurations

- **Continuous monitoring**: This involves regular monitoring and evaluation of the security design to identify potential vulnerabilities and update the security architecture as required

## Risk assessment in cybersecurity architecture

Risk assessment forms the foundation of a robust cybersecurity architecture. It's a systematic process that involves identifying key business objectives, determining which IT assets are critical for realizing these objectives, and assessing the likelihood and impact of potential cyber threats to these assets.

## Key steps in cybersecurity risk assessment

A cybersecurity risk assessment typically involves the following steps:

- **Scope definition**: Determine the scope of the risk assessment. It could encompass the entire organization or specific **business units** (**BUs**), locations, or processes.

- **Risk identification**: Identify potential risks that could impact in-scope assets. This includes understanding the threat landscape and pinpointing possible vulnerabilities.

- **Risk analysis**: Determine the likelihood of identified risks materializing and the potential impact on the organization.

- **Risk evaluation**: Evaluate the risks based on their likelihood and impact to prioritize mitigation efforts.

- **Documentation**: Document the risk assessment process and outcomes for future reference and continuous improvement.

## Implementing a risk cybersecurity architecture

Implementing a risk cybersecurity architecture involves creating a security blueprint that outlines the organization's approach to managing cybersecurity risks.

### Secure-by-design implementation

A secure-by-design implementation requires a shift in mindset toward viewing security as an integral aspect of the design and development process. This includes performing risk assessments during the design phase, adhering to security best practices, and incorporating multi-layered defense mechanisms.

### Secure-by-default implementation

In a secure-by-default setup, the most critical security controls are automatically enabled, providing robust protection against prevalent threats and vulnerabilities. This approach minimizes the chance of misconfigurations and reduces the burden on end users to configure security settings.

## Managing risk with cybersecurity engineering

Cybersecurity engineering forms a crucial part of a risk cybersecurity architecture. It involves designing, developing, and implementing secure systems and applications to mitigate potential cyber threats. Key considerations in cybersecurity engineering include understanding the business context, balancing trade-offs between security and functionality, and adopting a proactive approach to threat identification and mitigation.

## Role of continuous monitoring in risk management

Continuous monitoring plays a crucial role in managing cybersecurity risks. Regular monitoring and evaluation of the security design can help identify potential vulnerabilities, assess the effectiveness of security controls, and update the security architecture as needed.

### A proactive approach to cybersecurity risk management

Managing risk in cybersecurity architecture is not a one-time exercise. Instead, it requires a proactive, continuous approach to identifying, assessing, and mitigating potential cyber threats. By implementing a risk cybersecurity architecture, organizations can better safeguard their IT infrastructure, protect their digital assets, and foster a risk-aware culture. As cyber threats continue to evolve, so should the strategies to combat them. Adopting a risk-based approach to cybersecurity can help organizations stay one step ahead in the ongoing battle against cyber threats.

## Risk considerations – an in-depth analysis with practical exercises

An effective cybersecurity architecture begins with a candid appraisal of risks. Architects employ assessments, threat models, and mitigation plans to translate risks into resilient designs. Structured risk analysis identifies vulnerable assets, quantifies potential impacts, and prioritizes response efforts. Creative exercises bring risks into focus, from simulating data theft to evaluating vendor partnerships. Architects also consider emerging risks introduced by new technologies through continuous monitoring and experimental mitigation testing. Just as regular disaster preparedness drills harden infrastructure, risk modeling exercises hone instinct and skills. They forge the risk-aware mindset underpinning robust architectures. The following practical exercises will empower architects at any career stage to dissect and course-correct risks, transforming threat specters into informed decisions through experience. Well-architected security emerges from deep familiarity with risks in all forms.

### Performing risk assessments to identify and prioritize risks

Risk assessments serve as the cornerstone for identifying cybersecurity risks by considering both their likelihood and impact. These assessments often use matrices or qualitative labels for risk evaluation and prioritization.

Examples include the following:

- **Quantitative analysis**: Utilizing financial metrics to assess potential loss from cybersecurity incidents
- **Qualitative analysis**: Applying labels such as *High*, *Medium*, or *Low* to rank risks based on expert judgment

Let us look at an exercise regarding risk assessment:

- Use tools such as **Factor Analysis of Information Risk** (**FAIR**) or the NIST RMF to conduct a structured risk assessment
- Develop a risk matrix to prioritize risks based on their likelihood and potential impact
- Validate the matrix through red or blue teaming exercises to simulate attacks and assess preparedness

### Consideration of specific types of risks

Some risks require specialized attention due to their specific nature, such as data breaches, ransomware, insider threats, and third-party vendor vulnerabilities.

Examples include the following:

- **Data breaches**: Involves unauthorized access to sensitive information
- **Ransomware**: Malware that encrypts files and demands payment for their release
- **Insider threats**: Risks arising from disaffected or negligent employees
- **Third-party vendors**: Risks associated with outsourced services or products

Let us look at an exercise regarding specialized risk analysis:

- Use DLP tools to simulate data exfiltration scenarios
- Conduct phishing simulations to assess susceptibility to ransomware
- Use UEBA to model and identify anomalous behavior indicative of insider threats
- Conduct third-party risk assessments using standardized questionnaires such as **Standardized Information Gathering** (**SIG**) or **Vendor Security Alliance** (**VSA**)

### Evaluating risks associated with new projects and initiatives

New projects and initiatives often introduce new risk vectors. A thorough risk evaluation is essential during the planning and implementation phases.

Examples include the following:

- **Cloud migration**: Risks related to data sovereignty and multi-tenancy
- **Internet of Things (IoT) deployments**: Risks related to device security and data integrity
- **ML initiatives**: Risks related to biased algorithms and data poisoning

Let us look at an exercise regarding project-specific risk evaluation:

- Conduct a **preliminary hazard analysis** (**PHA**) during the project's conceptual phase
- Utilize tools such as **BowTieXP** or OWASP Threat Dragon for visual risk modeling in new projects
- Implement a continuous monitoring program to identify risks throughout the project life cycle

### Developing risk treatment plans to mitigate unacceptable risks

Once risks are identified and assessed, a risk treatment plan should be developed to outline strategies to mitigate unacceptable risks. This can involve risk transfer, avoidance, reduction, or acceptance.

Examples include the following:

- **Risk transfer**: Purchasing cybersecurity insurance for financial liability
- **Risk avoidance**: Discontinuing a service or product line that presents an unacceptable risk
- **Risk reduction**: Implementing additional security controls such as firewalls or IDS

Let us look at an exercise regarding risk treatment planning:

- Develop risk mitigation strategies mapped to each high-priority risk

- Utilize **decision trees** or **cost-benefit analysis** (**CBA**) to evaluate the effectiveness of mitigating controls or risk analysis

- Implement chosen risk treatments in a controlled environment and evaluate their effectiveness before full-scale deployment

## Summarizing risks

Managing risks is a dynamic and multifaceted endeavor that encompasses the identification and prioritization of potential vulnerabilities, specialized consideration of unique risk types, proactive evaluation of new projects, and the formulation of risk treatment plans. Through methodical risk assessments and the utilization of advanced cybersecurity tools, organizations can comprehensively address these considerations. In doing so, they substantially bolster their cybersecurity posture, thereby reducing the likelihood of successful cyber attacks and minimizing the impact of any that may occur.

# Governance

Governance in cybersecurity serves as the governing framework incorporating policies, processes, and roles that orchestrate the management of cybersecurity risks within an organization. CSAs are pivotal agents in this governance paradigm, contributing to policy development, secure system architecture, and holistic business integration of cybersecurity measures. This exposition articulates salient governance considerations and outlines practical approaches that CSAs should implement.

In the realm of information assurance, cybersecurity governance delineates the structural and procedural architecture that synchronizes an organization's cybersecurity endeavors. It fuses components such as risk assessment, regulatory compliance, and organizational roles, harmonizing them into a cohesive framework. CSAs, who serve as the vanguard of this framework, are responsible for the articulation of secure systems, policy development, and the procedural alignment of security initiatives with business processes.

## The imperative of cybersecurity governance

Cybersecurity governance serves multiple cardinal purposes:

- **Risk identification and assessment**: Utilizing methodologies such as the FAIR model, governance allows the organization to quantify and prioritize cybersecurity risks.

- **Risk mitigation strategies**: Governance frameworks guide the development and implementation of strategies including, but not limited to, **defense in depth** (**DiD**), ZTA, and threat modeling, that are aimed at mitigating identified risks.

- **Continuous monitoring and improvement**: Leveraging technologies such as SIEM, governance enables real-time monitoring of an organization's cybersecurity posture. The following screenshot represents a dashboard of known attackers:

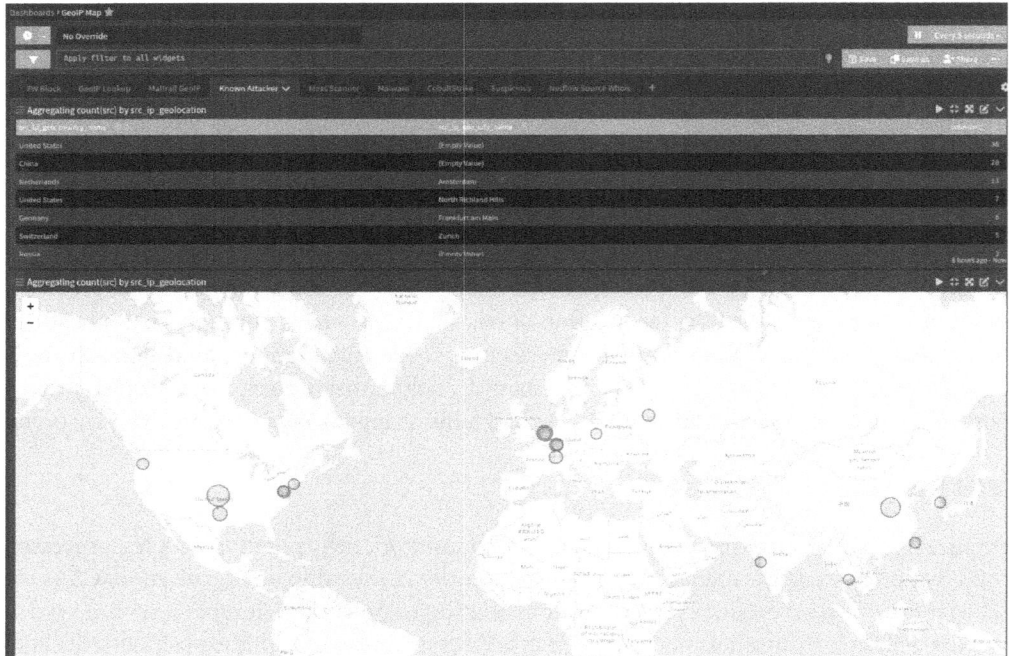

Figure 5.2 – Dashboard depicting known attackers' geolocation

The following screenshot represents a dashboard of **domain name system** (**DNS**) intel:

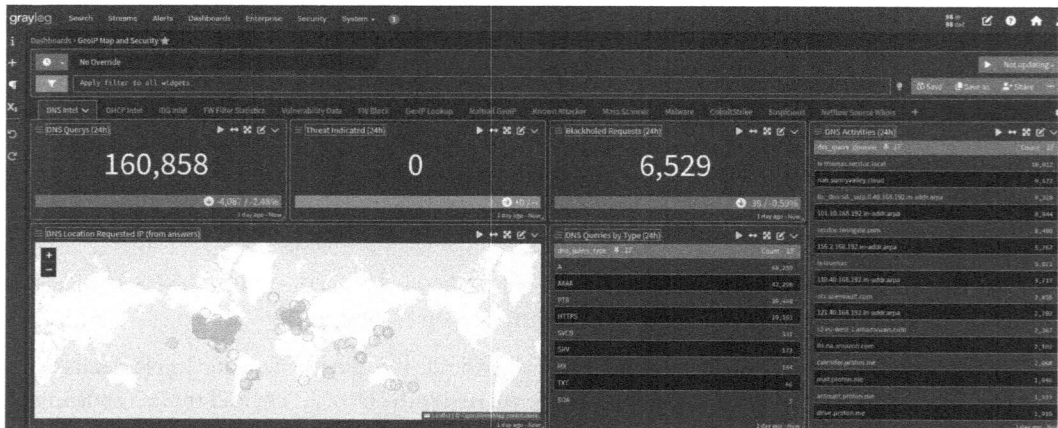

Figure 5.3 – Dashboard depicting DNS intel

- **Regulatory compliance**: With increasing legislative measures such as GDPR and CCPA, governance ensures that the organization adheres to pertinent laws and regulations, thereby mitigating legal repercussions.

- **Organizational reputation and customer trust**: Robust governance structures augment an organization's ability to thwart cyber attacks, thereby preserving brand reputation and customer trust.

## The multifaceted components of a cybersecurity governance framework

Any cybersecurity governance framework must be multifaceted, in that it must be built with multiple aspects or components to cover the full scope of visibility and understanding of what the framework is meant to govern. With this in mind, these are the common starting points within any given framework:

- **Policies and procedures**: Documents such as an **information security policy (ISP)** and **IR plans (IRPs)** define the organization's cybersecurity expectations, adherence protocols, and procedural guidelines. They usually adhere to international standards such as *ISO/IEC 27001*.

- **Roles and responsibilities**: Utilizing frameworks such as **Responsible, Accountable, Consulted, Informed (RACI)**, a clear delineation of cybersecurity roles and responsibilities is made across **organizational units (OUs)**.

- **Risk assessment and management**: Adopting frameworks such as *NIST SP 800-30*, the organization identifies, assesses, and prioritizes cybersecurity risks before formulating mitigation strategies.

- **Compliance management**: Leveraging tools such as **governance, risk, and compliance (GRC)** platforms, the organization maintains its compliance posture aligned with laws and industry regulations.

- **Security awareness and training**: Programs, often deployed through **learning management systems (LMS)**, impart employees with a detailed understanding of policies, as well as skills to recognize and report anomalies.

### The pivotal role of CSAs in governance

While we have touched upon this in previous chapters, we cannot stress enough the importance and fulcrum that the CSA plays within a successful governance program. With this stated, here is a reminder of some of those aspects:

- **System and network architecture**: CSAs design secure infrastructure, adhering to principles such as **least privilege** and **secure-by-design**. This includes the selection and configuration of security controls such as firewalls, IDS/IPS, and encryption solutions.

- **Policy and procedure formulation**: CSAs lead the development of technical guidelines, aligned with governance policies, for the configuration and operation of security solutions.

- **Cross-functional collaboration**: They often serve as the liaison between the technical teams and BUs, translating technical risks and countermeasures into business implications.

- **Risk assessment leadership**: Cybersecurity architects usually spearhead technical risk assessments, often leveraging specialized tools and methodologies to quantify risk metrics.

## Best practices for implementing and augmenting cybersecurity governance

Best practices are the bread and butter of a CSA's process. It is even more important to establish best practices to support the governance programs and frameworks implemented within your organization. We have touched upon this in previous chapters, but by now you should see a common thread being weaved within each section, in that this is not new but instead needs consistent application. With regard to governance, this also applies. The following are common considerations when implementing and/or modifying aspects of your governance process:

- **C-suite engagement**: Effective governance mandates commitment from senior executives, to allocate necessary resources and infuse a culture of cybersecurity awareness

- **Organizational tailoring**: Governance frameworks should be adapted to reflect the unique cyber-risk landscape, industrial sector, and organizational size

- **Business process integration**: Cybersecurity should intersect with various business functions, from procurement to customer relations, ensuring a uniform risk management approach

- **Adaptive evolution**: Given the dynamic threat landscape, governance frameworks should embrace iterative updates, possibly bi-annual revisions, informed by ongoing risk assessments and technological advancements

## Supplementary considerations

As always, there will be times when something does not fit the profile or does not align completely with the framework. It is also important that the framework not be so rigid or isolated that it dooms the organization to failure before implementation. With that in mind, there are some additional considerations when establishing a framework and socializing with stakeholders:

- **Business alignment**: Governance should not be an isolated entity but should be in sync with organizational objectives, portraying cybersecurity as an enabler rather than a hindrance.

- **Transparent risk communication**: The ability to articulate complex cyber risks to stakeholders in a comprehensible manner is crucial for gaining organizational buy-in.

- **Performance metrics**: KPIs and **key risk indicators** (**KRIs**) should be established to continually assess the efficacy of the governance structure.

- **Pragmatic implementation approaches by CSAs**: With pragmatism, architects avoid the pitfalls of *checklist security* and instill frameworks flexible enough to fulfill an organization's unique priorities and constraints. The art is in synthesizing guidance with business realities.

- **Policy life-cycle management**: CSAs can oversee the creation, review, and iteration of security policies and protocols.

- **Quantitative risk assessments**: Utilizing risk assessment platforms integrated with TI feeds, they can quantify and prioritize risks.

## Governance considerations – practical scenarios and exercises

A strong cybersecurity posture stems from comprehensive governance integrating people, processes, and technology. Architects must translate governance concepts such as policies, compliance, auditing, and training into operational realities. Immersive exercises bridge the gaps between frameworks and execution. Hands-on governance exercises allow architects to workshop critical IR planning, simulate staff phishing susceptibility, develop security metrics dashboards, and conduct mock audits. Just as fire drills embed preparedness, these simulations imprint governance proficiencies. Exercises provide low-risk environments to validate documentation, uncover oversights, and refine strategies. The following practical exercises will empower security architects at all levels to implement living governance regimes resilient to both digital threats and human factors. Architecting robust governance requires going beyond paper policies to skillfully direct an organization's ongoing cybersecurity activities.

### *Establishing cybersecurity policies, standards, and procedures*

Crafting a robust cybersecurity governance framework begins with the establishment of cybersecurity policies, standards, and procedures. These documents provide a formalized structure around which an organization's security controls and behaviors are built.

Examples include the following:

- **Policy creation**: Develop an **acceptable use policy** (**AUP**) to govern employee behavior related to information system usage

- **Standards specification**: Define **Secure Sockets Layer** (**SSL**)/TLS standards for secure data transmission

- **Procedure manuals**: Develop step-by-step guides for firewall configuration, among other procedures

Let us look at an exercise regarding policy and standards development:

- Use templates or policy generators to craft sample cybersecurity policies

- Create standards for secure data transmission, employing guidelines from NIST or similar organizations

- Develop a procedures manual and validate it through practical implementation

## Ensuring compliance with regulations

Regulatory compliance, including GDPR for data protection in the EU and HIPAA for health information in the US, is a critical governance consideration.

Examples include the following:

- **Data protection officers**: Appointment under GDPR regulations
- **HIPAA compliance checklist**: Regularly updated inventory of controls

Let us look at an exercise regarding compliance auditing:

- Conduct a GDPR or HIPAA mock audit using tools such as **ComplianceForge** or **StandardFusion**
- Correct non-compliance issues and rerun the audit to validate improvements

## Defining security roles and responsibilities

Clear definition and assignment of security roles and responsibilities are critical for accountability and operational effectiveness.

Examples include the following:

- **Chief information security officer (CISO)**: Oversees organization-wide cybersecurity
- **Security analyst**: Responsible for IR and threat analysis

Let us look at an exercise regarding role mapping:

- Develop a RACI matrix for cybersecurity tasks
- Validate role definitions through tabletop exercises

## Developing IR, disaster recovery, and contingency plans

IRPs and **disaster recovery plans** (**DRPs**) are essential governance components for reacting to and recovering from adverse events.

Examples include the following:

- **IR**: Phases including identification, containment, eradication, recovery, and lessons learned
- **DR**: Steps including initial response, recovery, and restoration
- **Contingency**: Formal documentation containing policies, procedures, and technical controls involved in the ability to sustain organizational operations and assets in response to abnormal disruptions or processing interruptions

Let us look at an exercise regarding IRP and DRP simulation:

- Conduct a simulated cyber incident to test your IRP
- Perform a DRP drill, simulating a data center outage

### Implementing security awareness training

Human factors often represent a significant security vulnerability. Regular security awareness training can mitigate this risk.

Examples include the following:

- **Phishing simulations**: Test staff susceptibility to phishing
- **Password hygiene**: Educate on creating and maintaining strong passwords

Let us look at an exercise regarding training assessment:

- Utilize platforms such as KnowBe4 or Wombat Security to simulate phishing attacks
- Assess the efficacy of training modules through pre- and post-training quizzes

### Conducting audits and assessments

Regular audits and assessments are essential for validating the effectiveness of governance structures and security controls.

Examples include the following:

- **Internal audits**: Regularly scheduled security audits
- **Third-party assessments**: External audits for unbiased evaluation

Let us look at an exercise regarding audit simulation:

- Conduct an internal audit using tools such as **Nessus** or **OpenVAS**
- Review findings and make necessary adjustments to governance policies and controls

### Providing security metrics and reports to leadership

Metrics and KPIs provide data-driven insights into the effectiveness of the cybersecurity governance program.

Examples include the following:

- **Mean Time to Detect (MTTD)**: Time taken to detect an incident
- **Patch management metrics**: Percentage of systems up to date

Let us look at an exercise regarding a metrics dashboard:

- Use SIEM solutions such as **Splunk, Elasticsearch, Logstash, Kibana (ELK)**, or **Graylog** to develop a metrics dashboard
- Generate monthly or quarterly reports and present them to leadership for strategic decision-making

## Summarizing governance

Effective governance in cybersecurity is a multifaceted undertaking that necessitates methodical planning, robust implementation, and continuous monitoring. With a coherent approach to policy formulation, compliance, role definition, IR, staff training, auditing, and metrics, organizations can substantially elevate their cybersecurity posture. This comprehensive approach enables organizations not only to protect their vital assets but also to meet regulatory demands, ultimately ensuring long-term resilience against an increasingly perilous threat landscape.

# How it all relates to the business

In today's complex and rapidly evolving global business environment, organizations face a myriad of threats and risks that can significantly impact their operations, reputation, and bottom line. At the same time, effective governance is crucial to ensure that these organizations not only comply with regulatory requirements but also align their strategies, resources, and processes with their overall business goals.

This section delves into critical considerations of threats, risks, and governance in the business landscape, offering insights and practical advice to help organizations navigate these challenges effectively.

## Understanding the concepts – threats, risks, and governance

This chapter has been detailing the aspects of threats, risks, and governance from the perspective of the CSA. It is also important to understand these same areas from a business perspective. It is important to remember that the CSA and the business may not always align or have the same definition or understanding regarding certain areas as they relate to threats, risks, and governance. In order to excel as a CSA, it is important to understand the business perspective to identify the middle ground between those differences when they arise, because they will.

### Threats in business

A threat in the business context refers to any potential event, action, or situation that could harm an organization. These threats could be internal, such as employee misconduct or system failures, or external, such as cyber attacks, market volatility, or regulatory changes.

## *Risk management*

Risk management entails proactively identifying, evaluating, and addressing threats that could impede an organization from accomplishing its goals. This involves the following:

- **Risk identification**: Cataloging key cybersecurity risks through methods such as threat modeling, vulnerability scans, and audits
- **Risk analysis**: Quantitatively and qualitatively assessing the likelihood and potential impact of identified risks using techniques such as risk matrices
- **Risk prioritization**: Determining risk severity and criticality to guide strategic mitigation efforts
- **Risk mitigation**: Implementing controls and safeguards to reduce unacceptable risks through avoidance, transfer, reduction, or acceptance
- **Risk monitoring**: Continuously tracking risks and controlling effectiveness through metrics and audits

Effective risk management takes a forward-looking, life-cycle view of risks rather than a reactive stance. By embedding risk considerations into key processes from planning to operations, organizations can cost-effectively allocate resources toward cyber defenses while enabling business objectives. This involves understanding the organization's risk tolerance, implementing appropriate controls, and monitoring and adjusting these controls as necessary.

## *Governance*

Governance in business refers to the system of rules, practices, and processes by which an organization is directed and controlled. This includes defining the roles and responsibilities of the board, management, and other stakeholders, setting strategic objectives, and monitoring performance against these objectives.

# The interplay of threats, risks, and governance

The concepts of threats, risks, and governance are intricately linked in the business context. Threats give rise to risks, which organizations must manage through effective governance. Here's a look at how these elements interact from the business perspective.

## *The origin of risk – threats*

Threats are the genesis of risks in business. Whether it's a cyber attack that could compromise sensitive data or a market downturn that could erode profits, threats create uncertainties that businesses must manage.

## *Mitigating risks through governance*

Effective governance is crucial for risk mitigation. By establishing clear policies, procedures, and controls, organizations can manage risks proactively, reducing their potential impact.

### The role of governance in identifying threats

Governance also plays a vital role in identifying potential threats. Through regular risk assessments and audits, organizations can uncover hidden threats and take appropriate action before they materialize into significant risks.

## Identifying and classifying risks

**Risk identification** and **risk classification** are crucial steps in the risk management process. This involves understanding the organization's risk landscape, pinpointing potential risks, and categorizing them based on their potential impact and likelihood.

### Risk identification

Risk identification involves recognizing potential events or situations that could negatively affect an organization's ability to achieve its objectives. This could be achieved through various methods, such as security audits, risk assessments, and benchmarking.

### Risk classification

Classifying risks involves assessing their potential impact and likelihood. This helps organizations prioritize their risk management efforts, focusing on the most significant risks first.

## Initial and residual risk assessment

**Risk assessment** involves determining the potential impact and likelihood of identified risks. From the business perspective, this could mean identifying the likelihood of a bank run, civil unrest, or a ransomware attack. Regardless of the scope or whether it is a technical, business, or social risk, this includes conducting an initial risk assessment before implementing risk mitigation measures and a residual risk assessment after these measures are in place.

### Initial risk assessment

The initial risk assessment involves estimating the potential impact and likelihood of each identified risk before any risk mitigation measures are implemented. This provides a baseline for measuring the effectiveness of these measures.

### Residual risk assessment

After risk mitigation measures are implemented, a residual risk assessment is conducted to determine the remaining risk. If the residual risk is still too high, further mitigation measures may be necessary.

# Risk mitigation strategies

**Risk mitigation** involves implementing strategies to reduce the potential impact and likelihood of risks. This could involve a range of measures, from implementing security controls to changing business processes.

## Implementing security controls

Security controls are measures that are put in place to protect an organization's assets. This could include physical controls such as locks and alarms, technical controls such as firewalls and encryption, and administrative controls such as policies and procedures.

## Changing business processes

In some cases, mitigating risks may involve changing business processes. For example, an organization may need to change its data handling procedures to mitigate the risk of data breaches.

# Monitoring and reviewing risks

Ongoing monitoring and review are crucial for effective risk management. This allows organizations to track the effectiveness of their risk mitigation measures, identify new risks, and adjust their strategies as necessary.

## Risk monitoring

**Risk monitoring** involves regularly checking the organization's risk landscape to ensure that risk mitigation measures are working and to identify new risks. This could involve regular risk assessments, audits, and reviews.

## Risk review

**Risk review** involves evaluating the organization's risk management strategies to ensure they are still relevant and effective. This could involve reviewing risk assessments, checking the effectiveness of risk mitigation measures, and reassessing the organization's risk tolerance.

# The role of enterprise architecture in risk management

Enterprise architecture plays a crucial role in risk management. By providing a holistic view of an organization's business processes, information systems, and technology infrastructure, enterprise architects can help identify potential risks and develop effective mitigation strategies.

## Identifying risks

Enterprise architects can use their understanding of the organization's systems and processes to identify potential risks. This could involve identifying vulnerabilities in the organization's IT systems, potential weaknesses in its business processes, or risks associated with its strategic objectives.

## *Developing mitigation strategies*

Once risks are identified, enterprise architects can help develop mitigation strategies. This could involve designing security controls, recommending changes to business processes, or helping to develop a risk management plan.

## The role of governance in risk management

Governance plays a crucial role in risk management. By establishing clear policies, procedures, and controls, governance frameworks help organizations manage their risks effectively.

### *Setting policies and procedures*

Policies and procedures provide guidelines for how an organization should manage its risks. This could involve defining roles and responsibilities for risk management, setting risk tolerance levels, and establishing procedures for identifying, assessing, and mitigating risks.

### *Establishing controls*

Controls are measures that help ensure that an organization's policies and procedures are followed. This could involve physical controls such as locks and alarms, technical controls such as firewalls and encryption, and administrative controls such as audits and reviews.

## Navigating regulatory and compliance risks

Regulatory and compliance risks are a significant concern for many organizations. These risks arise from the need to comply with various laws, regulations, and industry standards, which can be complex and continually changing.

### *Understanding regulatory requirements*

Organizations need to understand the regulatory requirements that apply to them. This could involve staying up to date with changes to laws and regulations, interpreting these requirements, and understanding how they apply to the organization's operations.

### *Implementing compliance measures*

Once regulatory requirements are understood, organizations need to implement measures to ensure compliance. This could involve changing business processes, implementing new controls, or training staff.

## Summarizing the business perspective

While at first glance, the steps are similar, the perspective and implications can be very different between that of a business and a CSA. Navigating the landscape of threats, risks, and governance is a complex but crucial task for businesses in today's dynamic and volatile environment. By understanding these elements, identifying and classifying risks, assessing and mitigating them, and continuously monitoring and reviewing their risk landscape, organizations can not only protect themselves from potential threats but also seize opportunities for growth and success.

# CSAs' balancing act

Balancing the scales of innovation and security has always been a tightrope walk for CSAs. Adding in potential business implications can be as equally challenging. The key lies in striking the right balance between enabling business innovation and ensuring robust security measures. This section aims to provide an overview of how CSAs can effectively manage GRC while avoiding potential risks.

With this in mind, this is a repetition of many of the concepts covered thus far within this book, but a repetition that provides context to the needed flexibility and creativity required for a CSA.

## Understanding the role of CSA

A CSA plays a crucial role in designing, implementing, and monitoring the security framework of an organization. Their expertise lies in developing strategies that align with the organization's business objectives while mitigating potential security risks. They need to keep abreast of the latest security trends and regulatory requirements to ensure the organization's security architecture is adaptable to changing threats and needs. The role of CSA is continually evolving in response to changes in the threat landscape and the emergence of new technologies.

### Key responsibilities

To summarize the previous chapter's discussion, the key responsibilities of a CSA include the following:

- **Risk assessment**: Identifying potential risks and threats to the organization's information system and formulating strategies to mitigate them
- **Policy development**: Developing and implementing security policies, standards, and guidelines that align with the organization's business objectives
- **Security architecture design**: Designing a robust security architecture that can effectively protect the organization's data and IT infrastructure from potential threats
- **Compliance management**: Ensuring the organization's security policies and practices comply with regulatory requirements
- **Monitoring and reporting**: Regularly monitoring the effectiveness of security measures in place and providing reports to management

### Keeping up with the changing landscape

CSAs must stay abreast of the latest threats and trends in cybersecurity. This requires ongoing learning and professional development, as well as a willingness to adapt and innovate.

### Preparing for the future

As technologies continue to evolve, the role of CSA will become increasingly complex and challenging. CSAs must be prepared to adapt and evolve their skills and approaches to meet these challenges.

## The art of risk management in cybersecurity

Risk management is an integral part of any cybersecurity strategy. It involves identifying, assessing, and managing risks associated with the organization's information system.

### Risk classification

Risks can be classified in various ways, depending on their impact on the organization, their probability of occurrence, and their potential severity.

### Risk identification and assessment

Identifying and assessing risks is a critical step in risk management. It involves evaluating the existing security architecture, identifying potential vulnerabilities, and assessing the likelihood and potential impact of these risks.

### Risk mitigation

Risk mitigation involves implementing measures to reduce the likelihood of risks or lessen their impact. This could include implementing security controls, developing contingency plans, or improving security awareness among employees.

## The framework of governance in cybersecurity

As discussed earlier in this chapter and in previous chapters, governance in cybersecurity refers to the processes and structures used to oversee and manage the organization's cybersecurity activities.

### Importance of governance

Good governance is essential for ensuring that the organization's cybersecurity activities align with its business objectives and comply with regulatory requirements. It also helps to ensure accountability and transparency in the organization's cybersecurity practices.

## Implementing a governance framework

Implementing a governance framework involves defining clear roles and responsibilities, establishing decision-making processes, and setting up mechanisms for monitoring and reporting on cybersecurity activities.

# The role of compliance in cybersecurity

Compliance in cybersecurity refers to adherence to laws, regulations, and standards related to information security.

## Understanding the role of CSA

A CSA plays a crucial role in designing and implementing an organization's security systems. They are responsible for creating a robust security architecture that can withstand potential cyber threats while supporting the organization's goals. The CSA's role is two-fold:

- **Innovation catalyst**: They must be innovative, staying abreast of the latest technologies and trends to ensure the organization's security measures are up to date and effective

- **Risk mitigator**: They must also be risk-averse, identifying potential threats and vulnerabilities and implementing measures to mitigate these risks

Balancing these two sides of their role is a complex task, requiring a deep understanding of both the organization's strategic objectives and the ever-evolving cybersecurity landscape. Creating a secure yet innovative cybersecurity architecture is a complex balancing act. This involves ensuring the security of an organization's data and systems, while also promoting innovation and efficiency.

## Importance of compliance

Compliance is crucial for ensuring that the organization's cybersecurity practices meet established standards and regulatory requirements. It can also help to prevent security breaches and protect the organization's reputation. Compliance with data protection and cybersecurity regulations is a vital aspect of cybersecurity architecture. Non-compliance can result in significant penalties, reputational damage, and loss of customer trust. CSAs must have a deep understanding of the compliance requirements that apply to their organization. They must ensure that the organization's cybersecurity measures meet these requirements and that compliance is maintained as regulations evolve.

## Integrating compliance into the cybersecurity architecture

Compliance should be integrated into the cybersecurity architecture from the outset. This can help to ensure that compliance is not an afterthought but is embedded into every aspect of the organization's cybersecurity strategy.

### The role of CSAs in IR

When a security incident occurs, the CSA plays a crucial role in responding to the incident and minimizing its impact.

### Preparing for incidents

Effective IR starts with preparation. CSAs must develop and implement IRPs that outline how the organization will respond to different types of security incidents.

### Responding to incidents

When an incident occurs, the CSA must coordinate the response, ensuring that the incident is contained, the impact is minimized, and the root cause is identified and addressed.

### Managing compliance

Managing compliance involves regularly reviewing and updating the organization's cybersecurity policies and practices to ensure they comply with changing regulations and standards. It also involves conducting regular audits and assessments to verify compliance.

## Striking a balance – security versus innovation

Striking a balance between security and innovation is challenging. Increasing security measures can often slow down processes, hinder innovation, and impact the user experience. On the other hand, prioritizing innovation over security can expose the organization to significant risks.

Therefore, a CSA must find a way to integrate security measures seamlessly into the organization's operations, promote secure practices, and foster a culture of security awareness, all while supporting innovation.

Striking the right balance between security and innovation is one of the biggest challenges faced by CSAs. While security measures are essential for protecting the organization's data and IT infrastructure, they should not hinder the organization's ability to innovate and adapt to changing business needs.

### The challenge

The challenge lies in ensuring that security measures do not become a barrier to innovation. This requires a flexible and adaptable approach to security that can accommodate the rapid pace of technological change while still providing robust protection against threats.

### The solution

The solution involves adopting a risk-based approach to security that focuses on managing risks rather than eliminating them entirely. This allows for greater flexibility and innovation while still maintaining a high level of security.

## Understanding cyber threats and vulnerabilities

Understanding the various types of cyber threats and vulnerabilities is crucial for effective risk management and the design of a robust security architecture.

Threat modeling is a proactive approach to identifying potential threats, assessing their impact, and implementing mitigation measures. It's a critical tool for CSAs, enabling them to balance the demands of innovation and risk management.

## The role of developers in threat modeling

Developers play a vital role in threat modeling. They have a deep understanding of the software they're building and can provide valuable insights into potential vulnerabilities. However, threat modeling is often seen as a burden by developers, who are under pressure to deliver software quickly and efficiently.

## DevSecOps

DevSecOps represents a culture shift, aiming to automate and integrate security across the entire **software development life cycle** (**SDLC**) rather than just tacking it on at the end. Core concepts include embedding security testing and compliance checks directly within **continuous integration** (**CI**) and **continuous delivery** (**CD**) pipelines. This enables developers to build and release software rapidly while still upholding robust protection. Techniques such as **static application security testing** (**SAST**) analyze source code for vulnerabilities early, while **dynamic application security testing** (**DAST**) scans running applications for risks. By leveraging automation through rigorous testing baked into CI/CD processes, security occurs continuously rather than as a one-time gate. This empowers sustainable velocity, securing software innovation and delivery natively across the pipeline. With security tooling and automation woven into their standard workflows, developers can remain laser-focused on building features quickly and safely. DevSecOps marries speed with protection through practices automating security fundamentals intrinsically across the life cycle.

## Making threat modeling appealing to developers

To make threat modeling more appealing to developers, CSAs must demonstrate its value. This could involve showing how threat modeling can streamline the development process, reduce the risk of security incidents, and ultimately save time and resources.

## Common types of cyber threats

Some of the most common types of cyber threats include malware, phishing attacks, MitM attacks, DDoS attacks, and APTs.

## Identifying and assessing vulnerabilities

Identifying and assessing vulnerabilities involves evaluating the organization's IT infrastructure and applications to identify potential weak points that could be exploited by cyber threats.

# Security architecture – design and implementation

The design and implementation of a robust security architecture is a key aspect of a CSA's role.

## Designing a security architecture

Designing a security architecture involves developing a detailed plan for how the organization's IT infrastructure and applications will be protected against cyber threats.

## Implementing a security architecture

Implementing a security architecture involves putting the plan into action, which may include installing security controls, configuring systems and networks, and training staff.

Risk management is a key aspect of cybersecurity architecture. It involves identifying, assessing, and responding to risks that could impact an organization's information security.

## Identifying and assessing risks

Risk identification and assessment involve pinpointing potential threats and vulnerabilities and evaluating their potential impact on the organization. This can be achieved through various methods, such as SWOT analysis, PESTEL analysis, and stakeholder analysis.

## Managing risks

Managing risks involves implementing measures to mitigate identified risks. This could involve reducing the risk, transferring it (for example, through insurance), avoiding it, or accepting it. The chosen approach will depend on the organization's risk appetite and the potential impact of the risk.

# The importance of continuous monitoring and improvement

Continuous monitoring and improvement are crucial for maintaining the effectiveness of the organization's security measures and adapting to changing threats and needs. A range of tools can support CSAs in their role. These tools can help to automate and streamline various aspects of cybersecurity, from threat modeling and risk assessment to IR and compliance management.

## The importance of choosing the right tools

Choosing the right tools is crucial to the success of a cybersecurity architecture. The chosen tools should align with the organization's strategic objectives, be easy to use, and provide meaningful and actionable insights.

## Leveraging AI and ML

Advanced technologies such as **artificial intelligence** (**AI**) and ML can bring significant cybersecurity benefits. These technologies can help to automate the identification and mitigation of threats, reducing the workload for CSAs and enabling them to focus on strategic tasks.

### Monitoring security measures

Monitoring security measures involves regularly reviewing and assessing the effectiveness of the organization's security controls, policies, and practices.

### Continuous improvement

Continuous improvement involves regularly updating and enhancing the organization's security measures based on findings from monitoring activities and changes in the threat landscape.

## The role of training and awareness in cybersecurity

Training and awareness play a crucial role in strengthening an organization's cybersecurity posture. Training is a vital component of a successful cybersecurity architecture. It ensures that employees understand the importance of cybersecurity, are aware of potential threats, and know how to respond to incidents.

### The importance of training

Training is essential for ensuring that staff understand their roles and responsibilities in relation to cybersecurity and are equipped with the knowledge and skills to perform these roles effectively. Regular training is critical to ensuring that employees stay up to date with the latest threats and security best practices. Training should be tailored to the needs of different roles and departments within the organization.

### Raising awareness

Raising awareness involves educating staff about the importance of cybersecurity, the types of threats they may face, and the steps they can take to protect themselves and the organization. Training can also play a crucial role in fostering a security-conscious culture within the organization. By highlighting the importance of cybersecurity and the potential consequences of security incidents, training can help motivate employees to take cybersecurity seriously.

## The future of cybersecurity architecture and GRC

The future of cybersecurity architecture and GRC is expected to be shaped by various trends, including the increasing use of AI and ML in cybersecurity, the growing importance of privacy and data protection, and the continuing evolution of cyber threats and regulations.

### Emerging trends

Emerging trends in cybersecurity include the increasing use of AI and ML for threat detection and response, the growing focus on privacy and data protection, and the evolution of cyber threats and attack techniques.

### *The future of GRC*

The future of GRC is likely to involve a greater focus on integrating GRC activities and the use of technology to automate and streamline GRC processes.

As the digital landscape evolves, the role of CSA is increasingly becoming a balancing act. A comprehensive cybersecurity strategy requires a delicate equilibrium between innovation and risk management. In this book, we delve into the intricate world of cybersecurity architecture, highlighting the challenging balancing act that professionals in this field must master.

The role of CSA is a complex balancing act. It involves promoting innovation while managing risks, fostering a security-conscious culture while navigating compliance requirements, and staying ahead of the ever-evolving threat landscape. Despite these challenges, the role of CSA is crucial in ensuring the security and success of an organization in today's digital world. As the digital landscape evolves, the role of CSA is increasingly becoming a balancing act. A comprehensive cybersecurity strategy requires a delicate equilibrium between innovation and risk management.

## Summary

This chapter provided an overview of key threats, risks, and governance factors that CSAs must consider when designing security architectures and programs. This included the following:

- Threat landscape:

  - Architects must have in-depth knowledge of threat actors, their motivations, and TTPs. Staying current on emerging threats through TI is critical.

  - Threat modeling using approaches such as STRIDE provides a systematic way to identify vulnerabilities and attack vectors.

- Risk management:

  - Risk assessments, both initial and residual, are essential to identify, analyze, and prioritize risks. Special consideration should be given to risks such as data breaches, ransomware, and third-party vendors.

  - Risk treatment involves selecting mitigation strategies to reduce unacceptable risks. This may include controls, process changes, or risk transfer.

- Governance:

  - Policies, standards, and procedures form the foundation of cybersecurity governance. Compliance with regulations such as GDPR must be ensured.

  - Clear roles and responsibilities provide accountability. IRPs and DRPs enable reacting to events.

  - Ongoing audits, metrics reporting, and security awareness training foster a culture of security.

In conclusion, CSAs must balance enabling business innovation and growth with managing cyber risks through their architecture designs and security programs. Threat modeling, risk assessments, and governance processes are essential tools to achieve this balance. Architects must balance security with innovation, using threat modeling, risk management, and governance processes. They serve as liaisons between technical units and BUs. Continued learning and adaptation are key in this evolving role responsible for creating secure yet agile architectures. CSAs have a crucial role in navigating the delicate balance between security and innovation through integrated threat, risk, and governance approaches tailored to the organization.

Architects must have a deep understanding of potential threat actors, their motivations, and tactics, as well as keep pace with the changing threat landscape. Conducting regular risk evaluations and implementing controls to mitigate unacceptable risks is also critical.

Strong governance with defined policies, procedures, compliance management, and security training fosters a culture of security and accountability across the organization. However, governance should ultimately align with business objectives and not hinder agility.

As cyber threats continue to evolve, architects must continuously monitor the effectiveness of controls and adjust their programs accordingly. Adopting flexible designs that allow for modular upgrades can help balance innovation and risk management.

Ultimately, managing this delicate balancing act requires CSAs to take a holistic, systems-based approach. By bringing together people, processes, and technology, they can enable organizations to securely innovate while creating cyber resilience and intuition based on organizational realities. This allows customization to your environment since organizational structures vary.

While technical expertise is core to architecture, clearly conveying designs is equally crucial. The next chapter explores how purposeful documentation benefits architects, teams, and organizations.

It examines documentation disciplines architects should master, from network diagrams to security control matrices. Best practices for style, formatting, versioning, and maintenance are discussed to ensure usability.

Powerful collaborative tools that integrate documentation deeper into architectures and workflows are highlighted, such as Microsoft Visio, Lucidchart, and Confluence. Guidance on building team knowledge management rituals using wikis, repositories, and documentation as code is provided.

By outlining documentation's critical role in architecture ideation, stakeholder communication, and institutional memory, the chapter aims to elevate documentation proficiency as a fundamental architectural skill set. You will gain insights into transforming documentation from obligation to strategic advantage.

With comprehensive documentation capabilities, architects produce not only robust technical architectures but high-impact knowledge repositories supporting execution, governance, and improved resilience.

# 6

# Documentation as a Cybersecurity Architect – Valuable Resources and Guidance for a Cybersecurity Architect Role

*"If words of command are not clear and distinct, if orders are not thoroughly understood, then the general is to blame. But, if orders are clear and the soldiers nevertheless disobey, then it is the fault of their officers."*

*– Sun Tzu*

In the previous chapter, we covered the potential challenges a cybersecurity architect may face in mitigating or level-setting controls against the threat/risk/governance of an organization. We also discussed how a cybersecurity architect accomplishes or manages the delicate balancing act required to take a holistic, systems-based approach to protecting the enterprise. By bringing together people, processes, and technology, they can enable organizations to securely innovate while creating intuitive cyber resilience based on organizational realities.

As has become the standard, I am using Sun Tzu's *Art of War* quotes to start this chapter. The preceding quote highlights the critical need for clear and thorough communication and understanding when conveying orders and instructions. The same applies to cybersecurity documentation. Effective cybersecurity relies on precise, well-structured documentation to align security teams and stakeholders. Unclear, disorganized documentation leads to confusion and disjointed security efforts, like an army that disobeys orders. Conversely, distinct policies, diagrams, and instructions promote compliance and coordinated security, like soldiers who thoroughly understand their marching orders. Cybersecurity architects bear the responsibility for producing comprehensible documentation. But security also depends on *officers* across the organization accurately implementing documented policies, models, assessments, and configurations. In summary, high-quality documentation is essential for cybersecurity architects to communicate security designs and requirements precisely. This clarity enables organizational alignment and action, avoiding obfuscation and missteps that can shatter security efforts.

Effective documentation is a critical, yet often overlooked, element of cybersecurity architecture. This chapter covers the following topics:

- Why document?

- Types of documentation

- Documentation tools

- Team approaches to documentation

As laid out in the preceding list of topics, this chapter explores best practices for documentation to enhance visibility, align security initiatives, and bolster compliance. First, we will examine why comprehensive documentation is imperative for cybersecurity architects. Next, we will provide an overview of the types of documents that are leveraged, spanning policies, diagrams, models, assessments, and configurations. Then, we will discuss documentation tools and team approaches to optimize creation and consumption. Throughout, the emphasis will be on pragmatic steps to elevate documentation from a checkbox activity to a value-adding endeavor. By adopting the methodologies in this chapter, cybersecurity architects can produce documentation that acts as a valuable organizational asset, rather than an afterthought. The documentation framework presented aims to balance detail and clarity, structure and flexibility. Adhering to these principles enables documentation that is comprehensive yet comprehensible – strengthening communication and collaboration in service of more robust security architectures.

The role of documentation within the purview of a cybersecurity architect is critically understated, yet fundamentally paramount. *It is also for this reason this chapter is one of the longest chapters in this book.* As cybersecurity architects, we are responsible for designing, implementing, and overseeing an organization's cybersecurity framework, and documentation serves as both the blueprint and the historical record of the organization's cybersecurity posture. These documents are not merely administrative formalities; rather, they embody a systematic knowledge base, providing a structured mechanism to capture complex configurations, policies, and procedures. They play a seminal role in facilitating institutional memory, thereby enhancing operational continuity, especially in scenarios that involve personnel changes or rapidly scaling cybersecurity infrastructure.

Intricate cybersecurity architectures often consist of multiple layers of defense, integrated through a variety of hardware and software solutions, each with its configuration nuances, inter-dependencies, and impact vectors. Proper documentation offers a distilled view of this complexity, acting as a guide for system administrators, security analysts, and decision-makers. It aids in diagnostics and troubleshooting, offers a basis for compliance audits, and serves as a cornerstone for training and awareness programs. Thorough and up-to-date records can expedite the processes of incident response and disaster recovery by providing accurate information when time is of the essence.

In the realm of regulatory compliance, be it the **General Data Protection Regulation (GDPR)**, **Health Insurance Portability and Accountability Act (HIPAA)**, or sector-specific regulations, documentation assumes a non-negotiable role. Failure to maintain accurate records can result in non-compliance, leading to hefty fines and reputational damage. Beyond its compliance utility, well-maintained documentation also provides an empirical foundation for risk assessments and threat modeling exercises. By offering a snapshot of the current security controls and configurations, it aids in identifying potential vulnerabilities and strategizing subsequent layers of defense, thus contributing directly to the robustness of the cybersecurity architecture.

Moreover, documentation serves as a communication bridge between technical and non-technical stakeholders, from engineers to executive leadership. By translating the intricacies of cybersecurity architecture into comprehensible terms, these records facilitate informed decision-making. Therefore, a well-documented cybersecurity architecture becomes a dynamic entity, offering the organization the flexibility to adapt to emerging threats while sustaining operational efficacy. The value proposition of documentation, therefore, extends beyond mere record-keeping to become an integral part of strategic cybersecurity governance. That is to say, you will be spending as much time documenting or working on documentation as you do designing, implementing, or evaluating the technology the documentation represents.

# Why document?

Documentation, an integral aspect of any organization, is often underestimated. However, its significance transcends diverse sectors, including IT, healthcare, finance, and government. Documentation is the backbone that supports the seamless functioning of systems, thus enhancing efficiency and promoting accountability. This section delves into the importance of documentation, exploring its various aspects and how it contributes to organizational success.

## What is documentation?

**Documentation** refers to the systematic process of organizing information in a structured manner to serve multiple purposes. It can range from user guides and manuals to reports, proposals, and regulatory submissions. The primary objective of documentation is to provide a tangible and enduring record of information that can be easily accessed and utilized when needed.

### Categories of documentation

Documentation can be classified into several categories, each serving a unique purpose:

- **Informative documentation**: These documents aim to elucidate a topic or concept. They provide comprehensive details and context to facilitate understanding.

- **Instructional documentation**: As the name suggests, these documents guide users on how to execute a task. They enumerate the necessary steps to accomplish a task efficiently.

- **Communication documentation**: These documents facilitate information flow between different entities. They play a crucial role in transmitting necessary documents or information to the intended recipients.

- **Plans**: Plans outline a project or initiative's development. They detail the objectives, significance, and strategies for achieving the goals of the plan.

## Why is documentation important?

Understanding the importance of documentation can provide your organization with a competitive edge. Here are some reasons why documentation is vital:

- **Ensures consistency and efficiency**: Documentation can serve as a roadmap, guiding employees on their roles and responsibilities. It ensures a standardized approach to tasks, leading to consistency in operations.

- **Mitigates risk from employee turnover**: Employee turnover can disrupt an organization's operations. Documentation helps in mitigating this risk by preserving the knowledge and expertise of departing employees.

- **Facilitates communication**: Documentation aids communication within the organization. It can also serve as a means of conveying information to external stakeholders, such as customers, vendors, or regulatory authorities.

- **Tracks progress**: Documentation is crucial for monitoring an organization's progress. It provides a historical record of activities and decisions, making it easier to track developments and measure growth.

- **Enhances professional image**: Proper documentation reflects an organization's professionalism. It gives the impression of a well-managed, organized, and accountable entity, thereby enhancing its reputation.

- **Effective documentation practices**: Creating effective documentation requires meticulous planning and execution. Here are some practices that can help you create high-quality documents:

  - **Create an outline**: Before writing a document, create an outline that delineates the document's structure and key points.

  - **Consider the audience**: Always keep the document's intended audience in mind. This will help you present the information in a manner that is easily comprehensible to them.

  - **Maintain clarity**: Make sure your document is clear and concise. Avoid jargon and complicated phrases as much as possible.

  - **Update regularly**: Documentation should always be up-to-date. Regular updates ensure that the information remains relevant.

  - **Review and proofread**: Always review and proofread your document before finalizing it. This helps eliminate errors and improves the overall quality of the document.

### The role of documentation in various industries

Documentation plays a pivotal role in various industries. Here are a few examples:

- **IT sector**: In the IT sector, documentation is crucial in software development. It provides a detailed account of the software's functionality and guides users on how to use the software effectively.

- **Healthcare sector**: In the healthcare sector, medical records act as essential documents. They provide a complete record of a patient's medical history, helping healthcare professionals make informed decisions about the patient's care.

- **Government sector**: In government agencies, documents such as official correspondences and records ensure that administrative processes are carried out as per government policies.

Documentation is a crucial aspect of any organization. It enhances efficiency, promotes accountability, and serves as a valuable resource for reference. Thus, organizations should invest time and effort in creating high-quality documents and maintaining a robust documentation system.

## Additional information

To further improve your organization's documentation process, consider using documentation software or tools. These tools can simplify the process of creating and managing documents, making it easier for your organization to maintain an effective documentation system.

Remember, good documentation practices contribute to an organization's success by providing a solid foundation for information management. Therefore, it's crucial to understand the importance of documentation and implement best practices in your organization.

# Types of documentation

In the complex world of cybersecurity architecture, documentation serves as both the roadmap and the rulebook, articulating both the what and the how of security controls. This chapter delves into the main categories of documentation that underpin a resilient cybersecurity architecture, serving as foundational elements for governance, design, risk management, and operational consistency.

The first critical category is *Policies and procedures*, which are high-level documents that establish the cybersecurity governance framework.

The second category zooms into architectural visualization. *System architecture diagrams* offer a bird's-eye view of the IT environment, illuminating the interplay between networks, systems, applications, and data flows.

The third category centers on risk-oriented documentation, such as threat models and risk assessments.

Finally, the fourth category addresses implementation and technical specifications, such as security requirements, solution design documents, and configuration documents. As this section progresses, we will explore each of these documentation types in detail, offering insight into their structure, purpose, and role within the broader context of cybersecurity architecture. By understanding the distinct functions of each document, cybersecurity professionals can better design, implement, and manage architectures that are not only secure but also scalable and compliant.

## Policies and procedures

In cybersecurity architecture, policies and procedures serve as the governing documents that establish a framework for organizational behavior and technical implementations. These documents are cardinal for aligning an organization's security stance with its business objectives and compliance requirements. They operationalize abstract security objectives into actionable directives, thereby serving as the cornerstone for planning, implementing, and assessing an organization's security posture.

### Types of policies and procedures

Let's look at the different types of policies and procedures:

- **Acceptable use policy (AUP)**: This policy outlines the acceptable ways in which organizational resources, such as networks and computing devices, can be used by employees. It often includes clauses relating to the use of external storage devices, browsing restrictions, and prohibitions against using organizational resources for illegal or unauthorized activities. For example, an AUP may explicitly forbid the use of peer-to-peer file-sharing software, thus mitigating the risk of software piracy and potential malware infections.

- **Data classification policy**: This policy governs how data is classified, stored, and handled within the organization. It usually divides data into categories such as *Public*, *Internal*, *Confidential*, and *Restricted*, each requiring varying levels of protection. For instance, data labeled as *Restricted* may necessitate encryption both at rest and in transit, as well as stringent access controls that are regularly audited.

- **Incident response procedure**: This is a specialized procedural document that delineates the specific actions to be taken when a security incident occurs. It often employs a phase-based approach encompassing identification, containment, eradication, recovery, and lessons learned. Each phase has its specific set of procedures; for example, containment could involve isolating affected systems from the network to prevent lateral movement by an attacker.

### Structural elements

Typically, a well-designed policy or procedure document will contain the following structural elements:

- **Scope**: This specifies the applicability of the policy, often identifying the organizational units or geographic locations it pertains to

- **Roles and responsibilities**: These define who is responsible for various aspects of policy enforcement and compliance

- **Compliance requirements**: These enumerate any relevant statutory, regulatory, or contractual obligations

- **Enforcement and sanctions**: These describe how the policy will be enforced and what penalties may be incurred for violations

## Technical embedding and implementation

The actualization of these policies often involves configuring various security controls. For instance, an acceptable use policy might be enforced through web content filtering and **data loss prevention (DLP)** solutions. Data classification policies often require technical mechanisms for tagging and access control, often integrated into an organization's **identity and access management (IAM)** system. Incident response procedures, on the other hand, may necessitate specialized tools for system forensics, network monitoring, and automated alerting.

## Auditing and revision

To ensure continued relevance and effectiveness, policies and procedures should be subject to periodic review and auditing. This involves both automated compliance checking – perhaps via configuration management tools – as well as manual assessments such as internal audits or third-party assessments.

In the context of **governance, regulatory, and compliance (GRC)** needs within a business, the documentation of policies and procedures assumes a paramount role. These high-level documents bridge the gap between an organization's strategic governance goals and the tactical implementation of cybersecurity measures. They not only provide a formalized structure for internal governance but also serve as evidence of compliance for external regulatory bodies and auditors. The following are specific ways in which such documentation contributes to GRC initiatives:

- **Governance support**:

  - **Strategic alignment**: Policies and procedures help translate the strategic goals set by governance bodies into executable, operational directives. This ensures that all cybersecurity activities are aligned with the organization's broader mission and objectives.

  - **Resource allocation**: By defining the scope and requirements of cybersecurity activities, these documents provide a foundation for budgetary decisions, helping to allocate resources where they are most needed.

  - **Accountability and oversight**: These documents establish roles and responsibilities, thus clarifying who is accountable for various aspects of cybersecurity. This facilitates better oversight and governance.

- **Regulatory compliance**:

  - **Evidence of due diligence**: Well-crafted policies and procedures act as evidence that an organization is taking cybersecurity seriously, often a requirement under laws such as GDPR or HIPAA.

  - **Audit readiness**: These documents form the basis for audit checks, whether they're self-assessments or external audits. They establish what controls should be in place, thus making it easier to demonstrate compliance during audits.

  - **Legal safeguards**: In the event of a security incident, having robust and up-to-date policies can act as a legal safeguard, potentially mitigating liabilities.

- **Compliance monitoring and reporting**:

  - **Metrics and key performance indicators (KPIs)**: Policies often stipulate performance metrics or KPIs that serve as quantitative measures of compliance. For instance, an incident response policy might specify that all incidents must be contained within 24 hours of detection.

  - **Continuous monitoring**: Compliance with these policies is often assured through continuous monitoring and enabled by **security information and event management (SIEM)** systems that alert administrators to any deviations.

  - **Reporting mechanisms**: These documents often define the structures for regular reporting to senior management or a governance body, thereby providing a structured approach for compliance reporting.

### An inter-relationship with technical controls

Moreover, these documents often specify the technical controls required for regulatory compliance. For instance, a data classification policy may prescribe the use of encryption technologies in compliance with GDPR's mandates for data protection. An acceptable use policy may also dictate the employment of firewalls and intrusion detection systems that align with specific industry regulations, such as the **Payment Card Industry Data Security Standard (PCI DSS)**.

The role of policies and procedures extends beyond merely dictating organizational behavior; they serve as a lynchpin in a comprehensive GRC strategy. They provide the structure and rigor that enable governance bodies to exert effective oversight, fulfill regulatory requirements, and continuously monitor compliance, thereby contributing to an organization's cyber resilience and risk management posture.

Policies and procedures are not merely administrative constructs; they have significant technical ramifications. By clearly defining rules, roles, and responsibilities, these high-level documents lay the groundwork for a security architecture that is both robust and aligned with an organization's overarching objectives.

# System architecture diagrams

System architecture diagrams serve as vital artifacts in the cybersecurity domain, offering visual representations that encapsulate various facets of an organization's IT environment. These diagrams can range from high-level overviews to highly detailed schematics that incorporate elements such as networks, computing resources, applications, data storage components, data flows, and trust boundaries. Their relevance stems from their ability to facilitate threat modeling, vulnerability identification, compliance validation, and even incident response. Ahead, we delve into specific technical details that highlight the critical role of system architecture diagrams in cybersecurity.

## Taxonomy and granularity

System architecture diagrams can be partitioned into several categories based on their focus and granularity:

- **Network topology diagrams**: These focus primarily on the networking infrastructure, showcasing how network segments and subnets are connected through switches, routers, and firewalls
- **Application architecture diagrams**: These concentrate on how application components interact with each other and with the underlying infrastructure
- **Data flow diagrams** (DFDs): These illustrate how data moves through the system, identifying points where data is at rest, in transit, or being processed

The choice of diagram depends on the specific use case. For instance, a DFD would be instrumental in complying with GDPR's data protection mandates, while a network topology diagram could be used to ensure that firewall rules meet organizational policies.

## Vulnerability identification

Diagrams are invaluable for performing vulnerability assessments and identifying potential attack vectors. By illustrating how components are interconnected, they allow security analysts to do the following:

- **Spot unsecure data transmissions**: Identify areas where data is transmitted without adequate encryption
- **Recognize unauthenticated access points**: Locate interfaces where authentication mechanisms are either weak or non-existent
- **Detect unnecessary trust relationships**: Examine trust boundaries to ensure that they adhere to the principle of least privilege

## Integration with security tools and processes

System architecture diagrams are often ingested into various cybersecurity tools and platforms:

- **SIEM**: Diagrams can help configure SIEM solutions more effectively by identifying which components and data flows to monitor

- **Intrusion detection systems (IDSs)/intrusion prevention systems (IPSs):** Knowing the architecture aids in strategically placing IDS/IPS sensors

- **Threat modeling tools:** Many threat modeling tools allow architecture diagrams to be imported to facilitate automated risk assessments

## Compliance and auditing

System architecture diagrams not only serve as foundational elements for security practices but also play a crucial role in governance, regulatory compliance, and business alignment. Their multifaceted utility ensures that they are not just technical documents but also governance artifacts that inform strategic decisions, resource allocation, and compliance verification. The following elucidation expands on how these diagrams are pivotal in satisfying GRC objectives:

- **Governance support:**

  - **Strategic planning and decision-making:** System architecture diagrams aid in the governance process by providing decision-makers with a clear view of the organization's IT landscape. This enables informed decisions around security investments and aligns security initiatives with business objectives.

  - **Resource allocation:** By visualizing the critical components and data flows, these diagrams assist in prioritizing resource allocation, ensuring that the most critical assets receive the highest level of security controls.

  - **Policy enforcement points:** Governance policies can be mapped directly onto the architecture, pinpointing where specific controls, such as DLP or access control mechanisms, need to be implemented.

- **Regulatory compliance:**

  - **Mapping regulatory requirements:** Regulations often mandate specific security controls for different types of data or components. Diagrams can help in mapping these requirements to specific parts of the architecture, making it easier to identify where compliance needs to be verified.

  - **Audit preparation and support:** During audits, system architecture diagrams serve as evidence for the implemented security controls and data flow mechanisms, thereby facilitating the verification process for compliance with standards such as GDPR, HIPAA, or **System and Organization Controls (SOC 2)**.

- **Risk management and compliance:**

  - **Compliance gap identification:** By integrating the architecture diagram with a compliance management tool, organizations can automate the process of compliance gap identification based on real-time architecture states and changes.

- **Risk assessments**: Diagrams are frequently used in risk assessment activities to quantify and prioritize risks based on the architecture. This is particularly useful in meeting the compliance requirements that mandate periodic risk assessments.

- **Documentation and record-keeping**:

  - **Change management**: In a governance framework, maintaining up-to-date system architecture diagrams is essential. These documents should be version-controlled and updated as part of the change management process.

  - **Legal preparedness**: In cases of breaches or legal disputes, having a well-documented system architecture can serve as part of the organization's due diligence and reasonable care records, which may be beneficial from a legal standpoint.

In essence, system architecture diagrams are instrumental in weaving cybersecurity considerations into the broader tapestry of organizational governance, regulatory obligations, and compliance activities. Their multi-tiered relevance makes them indispensable artifacts for businesses, transcending the boundaries between technical necessity and governance imperative. Through their capacity to illustrate, educate, validate, and guide, these diagrams constitute a keystone in a holistic cybersecurity and GRC program.

In the context of compliance, these diagrams serve multiple purposes:

- **Regulatory alignment**: Whether it's HIPAA's requirement for securing patient data or PCI DSS's mandate for safeguarding payment information, these diagrams help ensure that the architecture aligns with regulatory requirements.

- **Documentation for auditors**: During an audit, these diagrams serve as documentation that illustrates the organization's cybersecurity posture. They can be particularly effective in demonstrating the segregation of duties or the implementation of security zones.

System architecture diagrams act as a cornerstone in both strategic and tactical aspects of cybersecurity. They offer a visual map that guides not only the identification and mitigation of vulnerabilities but also facilitates compliance, monitoring, and incident response activities. Their integrative nature makes them indispensable for any robust cybersecurity program.

## Threat models

In the realm of cybersecurity, threat modeling is an analytical framework for systematically identifying, characterizing, and mitigating threats and vulnerabilities. It is generally recognized as an essential phase within the **software development life cycle (SDLC)** but is increasingly incorporated into broader risk management and governance procedures. One of the most widely adopted threat modeling methodologies is **STRIDE**, an acronym representing the different types of threats: **Spoofing, Tampering, Repudiation, Information Disclosure, Denial of Service, and Elevation of Privilege**.

## The technical specifics of threat models

Let's look at the technical specifics of threat models:

- **System decomposition**: Initially, threat modeling necessitates an exhaustive understanding of the system under scrutiny. This entails disentangling the architecture into its constituent components, such as servers, network links, data flows, and trust boundaries.

- **STRIDE categorization**: For each component or interaction that's identified, threats are characterized according to the STRIDE taxonomy. For instance, an API endpoint may be susceptible to *information disclosure* threats, while user authentication modules might be vulnerable to *spoofing*.

- **DFDs**: DFDs are employed to visually represent how data moves through the system, highlighting areas of potential vulnerability. Trust boundaries are particularly emphasized as crossing a trust boundary often entails a change in threat exposure.

- **Attack trees**: These offer a hierarchical visualization of potential attacks, delineating prerequisites and outcomes for each attack vector.

- **Mitigation strategies**: For each identified threat, possible mitigations are enumerated. These can range from code-level fixes to systemic changes such as incorporating additional layers of encryption or authorization.

## Utility in risk prioritization and mitigation

Threat models are used to quantify and prioritize risks. By associating severity levels and likelihood metrics with each threat, organizations can make data-driven decisions on where to allocate resources for mitigation activities. This not only meets compliance requirements for regular risk assessments but also aligns with governance objectives for risk management.

## Governance and regulatory compliance

Threat modeling serves as a nexus between cybersecurity implementations and broader governance and compliance obligations. Well-documented threat models directly strengthen governance frameworks by enabling policies aligned with actual risks. They also provide artifacts demonstrating due diligence for regulatory requirements such as GDPR and HIPAA. As auditable records of security evaluations, threat models can expedite external audits and certifications. By detailing risk analysis, threat models become strategic assets that allow organizations to implement cybersecurity fulfilling both technical effectiveness and legal/regulatory compliance. Savvy cybersecurity architects recognize threat modeling's immense value in securing organizations not only against digital threats but also potential legal and reputational damages. Integrating threat modeling deeply into security strategies allows architects to craft governance and compliance foundations as robust technical defenses:

- **Policy alignment**: Threat models help align security controls and policies with real-world risks, thereby strengthening governance frameworks.

- **Regulatory requirements**: Detailed threat models can serve as compliance artifacts to demonstrate due diligence in identifying and mitigating risks, as required by standards such as GDPR for data protection impact assessments or HIPAA for security risk analysis.

- **Audit trails**: The model serves as an auditable record, facilitating external audits or regulatory inspections. A well-documented threat model can expedite the audit process, substantiating the security posture of the organization.

Threat modeling, as an integral component of cybersecurity, extends beyond mere technical assessment to become a cornerstone of an organization's GRC strategies. The following are key ways that well-documented threat models contribute to fulfilling GRC objectives:

- **Governance**:

  - **Strategic alignment**: Effective threat models align with organizational objectives and governance frameworks, helping leadership make informed strategic decisions concerning cybersecurity investments and risk tolerance. Threat modeling serves as a critical tool for aligning an organization's security posture with governance, risk management, and compliance objectives. Threat models facilitate strategic governance by doing the following:

    - **Informing security policies and standards**: Models identify specific vulnerabilities such as SQL injection on web apps, enabling targeted policies like input validation requirements

    - **Driving budget/resource allocation**: Quantified risks allow leadership to prioritize spending on the highest-risk threats, such as patching critical servers first

    - **Assigning security responsibilities**: Models document responsible roles such as DBAs to enforce access controls

    - **Enabling risk-based decisions**: Executives can set risk tolerance and make calculated decisions by weighing threats against business needs

- **Risk management**: Threat modeling enables proactive risk management in the following ways:

  - **Allowing risk quantification**: Threats are assigned severity scores based on impact and likelihood, facilitating quantification.

  - **Prioritizing mitigations**: Threats are addressed in order of severity. For instance, blocking DDoS attacks would take priority over hardening public APIs.

  - **Defining metrics and key risk indicators (KRIs)**: Threat trends are monitored via metrics such as the frequency of SQL injection alerts falling below a threshold of $X$.

  - **Enabling continuous monitoring**: Models are updated regularly to account for new threats, systems, and mitigations.

- **Compliance**: Threat modeling helps demonstrate compliance by doing the following:

  - **Mapping to regulations**: Models can map threats to related compliance frameworks, such as the threat of unauthorized access to PCI DSS requirements

  - **Providing audit trails**: Threat models serve as evidence of due diligence during audits for standards such as ISO 27001

  - **Facilitating assessments**: Models contain the documentation required for assessments such as **Data Protection Impact Assessments** (**DPIAs**) under GDPR

Threat modeling forms the basis for building a robust GRC program by assessing risks, defining security controls, assigning responsibilities, and documenting due diligence. Integrating threat modeling strengthens alignment between security operations and broader organizational objectives:

  - **Policy development and implementation**: By identifying specific vulnerabilities and attack vectors, threat models can guide the formulation of targeted policies and procedures, fortifying an organization's overall governance structure

  - **Resource allocation**: Through systematic risk quantification and prioritization, threat models can inform resource allocation decisions, ensuring that security measures with the highest ROI are prioritized

  - **Accountability and roles**: Threat models explicitly list responsible parties for various security controls and mitigations, thereby promoting accountability and clarifying roles within the organization

- **Regulatory compliance**:

  - **Compliance artifacts**: Well-structured threat models serve as evidentiary material during regulatory audits. These documents prove that the organization has performed due diligence in identifying and planning to mitigate cybersecurity risks.

  - **Regulatory mapping**: Many regulations, such as GDPR or HIPAA, have specific requirements around data protection and risk assessment. Threat models can be designed to map directly to these requirements, making it easier to demonstrate compliance.

  - **DPIAs**: For regulations such as GDPR, a threat model can serve as a foundational document for conducting a DPIA, a mandatory exercise for data-intensive projects.

- **Risk management**:

  - **Quantification and prioritization**: Threat models aid in the quantification of risks by assigning severity and likelihood metrics to each identified threat, facilitating data-driven risk management strategies

  - **Risk treatment plans**: Leveraging the threat model, an organization can develop targeted risk treatment plans, complete with timelines and responsible parties

- **Risk metrics and KPIs**: Threat models can be used to define KRIs and KPIs, thereby creating metrics that can be tracked over time to measure the efficacy of risk management programs

- **Compliance reporting**:

  - **Board reporting**: A well-documented threat model can be summarized into high-level reports for board members and stakeholders, thereby aligning security operations with business governance

  - **Transparency and trust**: The rigor and detail encompassed in threat modeling demonstrate to stakeholders that the organization takes cybersecurity seriously, thereby boosting stakeholder trust and fulfilling transparency obligations mandated by various regulations

  - **Continuous compliance**: In dynamic regulatory environments, threat models offer a framework for continuous compliance by being easily updated to reflect new regulatory requirements, technological changes, or emerging threats

  The utility of threat modeling extends from tactical technical defenses to strategic governance and compliance requirements, serving as a unifying framework that satisfies a broad spectrum of organizational needs. Therefore, a well-implemented threat model becomes a linchpin in an organization's GRC arsenal, offering a multi-faceted approach to tackling complex cybersecurity challenges.

- **Integration with risk management programs**: Threat models should be dynamic and subject to updates in line with architectural changes, newly discovered vulnerabilities, or emerging threats. They become a part of an organization's ongoing risk management, designed to adapt to a continuously changing cyber threat landscape.

Thus, threat modeling serves as a potent tool that brings rigor and structure to the nebulous and often unpredictable domain of cybersecurity threats. When implemented and maintained correctly, it effectively serves not just technical needs but also satisfies an array of requirements concerning governance, risk management, and regulatory compliance.

## Cybersecurity threat modeling – examples and lab exercise

Cybersecurity threat modeling is an integral aspect of system design and development, enabling organizations to identify, assess, and remediate potential security threats. This section provides an in-depth exploration of cybersecurity threat modeling techniques, supported by practical examples and a lab exercise. Specifically, we'll focus on DFDs, STRIDE, and attack trees for elucidating threat scenarios and potential vulnerabilities.

### Methodologies

Threat modeling represents a proactive approach to identifying and addressing potential security vulnerabilities within applications and systems. Rather than waiting to be breached, organizations can get ahead of threats by methodically analyzing risks and architecting defenses prioritized to actual exposures. This section explores core methodologies that empower organizations to thoroughly assess threats and strategically strengthen protections.

By detailing techniques such as DFDs, STRIDE analysis, and attack trees, this section aims to equip you with actionable skills to model risks systematically. Hands-on practice through an example web application vulnerability assessment will reinforce how to apply threat modeling across diverse environments. Organizations often struggle to balance security with innovation amid rapid technological change. Threat modeling provides clarity amid the chaos, shining light on risks to forge defenses that enable fearless advancement. With these methodical approaches and practical experience, cybersecurity teams can work proactively, not reactively, securing organizations with their eyes wide open:

- **DFDs**: DFDs provide a graphical representation of how data flows within a system. The components include the following:

    - **Entities**: External actors (for example, users)

    - **Processes**: Actions or services that manipulate data

    - **Data stores**: Databases or storage mechanisms

    - **Data flows**: Movement of data

- **STRIDE methodology**: STRIDE is an acronym for various types of threats:

    - **Spoofing**: Impersonating another user or system.

    - **Tampering**: Manipulating data or code.

    - **Repudiation**: Denying the performance of an action. In terms of STRIDE, it is the inability to trace the attack back to a specific source/actor.

    - **Information disclosure**: Unauthorized access to information.

    - **Denial-of-service (DoS)**: Making a service unavailable.

    - **Elevation of privileges**: Gaining unauthorized access rights.

- **Attack trees**: Attack trees outline different attack vectors that achieve a particular malicious objective. Nodes represent conditions, while leaves symbolize attack vectors.

## Examples

Let's look at the first example – an online payment system using a DFD and STRIDE:

- **Entities**: Customer, payment gateway, bank

- **Processes**: Authenticate customer, validate payment, transfer funds

- **Data stores**: Customer database, transaction log

The following threats were identified via STRIDE:

- **Spoofing**: Fake customers can initiate transactions
- **Tampering**: Transaction logs can be manipulated
- **Information disclosure**: Sensitive payment information leaks

Now, let's look at the second example – cloud storage using attack trees:

- **Objective**: Unauthorized data access
- **Branch 1**: Exploit software vulnerability:
  - **Sub-branch**: SQL injection
  - **Sub-branch**: Buffer overflow
- **Branch 2**: Social engineering:
  - **Sub-branch**: Phishing
  - **Sub-branch**: Impersonation

## Lab exercise – threat modeling a web application

The objective of this exercise is to perform a threat modeling assessment for an e-commerce web application to identify security vulnerabilities:

- **System architecture**:
  - **Web server**: Apache Tomcat hosting a Java web application. Accessible from the internet.
  - **Database server**: MySQL database storing customer data. Resides on the internal network.
  - **Firewall with a DMZ**: Controls traffic between the web server in the DMZ and the database server.
- **Data flows**:
  - Internet | firewall | web server
  - Web server | firewall | database server
- **Tools required**: Microsoft Threat Modeling Tool or OWASP Threat Dragon

The steps are as follows:

1.  **Define the scope**: Identify the scope of the web application, the components involved, and the data flow. Decompose the application into components such as the web server, database, firewall, and so on.

2.  **Create a DFD**: Utilize a threat modeling tool to create a DFD representing entities, processes, data stores, and data flows. Draw a DFD showing how data moves between components. Highlight trust boundaries.

3.  **Apply STRIDE**: Use the STRIDE methodology to enumerate threats associated with each element in the DFD. Identify vulnerabilities using the STRIDE model:

    *   **Spoofing**: Weak authentication could allow a fake shop front

    *   **Tampering**: No input validation can allow attackers to modify the price/inventory

    *   **Repudiation**: Lack of logging may prevent auditing transactions

    *   **Information disclosure**: Plaintext passwords exposed in the database

    *   **Denial of service**: Unrestricted file uploads can fill the disk

    *   **Elevation of privilege**: SQL injection can allow privilege escalation

4.  **Generate attack trees**: For significant threats, create an attack tree outlining various attack vectors.

5.  **Recommend mitigations**: Propose security controls to mitigate identified threats. Develop mitigation strategies for each threat:

    *   **Spoofing**: Implement **multi-factor authentication (MFA)**

    *   **Tampering**: Validate all form inputs on the server side

    *   **Information disclosure**: Encrypt passwords using bcrypt

    *   **Denial of service**: Limit the file upload size and extension

    *   **Elevation of privilege**: Use prepared SQL statements

The expected outcomes are as follows:

*   A DFD representing the architecture of the web application

*   A list of identified threats categorized using STRIDE

*   Attack trees for critical threats

*   Recommendations for mitigating threats

This lab exercise demonstrates the core techniques involved in performing a threat modeling assessment for an application, such as STRIDE analysis, identifying security gaps, and defining mitigation strategies in priority order. You can build on this with actual code reviews, attack simulations, and designing controls aligned with the risks. Understanding the intricacies of cybersecurity threat modeling equips organizations with the knowledge to preemptively address vulnerabilities. By employing methodologies such as DFDs, STRIDE, and attack trees, organizations can cultivate a robust security posture.

## Risk assessments

Risk assessments in cybersecurity are a systematic methodology for evaluating the potential risks that could be involved in an IT system or infrastructure. These are comprehensive documents that record not just the vulnerabilities in a system, but also the various threats that could exploit these vulnerabilities and the potential impact such an exploit could have on the business. Assessments are pivotal in shaping an organization's cybersecurity strategy as they identify the key areas where mitigative action is most needed. The efficacy of a risk assessment is enhanced by its granularity and involves the following fundamental components:

- **Identification and scope definition**: The first step in risk assessment involves identifying the assets, systems, data repositories, and other critical components within the organization's IT landscape. This step may also encompass classifying data and assets based on their criticality to business functions.

- **Threat and vulnerability analysis**: Once the scope has been defined, the next step involves identifying known and potential threats and vulnerabilities. Threats could range from external actors, such as cybercriminals and nation-state actors, to internal threats such as disgruntled employees. Vulnerabilities could be software flaws, weak passwords, or even operational issues such as a lack of MFA.

- **Impact and likelihood estimation**: For each identified threat-vulnerability pair, the potential impact on the organization is estimated. The impact is often measured in financial terms, but it could also involve other types of harm, such as reputational damage or loss of competitive advantage. The likelihood of each threat exploiting its corresponding vulnerability is also estimated, often on a numerical scale or a qualitative measure such as low, medium, or high.

- **Risk quantification and prioritization**: Finally, the impact and likelihood estimates are used to calculate the level of risk using a formula such as *Risk = Impact x Likelihood*. These risks are then prioritized based on their severity, enabling organizations to focus their limited security resources on the most critical areas.

- **Risk treatment and documentation**: Based on the risk's prioritization, mitigative measures are prescribed for each risk. These could range from technical solutions such as patching a software vulnerability to administrative actions such as revising a security policy. The risk treatment steps, along with their cost, timelines, and responsible parties, are all documented, forming the basis for the risk treatment plan.

- **GRC support**: Risk assessments are also critical from a GRC perspective:

  - **Governance**: Risk assessments help in aligning security protocols with the organization's broader business objectives. They provide governance bodies such as the board of directors with quantitative metrics, facilitating data-driven decision-making.

  - **Regulatory compliance**: Compliance with standards such as PCI DSS, HIPAA, or GDPR often requires a comprehensive risk assessment. The documented risk assessment serves as evidence during compliance audits and can guide efforts to meet specific regulatory requirements around data security and privacy.

  - **Compliance reporting**: The findings from a risk assessment can be incorporated into compliance reports required by various regulations. They can also be used for internal compliance reporting to showcase risk management effectiveness.

Risk assessments serve as both critical tactical and strategic assets that are integrated into the broader cybersecurity architecture and GRC framework, thereby fortifying the organization's overall cybersecurity posture.

## Security requirements

Security requirements serve as the cornerstone for designing a resilient cybersecurity architecture, aimed at safeguarding an organization's assets while meeting operational objectives. These requirements explicitly enumerate the conditions, functionalities, and constraints that a system must satisfy to uphold the tenets of the **confidentiality, integrity, and availability** (**CIA**) triad. They are often detailed in **security requirements specifications** (**SRS**) documents and are derived through a methodological approach that usually includes stakeholder interviews, documentation reviews, and compliance guidelines. Here are some of the technicalities that underline the specification of security requirements:

- **Definition categories**:

  - **Confidentiality requirements**: These are designed to limit access to information to authorized users only. Requirements may include encryption algorithms to be used, access control measures, and secure transmission protocols.

  - **Integrity requirements**: These aim to ensure the accuracy and reliability of data and systems. They may specify the use of cryptographic hash functions, data validation measures, and digital signatures.

  - **Availability requirements**: These focus on ensuring that systems and data are accessible to authorized users when needed. This may involve specifying redundant systems, failover procedures, and backup strategies.

- **Techniques for eliciting requirements**:

  - **Use case analysis**: This defines how different types of users interact with the systems and what security constraints should be applied in each case.

  - **Threat modeling**: Techniques such as STRIDE or **Damage Reproducibility, Exploitability, Affected Users, and Discoverability (DREAD)** are used to understand potential threats and define countermeasures as requirements:

    - **Regulatory mapping**: Review relevant regulations such as GDPR, HIPAA, or PCI-DSS to extract specific security obligations that need to be met

- **Formal languages and notation**: Formal methods such as Z-notation, B-Method, and the Common Criteria's formal representation can be used to specify security requirements. These formalisms offer a mathematical basis for specifying and verifying the requirements, providing an unambiguous interpretation that mitigates risks associated with misunderstanding or miscommunication.

- **Verification and validation**: To ensure that the security requirements are adequately addressed in the system design, rigorous validation techniques are employed. This can include formal verification methods, testing against predefined security use cases, and performing code audits to verify that the implementation adheres to the specified requirements.

- **GRC support**: Security requirements are closely intertwined with GRC objectives:

  - **Governance**: They serve as actionable inputs in governance frameworks, ensuring that the organizational strategy integrates cybersecurity objectives effectively

  - **Regulatory compliance**: Security requirements are often derived from or mapped to regulatory mandates, ensuring that systems are designed to meet compliance from inception

  - **Compliance monitoring**: Automated compliance checks can be configured based on the security requirements, facilitating ongoing monitoring and reporting

Security requirements serve as a technical blueprint for implementing a robust cybersecurity framework. They assist in translating the often abstract principles of cybersecurity into concrete, actionable system specifications, and as such, they are fundamental in achieving GRC objectives.

## Logical architecture diagrams

Logical architecture diagrams are indispensable artifacts in cybersecurity planning, offering a coherent representation of the logical components, their interactions, and their dependencies within an organization's IT landscape. Unlike physical architecture diagrams, which focus on the physical connections and hardware specifications, logical architecture diagrams abstract away from hardware to focus on how data flows, how components communicate, and how services are orchestrated. These diagrams are formulated following specific modeling languages such as UML or ArchiMate, ensuring standardization and clarity.

## Components and granularity

System architecture diagrams provide invaluable blueprints for analyzing cybersecurity vulnerabilities when constructed with thoughtful components and appropriate granularity. By segmenting the network into zones to control, enumerating servers and services, detailing data stores and flows, and showing user access points, architects gain multidimensional visibility into risks. Meticulous diagrams enable targeted assessments focused on critical data conduits and hubs rather than generic evaluations. The component-based approach aids threat modeling by revealing where incidents may spread based on interconnections, while zonal segmentation facilitates designing security and access controls aligned to trust levels and data sensitivity. With proactive architectural diagrams mapping environments in nuanced ways, organizations can prioritize controls, authenticate risks, and respond effectively. Rather than reacting blindly to threats, meticulous diagrams shine a light on risks to forge resilient protections rooted in system comprehension. They provide foundations that enable organizations to innovate fearlessly by securing the unknown knowns:

- **Network zones**: These illustrate how to segment the network into zones based on trust levels and data sensitivity, such as DMZ, internal network, and restricted zones

- **Servers and services**: These highlight the servers responsible for specific services or applications, potentially divided by roles such as web servers, application servers, and database servers

- **Data stores**: These represent databases, file repositories, and other data storage mechanisms, detailing the data schema, if applicable

- **User locations and connections**: These depict how users or external systems connect to internal services, often visualizing VPN tunnels, public endpoints, and other access methods

- **Protocols and data flow**: This describes the protocols used for component interactions and how data moves within the system

- **Benefits in vulnerability identification**:

  - **Data flow analysis**: Helps in identifying potential weak links or bottlenecks where sensitive data is transmitted or stored

  - **Component-based risk assessment**: Facilitates the targeted assessment of specific components based on their role and exposure

  - **Incident response planning**: Provides valuable insights for developing incident response scenarios, allowing responders to understand where potential breaches might propagate

## Best practices

Thoughtfully constructed logical architecture diagrams provide immense strategic value but realize their full potential through prudent design choices. Savvy architects employ best practices such as judiciously layering diagrams, maintaining rigorous version control, and annotating with key attributes. By mindfully applying layering, versioning, and annotations, logical architecture diagrams evolve from mere maps into navigational charts that guide organizations securely into the future. Rather than static snapshots, they become living references that improve situational awareness as both threats and architectures inevitably evolve. By applying best practices, architects can craft logical diagrams that pay dividends over time with clarity and strategic insights to overcome future unknowns:

- **Layering**: Decomposing the diagram into multiple layers, each focusing on specific aspects such as data, services, or security components can offer a more in-depth view

- **Version control**: Keeping versions of logical architecture diagrams can help in tracking changes over time, which is crucial for analyzing past incidents and understanding the evolution of the architecture

- **Annotations**: Annotations that specify certain attributes such as data sensitivity, encryption standards, or compliance markers can add value to the diagram

## GRC support

Logical architecture diagrams provide immense value beyond technical representations by strengthening governance frameworks, evidencing regulatory compliance, and enabling at-a-glance security postures. With thoughtful application, logical architecture diagrams are transformed from isolated technical artifacts to interconnected governance mechanisms that enhance compliance, communication, and strategic vision. Rather than rigid retrospective assessments, they become active tools that align protections with priorities and steer organizations confidently through complex regulatory environments. Architects play a pivotal role in realizing this potential by linking diagrams holistically across governance, compliance, and strategy:

- **Governance**: Logical architecture diagrams can be a critical input for IT governance, giving stakeholders a high-level view of the current architecture and facilitating informed decision-making.

- **Regulatory audits**: The diagrams can be used as evidence during audits to prove that certain regulatory requirements are being met, such as data separation, network segmentation, or secure communication channels.

- **Compliance mapping**: Components in the diagram can be annotated or color-coded based on compliance status, enabling quick visual assessments of compliance posture.

- **Logical architecture diagrams**: These are visual models that depict the logical components and interactions in an IT system or application. They are documented using standard diagramming conventions and tools such as Visio, Lucidchart, DrawIO, and others.

## Key elements

The key elements that are represented in a logical architecture diagram are as follows:

- Boxes to represent logical components such as servers, databases, user devices, and so on. Components are labeled with their role/function.

- Lines and arrows between boxes to show data flows and connections and annotated with protocols such as HTTP, SSH, and others.

- Swimlanes or boundaries to segment components into logical zones such as public zones, private zones, and others based on trust levels.

- Annotations for additional attributes such as sensitivity, compliance status, or technology used. Color coding can also be used.

- A legend to define the meaning of shapes, lines, colors, and other symbols used in the diagram.

The level of granularity can vary based on the purpose of the diagram. High-level diagrams may just show core networks, whereas detailed diagrams may represent individual applications and data stores.

Logical architecture diagrams enable an understanding of how data and transactions flow in the system, highlighting trust boundaries and potential vulnerabilities. They provide the blueprint for cybersecurity architecture and are leveraged for risk assessments, incident response, and compliance audits. Periodically updated diagrams help track changes to the IT environment.

By distilling complex system interactions into an interpretable visual format, logical architecture diagrams serve as a critical tool for cybersecurity professionals. They not only aid in understanding the system and identifying vulnerabilities but also facilitate compliance, governance, and strategic planning.

## Physical architecture diagrams

Physical architecture diagrams serve as a foundational element in the pantheon of cybersecurity documentation, offering meticulously detailed mappings of the hardware infrastructure and the network topology within an organization. Unlike logical architecture diagrams, which abstract away from the physical layer to present a high-level view of data flow and system interaction, physical architecture diagrams are deeply concerned with the tangible constituents of an IT landscape – such as servers, switches, routers, firewalls, endpoints, and even the cabling that interconnects these components.

Physical architecture diagrams visually depict the actual hardware and infrastructure components in an IT environment. Like the previous types of architecture diagrams, they are documented using diagramming tools such as Visio, Lucidchart, DrawIO, and others.

### Components and specificity

Meticulous physical architecture diagrams illustrate the intricate details of infrastructure and environments, but thoughtful component specificity takes their value to the next level. Rather than generic placeholders, precise device models, hardware specifications, cabling routes with lengths, and spatial layouts within facilities enable architects to assess vulnerabilities and model threats with surgical precision. Detailed component inventories empower targeted hardening of critical servers based on risk profiles. Granular cabling diagrams facilitate incident response by mapping potential breach conduits and blast radii. Even the physical locations of racks identify potential physical access weaknesses. With nuanced specificity rather than abstract overviews, physical architecture diagrams become indispensable references informing policies, access controls, maintenance procedures, and disaster recovery. They enable architects to optimize controls precisely matched to unique environments, not theoretical templates. The adage rings true – the cybersecurity devil is in the details. With meticulous diagrams capturing those details, architects gain superpowers to secure environments proactively from the server rack-up:

- **Network devices**: Details specifications, makes, and models of switches, routers, and firewalls. It may also indicate port configurations and VLAN assignments.

- **Servers and hardware**: Describes server rack arrangements, server models, and hardware specifications such as CPU, RAM, and storage arrays.

- **Endpoints**: Depicts workstations, laptops, and other user devices, often categorized by department or role.

- **Cabling and connectors**: Enumerates the types of cables used (for example, Cat 6, fiber optic) and their routes, possibly including cable lengths and identifiers.

- **Physical locations**: The layout of the devices in actual physical space, often within data centers, may also be included, capturing elements such as rack numbers and room identifiers.

## Benefits for security considerations

While virtual threats dominate headlines, physical architecture diagrams spotlight the immense value of securing the tangible. By mapping the nitty gritty of devices, cabling, and facilities, architects gain powerful perspectives into physical risks and responses. Just as castle builders relied on architectural plans, cyber defenders benefit tremendously from blueprints that map the physical foundations underpinning logical layers:

- **Attack surface analysis**: Enables the identification of potential physical entry points in the network, aiding in hardening strategies

- **Incident response**: Acts as a reference during incident responses for locating affected hardware quickly, which is especially crucial when physical access to hardware is required

- **Resource optimization**: Helps in identifying underutilized resources, thus informing hardware consolidation strategies that minimize exposure to physical attack vectors

## Best practices

Though focused on the tangible, prudent architects enhance physical architecture diagrams by incorporating logical and security considerations. Just as fusing engineering and architecture creates enduring infrastructure, blending the tangible with the conceptual crafts resilient cybersecurity foundations:

- **Layering**: Though focused on physical components, these diagrams could be layered to indicate relationships with logical constructs such as subnets or VLANs

- **Versioning**: Physical architectures are often subject to change; version-controlled diagrams help in rollback and auditing tasks

- **Security markings**: Indicating the security features of each hardware component (for example, TPM chips in servers or firewall capabilities) can be beneficial

## GRC support

Though commonly treated as isolated technical documentation, physical architecture diagrams provide immense strategic value in enabling governance, regulatory compliance, and security policy alignment. Just as ancient maps guided explorers, modern physical architecture diagrams help organizations navigate complex regulatory terrain to securely chart their future:

- **Asset management**: Physical architecture diagrams serve as asset inventories, an essential requirement for various compliance standards such as ISO 27001

- **Regulatory audits**: For regulations that require physical security measures, such as FISMA or HIPAA, these diagrams can serve as corroborative evidence of implemented controls

- **Policy alignment**: By mapping physical resources, these diagrams aid in the formulation of policies regarding hardware security, disposal, and maintenance, thus streamlining governance

Physical architecture diagrams are not merely representational artifacts but serve as instrumental frameworks in the cybersecurity domain. They contribute to both the tactical and strategic aspects of cybersecurity, from immediate incident response to long-term governance and compliance planning.

## Key elements

The key elements that are represented in a physical architecture diagram are as follows:

- Symbols for hardware, such as servers, routers, switches, firewalls, endpoints, and others. Symbols are industry standard or legible.

- Connecting lines to represent the physical network cabling and connections between devices. Lines indicate cable types.

- Annotations with specifications such as model numbers, configurations, IP addresses, and so on.

- The layout of hardware components in their actual physical locations – for example, racks, rooms, and buildings.

- Boundaries to demarcate network zones, departments, geographical locations, and so on.

- A legend that specifies the meanings of the symbols, lines, and colors used in the diagram.

The level of detail can vary from high-level network overviews to comprehensive maps of data center layouts that include device placements. Physical diagrams evolve continuously as infrastructure changes.

These diagrams help identify physical vulnerabilities such as single points of failure, insecure rack access, unsupported hardware, and more. They provide valuable reference during incidents for locating devices. Physical diagrams also facilitate governance activities such as asset management and policy formulation.

## Solution design documents (SDDs)

In cybersecurity, SDDs act as technical roadmaps, delineating the architecture, components, modules, interfaces, and data for a particular security control or solution. Often developed post-requirements analysis, these documents embody the blueprint that links business, functional, and technical requirements to the specificities of the implementation. They serve as the primary artifacts during the SDLC for ensuring that the devised solution aligns with the intended security posture of an organization.

### Components and specificity

Robust cybersecurity solution designs transcend isolated technical specifications by interweaving architectural alignment, granular configurations, phased deployment plans, and fallback precautions. Just as meticulous blueprints enable complex engineering feats, thoughtful cybersecurity designs manifest rigorous protection tailored to unique environments:

- **Architectural overview**: Provides a high-level description of the system architecture, often including diagrams, and places the security control or solution within the broader organizational IT landscape

- **Technical specifications**: Outlines the hardware and software prerequisites, dependencies, configurations, and interface specifications in detail

- **DFDs**: These depict how data will traverse the system, indicating the points at which encryption, logging, or other security controls are applied

- **Implementation plan**: This enumerates the steps for installation, configuration, and deployment, often broken down into sprints or phases with associated timelines and resources

- **Validation criteria**: Describes the performance metrics, KPIs, and testing methods to validate the security controls post-implementation

- **Rollback plans**: Procedures for reverting the system to a prior state in the case of a failed deployment or unforeseen vulnerabilities

## Benefits for security considerations

Thoughtful cybersecurity solution designs deliver immense advantages even before implementation by enabling traceability, standardization, and quality assurance, all of which are built in by design. With clarity emerging from solution documents, security transcends from being an afterthought to becoming a competitive advantage that scales the heights of digital transformation:

- **Traceability**: This ensures that every business, functional, and technical requirement is mapped to a specific component or process in the solution

- **Standardization**: This facilitates standard approaches and best practices in implementation, thus avoiding "security through obscurity" or ad hoc, unverifiable security measures

- **Quality assurance**: This acts as a baseline for various forms of testing, such as unit, integration, and security tests, as well as for conducting code reviews and audits

## GRC support

Beyond technical implementations, meticulous solution designs provide immense value in aligning security architectures with governance policies, validating regulatory compliance, and assessing the impacts of changes. Just as detailed plans enabled ancient wonders such as the Parthenon, robust cybersecurity designs manifest protections on time, on budget, and on compliance amid relentless change:

- **Policy alignment**: The design specifications should be directly informed by the cybersecurity policies, standards, and guidelines of the organization, thereby ensuring governance alignment.

- **Compliance validation**: Detailed documentation regarding security features and controls facilitates easier compliance audits as it offers verifiable evidence that prescribed measures are implemented. This is particularly relevant for frameworks such as PCI-DSS, HIPAA, or GDPR, which mandate specific controls.

- **Impact assessment**: A well-documented solution design allows for structured impact assessments when changes to regulations or business processes occur, enabling efficient modifications to the security controls.

## Examples

Here are some examples of how solution design documents are represented in cybersecurity architecture:

- **Secure remote access solution**:

    - Provides remote employees secure access to the internal corporate network over the internet using a site-to-site IPsec VPN tunnel and MFA

    - Integrates with existing network topology comprising a border firewall, DMZ, and internal zones

- **Technical specifications**:

  - VPN concentrator model, throughput, redundancy, IPsec tunnel parameters, and encryption algorithms

  - MFA provider, token mechanisms such as RSA SecurID, and RADIUS integration specifications

  - Operating systems, routing protocols, and firewall rules for traffic segmentation

- **DFDs**:

  - Visualize the data flow from remote user devices over a VPN tunnel to application servers/databases in the internal zone

  - Highlight encryption points and logging by VPN server and firewall

- **Implementation plan**:

  I.    **Phase 1**: Procure and configure VPN hardware.

  II.   **Phase 2**: Integrate the MFA solution with Active Directory.

  III.  **Phase 3**: Provision remote access profiles and VPN client software to users.

- **Validation criteria**:

  - VPN tunnel throughput meets the expected load

  - Latency within the defined SLA

  - Penetration testing does not reveal exploitable vulnerabilities

  - Audit logging enabled on all devices

The preceding sections demonstrate how an SDD specifies technical, process, validation, and architectural details for implementing a remote access security solution. Similar SDDs are created for other solutions, such as IDS/IPS, SIEM, and endpoint security, to establish the cybersecurity architecture.

In essence, SDDs are imperative artifacts that codify the security principles, architectural decisions, and implementation specifics into a tangible plan. They serve multiple strategic functions, from serving as a communication medium among stakeholders to playing a critical role in governance and compliance endeavors.

## Configuration documents

Configuration documents serve as meticulous technical manuals that stipulate the necessary settings, parameters, and procedural steps for configuring security solutions and technologies within an organization's IT ecosystem. These documents are tailored to align with the organization's cybersecurity policies, operational requirements, and governance standards. They are critical for the accurate and consistent implementation of security controls, ensuring that the solutions perform as intended, both in isolation and as part of the integrated security architecture.

## Components and specificity

Effective cybersecurity configuration documentation transcends generic instructions by incorporating granular technical specificity, procedural clarity, change tracking, and contextual transparency. Just as checklists enabled the Apollo moon landings, meticulous configuration documentation empowers organizations to successfully traverse the intricate terrain of cybersecurity implementation with precision, transparency, and accountability:

- **Prerequisites**: These list hardware and software dependencies, including version numbers and compatibility requirements

- **Configuration settings**: These detail the specific parameters, values, and options to be set, often accompanied by code snippets, XML configurations, or GUI screenshots

- **Step-by-step procedures**: These are sequenced instructions for implementing the configuration that often include validation steps to ensure the settings are applied correctly

- **Rollback procedures**: These are guidelines for reverting to previous configurations in the event of incorrect implementation or operational issues

- **Change logs**: These provide a record of updates, modifications, and who performed them, providing an audit trail

- **Security considerations**: These explicitly identify how each configuration setting contributes to the overall security posture – for example, why a particular encryption algorithm was chosen or a specific port was closed

## Benefits for security considerations

Thoughtfully constructed configuration documentation delivers immense advantages by enabling consistent standardization, streamlining audits, and empowering users before ever being implemented. Just as checklists enabled complex achievements such as moon landings and skyscraper construction, meticulous configuration documents manifest cybersecurity excellence through clarity, accountability, and standardization:

- **Consistency**: Facilitates uniform application of security settings across various environments, thereby mitigating risks associated with inconsistent configurations

- **Auditing**: Serves as a reference point for internal and external audits, enabling efficient validation of the security infrastructure

- **Operational efficiency**: Provides a reliable guide for system administrators, network engineers, and cybersecurity professionals, thereby expediting the configuration process and reducing human error

## GRC support

Beyond technical guidance, meticulous configuration documentation provides immense strategic value that strengthens governance alignment, regulatory compliance, and organizational accountability. With clarity emerging from detailed documentation, security transcends being an obstacle to becoming a strategic driver of productivity and trust:

- **Policy adherence**: Configuration documents ensure that the technical settings align closely with high-level cybersecurity policies and standards, enabling effective governance.

- **Regulatory compliance**: This enables a more straightforward mapping of implemented configurations to specific regulatory requirements, such as the controls mandated by GDPR, HIPAA, or NIST frameworks. This simplifies the compliance verification process.

- **Documentation for accountability**: A well-maintained configuration document can serve as evidence in demonstrating due diligence in the event of legal scrutiny or regulatory audits.

## Examples

Here are some examples of how configuration documents are represented in cybersecurity architecture:

- **Firewall configuration document**: The document provides detailed instructions for configuring corporate next-generation firewalls to enforce security policies.

- **Prerequisites**:

  - Firewall model XYZ running firmware version 2.1

  - Management IP address reachable from admin workstations

  - Access to encryption keys for VPN and SSL inspection

- **Configuration settings**:

  - Ruleset for segmenting traffic between VLANs

  - Access control lists limiting traffic from the internet to DMZ services

  - Decryption policies for HTTPS traffic to enable content inspection

  - IPsec VPN parameters such as encryption algorithms, tunnels, and pre-shared keys

  - Syslog and SNMP settings for integration with monitoring tools

- **Procedures**:

  - Step-by-step instructions to configure rules, objects, and policies in the firewall GUI

  - Use of a command-line interface for certain advanced settings

  - Testing methodology for validation

The preceding sections demonstrated how a configuration document provides standardized details for implementing firewall settings across an enterprise. This drives consistency, enables auditing, and ensures alignment with security policies. Similar documents can be created for other solutions, such as IDS, web proxies, endpoint security, and others.

Configuration documents are integral assets in a cybersecurity framework that formalize the implementation details of security controls. They bridge the gap between high-level policies and on-the-ground technical actions, serving dual roles in both operationalizing security and satisfying governance, regulatory, and compliance imperatives.

The section provided an overview of several critical types of documentation that form the foundations of a cybersecurity architecture. In essence, the diverse documentation provides structure, alignment, and transparency across governance, risk management, and technical realms, thereby enabling organizations to build cybersecurity resiliency holistically. The interconnections between documentation types manifest coherent cybersecurity strategies rooted in principles yet tailored to unique operational landscapes.

# Documentation tools

Documentation is an integral component of effective cybersecurity governance. The choice of appropriate tools for documentation varies according to the specific needs of each organization. This section provides a technical overview of several classes of tools used for cybersecurity documentation, including diagramming tools, configuration documentation tools, collaborative platforms, compliance management tools, and office products for general document editing or spreadsheet management.

Cybersecurity governance is contingent upon robust, detailed, and easily accessible documentation. This extends across the spectrum, from policies and procedures to configurations and network topologies. The landscape of tools available for achieving this is vast and includes specialized software for specific documentation tasks and more general-purpose office products for creating and managing documents and spreadsheets.

## Categories of documentation tools

Let's look at the different categories of documentation tools.

### *Diagramming tools*

Effective diagramming relies on choosing purpose-built tools tailored to specific architectural visualization needs. Just as a craftsperson selects precise tools, cybersecurity architects must adopt fit-for-purpose platforms to diagram diverse complexities understandably. The axiom rings true: when the only tool you have is a hammer, everything looks like a nail. With thoughtful tool selection, architects can broaden perspectives and gain clarity amid complexity:

- **Microsoft Visio**:
  - **Link**: `https://www.microsoft.com/en-us/microsoft-365/visio/flowchart-software`

- **Features**: Object-oriented drawing, extensive shape libraries, layering, grouping, and hyperlinking
- **Advantages**: Integration with Microsoft Office and is customizable
- **Use cases**: Network topologies, DFDs, and **entity-relationship diagrams (ERDs)**

- **DrawIO**:

  - **Link**: `https://app.diagrams.net/`
  - **Features**: Cloud-based, real-time collaboration, and multiple export formats
  - **Advantages**: Open source and no installation required
  - **Use cases**: Logical mappings, system interconnections, and **Unified Modeling Language (UML)** diagrams

### *Configuration documentation tools*

Robust configuration documentation relies on versatile tools tailored to capturing technical intricacies alongside context. Just as a master chef curates quality utensils, prudent architects choose tools to potentiate productivity rather than present obstacles. With thoughtful tool selection, configuration documentation evolves from an operational chore into a strategic asset that enables security excellence:

- **CherryTree**:

  - **Link**: `https://www.giuspen.net/cherrytree/`
  - **Features**: Hierarchical notes, syntax highlighting, and rich-text editing
  - **Advantages**: Lightweight, cross-platform, and password protection
  - **Use cases**: Device configurations, hardening checklists, and procedural documentation

### *Collaborative platforms*

Effective cybersecurity documentation requires extensive collaboration across teams and functions. Just as an orchestra requires harmony across musicians, cybersecurity documentation necessitates tightly orchestrated collaboration. With thoughtful platform selection, organizations gain a force multiplier to realize the collaborative potential of comprehensive documentation:

- **Confluence**:

  - **Link**: `https://www.atlassian.com/software/confluence`
  - **Features**: Real-time editing, version history, and extensive integration capabilities
  - **Advantages**: Enterprise scalability and a robust plugin ecosystem
  - **Use cases**: Security architecture, incident response plans, and technical protocols

## Compliance management tools

Navigating complex compliance landscapes requires purpose-built tools centralizing control evidence, risk analysis, and audit artifacts. Just as GPS provides situational awareness when navigating, thoughtful tools give architects clarity on compliance status amid turbulent regulatory seas. With tailored solutions scaling from frameworks to controls, organizations can transform regulatory mandates from obstacle to opportunity:

- **Archer**:
  - **Link**: `https://www.archerirm.com/`
  - **Features**: Control mapping, risk assessment modules, and audit management
  - **Advantages**: A centralized GRC dashboard and automation
  - **Use-cases**: Regulatory compliance and audit reports
- **OneTrust**:
  - **Link**: `https://www.onetrust.com/`
  - **Features**: Data mapping, risk quantification, and compliance tracking
  - **Advantages**: Cloud-based with data privacy features
  - **Use cases**: GDPR compliance and privacy impact assessments

## Office products for document editing and spreadsheets

Effective documentation requires versatile tools for authoring, calculations, visualizations, and collaboration. Just as craftspeople choose materials suiting their medium, prudent cybersecurity teams adopt tools that optimize documentation workflows. With capabilities spanning authoring to analysis to sharing, office suites become force multipliers for realizing the potential of interconnected documentation. They provide essential raw materials to construct holistic cyber defenses:

- **Microsoft Office and Office 365**:
  - **Link**: `https://www.office.com/`
  - **Features**: Word processing, spreadsheets, presentations, and cloud storage (Office 365)
  - **Advantages**: Industry standard and integration with other Microsoft products
  - **Use cases**: Policy documents, risk assessment spreadsheets, and training presentations
- **OpenOffice and LibreOffice**:
  - **Link**: `https://www.openoffice.org/` and `https://www.libreoffice.org/`
  - **Features**: Word processing, spreadsheets, presentations, and drawing

- **Advantages**: Open source, cross-platform, and no licensing costs
- **Use cases**: General documentation, financial modeling, and presentations

- **OnlyOffice**:
  - **Link**: `https://www.onlyoffice.com/`
  - **Features**: Document editing, spreadsheets, presentations, and cloud-based collaboration
  - **Advantages**: Real-time collaboration and an API for customization
  - **Use cases**: Policy drafting, risk calculations, and team presentations

## Comparative analysis

Robust cybersecurity architectures rely on comprehensive documentation spanning policies, diagrams, configurations, and compliance evidence. This necessitates versatile tools tailored to specific documentation needs. Like artisans choosing mediums suited to their creations, prudent cybersecurity leaders adopt diverse tools that enable comprehensive documentation production. With capabilities spanning authoring to compliance, organizations gain strategic assets to help them realize the potential of interconnected documentation. The following table provides a feature comparison of the products discussed:

| Feature | Visio | DrawIO | CherryTree | Confluence | Archer | OneTrust | MS Office | OpenOffice/ LibreOffice | OnlyOffice |
|---|---|---|---|---|---|---|---|---|---|
| **Real-Time Collaboration** | No | Yes | No | Yes | No | Yes | Yes (Office 365) | No | Yes |
| **Cloud-Based** | No | Yes | No | Yes | Yes | Yes | Yes (Office 365) | No | Yes |
| **Customization** | High | High | Medium | High | High | Medium | Medium | High | High |
| **Platform Integration** | High | Medium | Low | High | High | Medium | High | Medium | Medium |
| **Cost** | High | Low | Low | Medium | High | High | High | Low | Medium |

Table 6.1 – Documentation product feature comparison

Effective cybersecurity architecture and governance rely heavily on comprehensive documentation across policies, processes, systems, and controls. To enable robust and standardized documentation, cybersecurity professionals employ a diverse set of tools. For creating and editing text-based documents, ubiquitous office suites such as Microsoft Office, Office 365, OpenOffice, LibreOffice, and OnlyOffice provide versatile word processors, slide decks, and spreadsheet editors that can be used to craft policies, protocols, training materials, reports, and more. For visualizing architectures, dedicated diagramming tools such as Microsoft Visio and open source DrawIO facilitate the creation of detailed network topology diagrams, logical mappings, data flows, and system interconnections. To codify configurations, tools such as CherryTree provide structured scratchpads for documenting device settings, hardening checklists, and procedural steps in a granular fashion. For collaborative documentation, the Confluence enterprise wiki allows multiple teams to collectively develop security architecture blueprints, incident response plans, and technical specifications. To manage compliance, purpose-built tools such as Archer help map controls to regulations, manage assessments, and generate audit-ready reports. With the optimal blend of both general-purpose and specialized documentation tools, organizations can enable transparency, continuity, accountability, and measurability across their cybersecurity apparatus.

# Team approaches to documentation

Effective cybersecurity documentation is a collaborative endeavor that requires the active participation of various stakeholders, ranging from security experts to compliance officers and system administrators. This section discusses how teams can employ a synergistic approach using a variety of tools to document cybersecurity aspects comprehensively. The focus will be on dividing responsibilities, using specialized and general-purpose tools, and managing documentation in a collaborative and dynamic environment.

In cybersecurity governance, documentation serves as the foundation upon which security postures are built, validated, and maintained. Given the complexity of modern information systems and the multifaceted nature of cybersecurity threats, a team approach is often requisite for effective documentation. This section aims to provide a technical framework that outlines how teams can collaboratively work on documenting different facets of cybersecurity using both specialized and general-purpose tools.

## Division of responsibilities

For example, security analysts can create network architecture diagrams on Visio to explain firewall placements and data flows. Compliance officers can maintain control repositories on Archer, linking them to regulations. Developers can collaboratively edit API documentation on Confluence.

This model enables organizations to tap into the specialized expertise of each team while aligning documentation efforts through collaboration platforms.

### *System architects and network administrators*

Let us look at the responsibility and tools:

- **Responsibility**: Network topology, system configurations, and access control policies
- **Tools**:
    - Microsoft Visio for creating detailed network topologies
    - CherryTree for documenting system configurations

### Security analysts and experts

Let us look at the responsibility and tools:

- **Responsibility**: Threat modeling, vulnerability assessments, and incident response plans
- **Tools**:
  - DrawIO for DFDs related to threat models
  - Confluence for collaborative editing of incident response plans

### Compliance officers

Let us look at the responsibility and tools:

- **Responsibility**: Regulatory compliance, control mapping, and risk assessments
- **Tools**:
  - Archer for control mapping and compliance tracking
  - Microsoft Office 365 for creating and sharing risk assessment matrices

### Developers and DevOps teams

Let us look at the responsibility and tools:

- **Responsibility**: Code base documentation, including API specifications, and system deployment procedures
- **Tools**:
  - Confluence for API documentation
  - Only Office for real-time collaborative coding and deployment checklists

### Project managers and coordinators

Let us look at the responsibility and tools:

- **Responsibility**: Project timelines, deliverables, and progress tracking
- **Tools**:
  - Microsoft Project for timeline management
  - LibreOffice Calc for budget and resource allocation

## Collaborative platforms for a team-based approach

Effective cybersecurity requires a collaborative, team-based approach that spans multiple departments and specializations. To enable seamless coordination for creating, managing, and operationalizing security documentation, organizations often leverage dedicated platforms tailored for cross-functional cooperation and transparent version control. This section explores leading solutions that provide the foundation for a vigorous documentation program through real-time co-authoring abilities, customizable templates, commenting tools, chat facilities, and tight integrations. By centralizing documentation workflows onto purpose-built collaboration platforms, security teams can mitigate risks stemming from scattered information silos and better synchronize complex activities across disparate stakeholders. The use of structured templates and standardized storage protocols also brings rigor to documentation efforts, facilitating subsequent retrieval, reporting, and auditing processes.

The following analysis highlights specialized platforms that empower seamless teaming to produce consistent, accessible, and cohesive security documents:

- **Confluence**:

  - **Features**: Version control, commenting, customizable templates

  - **Use case**: Centralized repository for all collaborative documentation

- **Microsoft Teams (integrated with Office 365)**:

  - **Features**: Seamless integration with Office 365, real-time collaboration, and file sharing

  - **Use-case**: Sharing and editing documents instantaneously within team channels

- **Slack**:

  - **Features**: Real-time messaging, file sharing, and integration with other tools via APIs

  - **Use-case**: Quick communication and file-sharing for agile teams; facilitates real-time discussions to support the documentation process

## Documentation life cycle management

Developing rigorous cybersecurity documentation requires extensive collaboration across distributed teams with diverse expertise. Like an orchestra blending disparate talents, prudent platform use synchronizes enterprise-wide contributions into cohesive narratives. With capabilities spanning real-time creation to governance, modern tools empower taking documentation from fragmented to fortified:

1. **Initiation phase**: Teams define the scope, objectives, and tools for documentation.

2. **Development phase**: Individual team members create drafts using specialized tools.

3. **Review phase**: Teams use collaborative platforms such as Confluence or Microsoft Teams for peer reviews.

4. **Approval phase**: Compliance officers and project managers review the documentation for completeness and accuracy.

5. **Maintenance phase**: Regular updates and revisions are made to keep the documents current, usually overseen by a combination of roles.

## Comparative analysis

Comprehensive cybersecurity documentation requires collaboration across diverse roles leveraging tools aligned to their specialized needs. Like musicians blending complementary talents, cross-functional collaboration enables harmonizing disparate documentation into unified narratives mapping controls enterprise-wide. With capabilities spanning authoring to compliance, integrative tools empower comprehensive documentation. The following table provides a feature-use-by-role comparison:

| Role | Specialized Tools | General-Purpose Tools | Collaborative Platforms |
|------|-------------------|------------------------|--------------------------|
| **System Architects** | Microsoft Visio | Microsoft Office | Confluence, Slack |
| **Security Analysts** | DrawIO | LibreOffice | Microsoft Teams, Slack |
| **Compliance Officers** | Archer | OpenOffice | Confluence, Slack |
| **Developers and DevOps** | N/A | OnlyOffice | Microsoft Teams, Slack |
| **Project Managers** | N/A | Microsoft Project | Confluence, Microsoft Teams, Slack |

Table 6.2 – Feature use by role

Robust cybersecurity documentation requires a collaborative approach, with cross-functional teams employing specialized and general-purpose tools aligned to their domains. This facilitates accuracy, comprehensiveness, and timeliness in documenting policies, procedures, architectures, and controls.

In today's complex and ever-evolving cybersecurity landscape, a team approach to documentation is essential for achieving comprehensive and up-to-date governance. By strategically dividing responsibilities and employing a diverse set of specialized and general-purpose tools, teams can collaborate effectively throughout the documentation life cycle. Utilizing centralized platforms for collaboration further enhances the efficiency and accuracy of the documentation process.

# Summary

In this chapter, effective documentation served as the cornerstone of a resilient cybersecurity architecture. The policies, diagrams, models, assessments, and configurations covered in this chapter provide a multidimensional view of an organization's security posture. By adopting pragmatic documentation practices, cybersecurity architects can enhance visibility, facilitate compliance, and enable organizational alignment. However, documentation is not simply an isolated governance activity. The methodical approaches outlined aim to make documentation an integrated, value-adding aspect of daily operations. Whether through streamlined creation workflows or easy-to-consume formats, the principles discussed help transform documentation from an obligation into an asset. Fundamentally, documentation is about communication – conveying policies, designs, and requirements with clarity. Organizations that embrace documentation as an enabler of transparency, not just a ceremonial necessity, are better positioned to evolve their security architectures in alignment with business objectives.

In summary, comprehensive and communicative documentation serves as the basis for effective cybersecurity architecture and governance. The combination of standards-based formats, purpose-built tools, and collaborative approaches enables organizations to create, manage, and consume documentation efficiently. By adopting the recommendations outlined in this chapter, cybersecurity teams can produce documentation that informs and educates, rather than obfuscates. The methodologies covered aim to make documentation comprehensive yet comprehensible – an organizational asset that cybersecurity architects can leverage for strengthening security postures through enhanced communication and transparency. With the exponential increase in technological complexity, documentation has become mission-critical. By transforming its role and value, organizations position themselves for security resilience and regulatory compliance.

In the next chapter, we'll discuss the journey to the top as a cybersecurity architect. It is not without mentioning that certain career paths are more direct than others for a cybersecurity architect. Like most things in technology, *it depends* can be a common answer. The upcoming chapter provides various approaches to gaining the experience or skillset required to become a cybersecurity architect.

# 7
# Entry-Level-to-Architect Roadmap

*"There are not more than five musical notes, yet the combinations of these five give rise to more melodies than can ever be heard. There are not more than five primary colours, yet in combination they produce more hues than can ever been seen. There are not more than five cardinal tastes, yet combinations of them yield more flavours than can ever be tasted."*

*–Sun Tzu*

In the previous chapter, we covered an understanding of how comprehensive and communicative documentation serves as a basis for effective cybersecurity architecture and governance. The combination of standards-based formats, purpose-built tools, and collaborative approaches enables organizations to create, manage, and consume documentation efficiently. By adopting the recommendations outlined in this chapter, cybersecurity teams can produce documentation that informs and educates rather than obfuscates. The methodologies aim to make documentation comprehensive yet comprehensible—an organizational asset that **cybersecurity architects (CSAs)** can leverage for strengthening security postures through enhanced communication and transparency. It is also important to note that a CSA should not be tied to specific products or vendors. Instead, they should strive for technology independence, recognizing diverse approaches to building secure and resilient IT systems. The development of a career path can help lead to this outcome.

Though the foundations of cybersecurity may seem simple on the surface, mastering the field requires creatively combining and applying core principles. This means that you need to be able to master the fundamentals. Just as only a few musical notes or primary colors yield endless permutations, the building blocks of cybersecurity—such as encryption, access controls, vulnerability management, and network segmentation—can be arranged and implemented in varying and numerous ways.

An entry-level position may understand these basic cybersecurity concepts, but translating that knowledge into an advanced architecture requires finesse. The most skilled architects don't just implement textbook security controls—they craft adaptable systems tailored to their company's specific threats and business needs. Their solutions blend art and science.

Like an artist mixing paints on a palette, great security architects leverage their versatile toolkit to produce something new. They select and configure the right controls for their organization while staying up to date on new techniques and technologies to incorporate. Their architecture reflects a wisdom deeper than any single brushstroke.

Over time, an analyst can gain this wisdom by continuously experimenting with and reflecting on how foundational cybersecurity techniques combine in different situations. There is always more to learn. Just as there are endless melodies to be heard, there are countless ways to arrange the notes of cybersecurity into an elegant masterpiece.

I would be remiss if I did not call out the fact that Packt has a great book on creating a cybersecurity career plan. *Cybersecurity Career Master Plan*, by Dr. Gerald Auger, Jaclyn "Jax" Scott, Jonathan Helmus, and Kim Nguyen, provides a great wealth of information to consider and should be the first book on your reading list to help you map out your career path. It also includes considerations beyond the scope of this book and, specifically, this chapter.

With this in mind, it does not specifically require a specific starting point or next step; as per the colloquialism, "*there are a thousand ways to skin a cat*," there are just as many ways to become a CSA. It is this wisdom and experience that is gained over time that we will look to explore within this chapter, using my own journey as an example. Being a CSA is not something you just *fall into* or a state you wake up in; it is a pathway that takes years to excel in and requires dedication and resolve to achieve. The pursuit of this mastery should excite any practitioner looking to advance in their career.

This chapter covers the following topics:

- The journey
- Where to start
- The cold open
- The transfer
- How to expand

# The journey

It is important to remember that the journey begins with the first step. In this case, it is deciding where to go. While it is possible to just travel along life with no direction or destination, this can lead to great excitement or utter stagnation. Like a boat that has no rudder or sail, you are left to tidal forces to take you from place to place. This can definitely provide adventure and excitement but also has the potential to leave you stranded in the middle of the ocean without resources and at the mercy of the destructive power of an ocean storm.

Your career can be as equally challenging, making the desired destination an important decision to make regardless of where you begin. Using Jeff Goldblum's character Ian Malcolm from *Jurassic Park* as an example, he explains chaos theory using drops of water. Ian takes a drop of water and places it on the hand of another character, and it flows down the hand in a specific direction. He then repeats what he did initially. When the water rolls off in another direction the second time, he explains, *"It changed, because tiny variations, the orientation of the hairs on your hands, the amount of blood distending your vessels, imperfections in the skin... never repeat and vastly affect the outcome."* These are decisions we make along our journey. While many may head for the same destination, the path we take can vary and is unique to each individual.

The journey from an entry-level position to a senior CSA is filled with crucial milestones. While rewarding, it requires strategic planning and avoidance of potential pitfalls to achieve career advancement. This guide serves as a roadmap highlighting core knowledge areas, necessary certifications, common job roles, and fundamental proficiencies at each stage of the cybersecurity career life cycle. It provides perspective on transitions between early technology jobs to mid-level security analyst roles, then specialist and engineer positions, and finally, the advanced architect level.

By understanding the incremental evolution required at each level, aspiring cybersecurity professionals can thoughtfully chart their career trajectories, set targeted goals, and ultimately attain leadership roles in this critical and ever-evolving field. Whether starting from IT support, software development, systems administration, or network engineering, this guide outlines domains to expand into, skills to hone, credentials to acquire, and pitfalls to sidestep at each step of the cybersecurity career journey.

The cybersecurity field offers a wide range of career growth opportunities, from entry-level roles to advanced architect positions. However, the path is not always linear and requires diligent planning, continuous skill-building, and avoiding potential pitfalls. This guide provides an overview of typical milestones and learning priorities at each stage, helping aspiring cybersecurity professionals chart out an optimal career progression strategy.

Before we begin the discussion on the various pathways from getting into cybersecurity to becoming a CSA, it would be helpful to have a more visual representation to understand the direction and steps:

| | |
|---|---|
| **Initial entry-level roles** | Help desk support |
| | Software developer |
| | Network administrator |
| **Key intermediate steps** | Systems administrator |
| | Application security engineer |
| | Security engineer (focus areas such as firewalls, **intrusion detection systems (IDSs)/intrusion prevention systems (IPSs)**, and so on) |

| Important certifications to obtain | CompTIA (A+, Network+, Security+) |
|---|---|
| | Cisco (**Cisco Certified Network Associate (CCNA)**, **Cisco Certified Network Professional (CCNP)** Security) |
| | **International Information System Security Certification Consortium (ISC2) Certified Information Systems Security Professional (CISSP)**, **Certified Cloud Security Professional (CCSP)**) |
| Critical skills to develop | Hands-on technical skills (networking, coding, systems, and so on) |
| | Communication and collaboration abilities |
| | Understanding of risk management frameworks |
| Years of experience before the architect role | Typically 7–10 years |
| | Deep expertise and well-rounded experience are key |
| Architect job responsibilities | Design and integrate security solutions |
| | Bridge technical capabilities and business needs |
| | Guide strategic roadmaps and governance |

Table 7.1 – Pathway to becoming a CSA

The preceding table provides a visual representation of items that will be discussed in this chapter.

## Entry level – starting in a technology field

For those just embarking on a technology career, early roles tend to focus on building core competencies such as networking, systems administration, and basic programming. It is crucial even at this stage to avoid overspecializing and to keep exploring adjacent domains. Continuously learning new skills, experimenting with projects outside work, and avoiding complacency are key. Certifications such as A+, Network+, and language-specific programming certs can help build credibility.

Obtaining critical certifications early validates core competencies. Study guides, practice tests, and online courses can prep for exams such as CompTIA A+, Network+, and Security+. Studying 10–15 hours weekly in the first two years to pass 3–4 foundational certs is recommended. Learning adjacent domains builds well-rounded abilities.

## Example pathways

Transitioning from entry-level technology roles to a CSA requires meticulous planning, diversifying skills, and staying updated with industry trends. While the journey may start in different tech domains, the ultimate convergence is toward a robust understanding of cybersecurity principles. Here's a deep dive into some example pathways, accompanied by tailored study and training schedules to become a CSA, starting from an entry-level technology role:

- **Starting in help desk support**: Progress to a systems administrator role to gain networking and systems expertise. Pursue cybersecurity certifications such as Security+, CISSP, and **Certified Ethical Hacker (CEH)** in your free time. After 3–5 years, attempt to transition into an information security analyst job. From there, earn certs such as CCSP and advance to leading security engineering projects. After 7-10 years total, you can achieve a security architecture role.

- For help desk techs, self-study for certs such as CCNA. Avoid overspecializing too early:

  - **Pathway**:

    - **Initial role**: Help desk support

    - **Intermediate steps**: Progress to systems administrator | information security analyst | security engineer

    - **Final destination**: CSA

  - **Study schedule**:

    - **Years 1–2**: Focus on foundational IT concepts and obtain certifications such as A+.

    - **Years 3–4**: Dive into networking with certifications such as Network+ and start exploring cybersecurity concepts. Prepare for and earn the Security+ certification.

    - **Years 5–6**: Dedicate considerable time to advanced cybersecurity studies. Aim for the CISSP and CEH certifications.

  - **Training**:

    - Engage in hands-on labs and real-world scenarios.

    - Join online forums and communities focused on systems administration and cybersecurity.

    - Attend workshops and conferences.

  - **Pitfalls**:

    - Becoming confined to non-technical support roles.

    - Not acquiring enough practical security experience early.

- **Starting as a software developer**: Look for opportunities to gain experience in secure coding practices and designing secure architectures. Learn system administration basics on the side. After a few years, try to switch to an application security engineer role. Obtain advanced certs such as **CompTIA Advanced Security Practitioner** (**CASP+**) and gain expertise in auditing and pen testing. After 5+ years, you can aim for a lead architect job focusing on application and **application programming interface** (**API**) security.

- **Creating a training plan focusing on next-career-step-tailored learning**: Those aiming for security analyst roles can pursue intermediate certs such as Security+ and CISSP while working. Studying 1–2 hours on weeknights and 4–6 hours on weekends can prepare for exams in 6–12 months per cert:

  - Pathway:

    - **Initial role**: Software developer

    - **Intermediate steps**: Master secure coding | application security engineer | lead in application/API security

    - **Final destination**: As a developer, you are able to pivot to any role, so there is no specific final destination as with other career paths.

  - Study schedule:

    - **Years 1–2**: While mastering coding, start gaining foundational knowledge in cybersecurity. Explore certifications that focus on secure coding practices.

    - **Years 3–4**: Transition focus to designing secure architectures and delve into system administration basics. Seek the CASP+ certification.

    - **Years 5–6**: Deepen expertise in application security and work on advanced certifications such as CISSP.

  - Training:

    - Participate in coding bootcamps with a focus on security.

    - Engage in secure coding challenges and **capture-the-flag** (**CTF**) events.

    - Regularly attend workshops and seminars on secure application design and development.

  - Pitfalls:

    - Not acquiring a broad foundation in networking or infrastructure.

    - Letting coding skills become obsolete.

- **Starting in network administration**: Obtain vendor certs such as CCNA and gain firewall configuration skills. Volunteer for security-related initiatives and policy planning. After 2–3 years, look to transition into a security engineering role managing firewalls/VPNs. Study for advanced certs such as CCNP Security and CISSP while seeking opportunities to gain experience with cloud and identity management systems. After 6+ years and with diverse hands-on skills, you can attain an architect position.

- For network admins, take online programming courses on nights/weekends:

  - **Pathway**:

    - **Initial role**: Network administrator

    - **Intermediate steps**: Master network security | security engineer focusing on firewalls/VPNs | lead in network security architecture

    - **Final destination**: CSA with a specialization in network security

  - **Study schedule**:

    - **Years 1–2**: Get foundational networking certifications such as CCNA. Begin studying firewall configurations and security protocols.

    - **Years 3–4**: Deepen knowledge of network security. Obtain certifications such as CCNP Security and broaden your horizons into cloud security principles.

    - **Years 5–6**: Focus on comprehensive cybersecurity principles and aim for the CISSP certification.

  - **Training**:

    - Join specialized training programs for network security.

    - Participate in simulated network attack and defense exercises.

    - Attend industry conferences focused on network security trends and innovations.

  - **Pitfalls**:

    - Remaining restricted to purely network operations roles.

    - Not diversifying into comprehensive security architecture and policy formulation.

Irrespective of the starting point in technology, the journey to becoming a CSA demands a multifaceted approach. Emphasizing continuous learning, acquiring diverse technical skills, and securing practical experiences are pivotal. By following tailored pathways and avoiding common pitfalls, professionals can streamline their journey to senior cybersecurity roles, ensuring they are well prepared for the challenges and responsibilities they entail.

This has been mentioned previously in previous chapters; in fact, several labs were featured to prompt you to create a lab-based environment, but maintaining an updated home lab to tinker with new technologies prevents stagnation while adding demonstrated initiative. Set aside 4–6 weekends per year for refreshing lab systems and software. The key is balancing focused credentials, hands-on experimentation, adjacent knowledge, forward-looking skills, and leveraging employer resources to maximize foundational learning and avoid entry-level pitfalls.

### Real-life example

My technology career did not follow a traditional linear path. I originally pursued medical training with aspirations of becoming a physician. Throughout high school and college, I took extensive science coursework and worked summer jobs in various healthcare settings—from a phlebotomist drawing blood to a medical assistant to an oncology lab technician. This immersion only solidified my passion for medicine.

However, I also nurtured a growing interest in the booming personal computer revolution in the 1990s. Using my own savings, I purchased a modest Pentium system in 1997, a significant upgrade from the Apple IIc of my youth. This gateway into IT led me to dabble in building Microsoft Access databases and helping an anesthesiologist digitize his invoicing system. These technology projects made me realize my natural analytical and troubleshooting abilities might translate better to a career in computers rather than medicine.

The computer I purchased was the first I purchased with my own money. It was around 1997 when I purchased an Intel Pentium MMX 200 MHz system with 64 MB of RAM, a 500 MB hard drive, and a ZIP drive, running Microsoft Windows 95. Thinking back on this now, the phone I use today has more system resources and capabilities than that computer. With this, I started my journey into IT and getting an understanding of computers in general. When working on the aforementioned project helping an anesthesiologist with his invoicing information, I realized I needed to make a change, even though I enjoyed working within the medical field.

I started working for a non-profit organization. While my role or job was not glamorous, it was a job that paved the way for my introduction to organizational IT. After getting a taste for technology, I realized that moving from medicine to computers was a natural progression and not really a step backward. The way I looked at it, it allowed me to use my skills in a similar way to diagnose issues; computers and networking technology just *did not complain* as much as a patient could. It was also at this time that I realized that I needed to get a more structured education related to information technology and computers in general. While working at the non-profit and going to school, I started working on web development and transitioning the organization's infrastructure from a Novell NetWare infrastructure to Windows NT 4.0. This was definitely an entry-level position and was not the best paying, but the work environment and the information and experience I was gaining made up for the pay.

After working for the organization for several years, changes were being made within the organization, and I made the decision to start looking for other opportunities with the skills I had gained. I was hired by a mid-sized California bank as an IT project administrator. It was here that I started getting a more in-depth understanding of processes, projects, and technology integration. During my time at the bank, I studied and became familiar with various technologies. Much of my time was spent becoming familiar with the systems within the environment. This included Cisco routers and switches, Cisco **Adaptive Security Appliance** (**ASA**) firewalls, Windows NT 4/2000/XP, Linux, Snort, and a new technology called VMware ESXi. With this, I started studying for the Cisco CCNA and Microsoft certifications.

I worked for the bank for over six years, moving from project administrator to database administrator, and then network engineer. I also graduated with a bachelor's degree in information technology at this time and started my master's degree.

At this point, you may be asking why you need to know this or why I am even providing this background. While an unconventional pivot from medicine to IT, this career detour allowed me to apply my analytical abilities in an environment where technology, not people, was the focus of diagnosis and treatment. Through a combination of self-study, entry-level exposure, advanced degrees, and a willingness to learn, I successfully charted a new professional course—proof that career changes into technology, with dedication, are achievable.

## Mid-level – transitioning to cybersecurity

The mid-level phase typically involves pivoting into specialized cybersecurity roles through either internal transfers or external job changes. Here, certifications become critical to validate expertise. Hands-on experience with vulnerability assessments, penetration testing, and **incident response** (**IR**) is invaluable. Soft skills such as communication and collaboration are equally crucial. Job hopping too frequently can be a red flag. At this stage, maintaining strong professional relationships and networks provides visibility to new opportunities.

Intermediate certifications such as CISSP and **Certified Information Security Manager** (**CISM**) validate core cyber capabilities. Using prep courses such as bootcamps, professionals should study 15–20 hours weekly over 2–3 months per cert. Passing one advanced certification annually displays continuous learning.

Immersing in hands-on work provides invaluable experience. Volunteering for projects such as performing penetration tests, running security tool evaluations, and building **proof-of-concept** (**PoC**) environments cements practical skills. Make time for 5–10 extra hands-on hours weekly.

### Example pathways

The transition from mid-level roles in cybersecurity to more advanced positions such as CSA requires meticulous planning, a broadening of the skill set, and a deep commitment to the craft. While certifications play an essential role in showcasing one's knowledge, it's the blend of hands-on experience, soft skills, and strategic networking that makes a difference.

Here are some detailed example pathways to becoming a CSA, starting from a mid-level cybersecurity role:

- **Starting as a security analyst**: Obtain certifications such as CISSP and CISM to validate your knowledge. Seek opportunities to gain experience with risk assessments, vulnerability management, and developing security roadmaps. After 2–3 years, attempt to transition into a security engineer role. Focus on gaining hands-on experience with firewalls, IDSs/IPSs, and **security information and event management** (**SIEM**). Study for advanced certs such as CCSP. After 5+ years, aim for a lead architect position:

  - Pathway:

    - **Initial role**: Security analyst
    - **Intermediate steps**: Master risk assessments and vulnerability management | security engineer focusing on IDS/IPS and SIEM | lead in security roadmap formulation
    - **Final destination**: CSA

  - Certifications:

    - **Years 1–2**: Focus on the CISSP and CISM certifications, which are fundamental for anyone serious about a long-term career in cybersecurity.
    - **Years 3–4**: Branch out to tools-specific certifications and consider an advanced cert such as CCSP.

  - Training:

    - Engage in workshops focused on threat modeling and vulnerability management.
    - Seek mentorship from seasoned professionals and architects to gain insights into the nuances of crafting security strategies.
    - Participate in industry conferences and webinars.

  - Pitfalls:

    - Becoming confined to only compliance and policy-driven roles.
    - Not continuously upgrading technical capabilities.

- **Starting as an ethical hacker**: Continue honing penetration testing and vulnerability research skills. Expand knowledge of networking, OS internals, and application security. Pursue certs such as **Offensive Security Certified Professional** (**OSCP**) and CCNA. After 3–4 years, try to move into a security engineering job focused on architecture and system hardening. Later, gain experience in cloud platforms and identity management.

After 7+ years, achieve a lead architect role:

- **Pathway**:

  - **Initial role**: Ethical hacker

  - **Intermediate steps**: Expand to comprehensive security research | security engineer focusing on system hardening | lead in application security

  - **Final destination**: CSA specializing in application security

- **Certifications**:

  - **Years 1–2**: Deepen penetration testing skills with the OSCP certification.

  - **Years 3–4**: Branch out to network-focused certifications such as CCNA, ensuring a well-rounded profile.

- **Training**:

  - Join penetration testing bootcamps and challenges.

  - Collaborate with software developers to understand the intricacies of secure coding practices.

  - Attend workshops and webinars focusing on the latest vulnerabilities and countermeasures.

- **Pitfalls**:

  - Remaining too focused on hacking without considering broader security strategies.

  - Neglecting the importance of effective communication and collaboration.

- **Starting as an incident responder**: Gain well-rounded experience responding to various types of security incidents. Improve skills across detection, analysis, containment, and recovery processes. Pursue certs such as **GIAC Certified Incident Handler (GCIH)** and **GIAC Certified Intrusion Analyst (GCIA)**, and study risk management frameworks. After 4+ years, attempt to transition into a security engineering job focused on detection and response capabilities. Later, lead projects to architect **security operations center** (**SOC**) and IR capabilities. After 8+ years, attain an architect job:

  - **Pathway**:

    - **Initial role**: Incident responder

    - **Intermediate steps**: Master comprehensive IR | security engineer with a focus on SOC and IR capabilities | lead in **threat intelligence** (**TI**)

    - **Final destination**: CSA with a specialization in IR

- **Certifications**:

    - **Years 1–2**: Gain certifications such as GCIH and GCIA, which validate expertise in incident handling and intrusion analysis.

    - **Years 3–4**: Explore certifications that delve into risk management frameworks, fortifying the bridge between technical and strategic roles.

- **Training**:

    - Engage in simulated cybersecurity incident scenarios.

    - Attend trainings that offer insights into emerging threat vectors and **advanced persistent threats (APTs)**.

    - Collaborate with teams responsible for network monitoring and threat detection to understand real-world challenges.

- **Pitfalls**:

    - Becoming restricted to only IR without exposure to overarching security strategies.

    - Overlooking the importance of understanding security solution design and enterprise risk management.

The keys are expanding technical breadth, not just depth in one specialty, gaining well-rounded hands-on experience, and developing risk management and communication skills before pursuing senior architect jobs. Continuous learning across the cybersecurity landscape is key. Create two-year plans balancing capabilities and exposure. Those pursuing architect roles can obtain specialized credentials in security design while leading retrospectives to improve team operations.

With tailored certifications, diversified hands-on skills, softened perspectives, strengthened relationships, and focused planning, mid-career professionals can set the stage for leading complex initiatives on the road to cybersecurity architecture.

## Real-life example

While working at the bank, I began getting more interested in the security of computer systems. During this time, I began learning more about technologies such as Cisco, Linux, and, of course, Windows. Making my transition to a network engineer within the bank afforded me more opportunities to configure firewalls and get into newer technologies at that time.

To strengthen my security posture, I pursued a master's degree in information assurance to formalize my knowledge. This *pre-cybersecurity* program provided crucial concepts around risk frameworks, access controls, and security operations.

While working as a network engineer at the bank in the early 2000s, I became increasingly interested in cybersecurity. This was an era of growing regulatory pressure, as guidelines such as the **Gramm-**

**Leach-Bliley Act (GLBA)**, the **Health Insurance Portability and Accountability Act (HIPAA)**, and the **Sarbanes-Oxley Act (SOX)** raised the bar on compliance and risk management.

During this time, I was understanding the impact of risk from a business perspective. New and more stringent compliance and regulatory requirements were being drafted and required. This meant understanding more how not only business risk can impact the security of an organization but also how cybersecurity risk can compound business goals or effectiveness.

As expected, this additional push on compliance and risk mitigation became a large portion of the work I began doing. It is important to understand that many of the tools and resources that we now take for granted were not completely realized or available in the early 2000s.

A good example is enterprise anti-virus or anti-malware. The bank used McAfee (now known as Trellix), but there was no McAfee **ePolicy Orchestrator (ePO)** to centrally manage and report on compliance. While we did image systems with default applications, which included McAfee, it was essentially a manual install. This meant that monitoring and reporting on these systems were manual as well.

New compliance and regulatory guidelines were moving toward the continuous monitoring standards we take for granted today. Without a central management mechanism, reporting and validation was a manual process that literally took days to accomplish across all of the bank's branches. With this becoming a much larger issue, and with me getting more understanding of Windows batch scripting along with my experience with web interfaces, I decided to implement a solution.

After some testing and initial research, I came up with the ability to write the information needed to a Microsoft SQL database. This was rather a simple and basic solution. This DIY approach provided continuous monitoring well before tools such as McAfee ePO existed.

By no means was this what McAfee ePO provides, but it generated real-time reporting based on the last login of the workstation or server. The solution was a batch script that was part of the login script of the workstations or a scheduled task on the servers. Once the script was run, it would check the McAfee install directory and look at the antivirus engine and signature files. This would provide version information and the last update. This information would be written to the database along with the date/time, device name, and IP address.

Now that the data was getting into the database, there was a need to provide simple reporting. For this, I created a simple **Active Server Pages (ASP)** web page that would connect to a database that allowed querying of data based on a specified date and time range. With this, you could query and sort the data by date-time, IP address, computer name, or McAfee information. This showed the systems getting updated and if a system was not up to date, this was not something that came up days later. Appropriate resources and remediation could be acted upon in a more immediate fashion, thereby reducing potential risk and exposure by poorly updated anti-malware signatures.

Although management hesitated to formally deploy my solution, I activated it regardless since it saved security teams countless manual hours. Sometimes, you have to bypass bureaucracy and implement what you know is right. In the end, my unauthorized initiative resulted in a major audit win for the bank.

This experience taught me creative problem-solving and perseverance can overcome organizational inertia. With the will to learn and drive to execute, security-minded technologists can construct their own solutions to enable risk reduction, even in the face of resistance.

## Advanced level – becoming a cybersecurity specialist

At the advanced stages, cybersecurity professionals start developing deep expertise in specific domains such as application security, TI, or cloud security. Highly specialized certifications demonstrate niche skills, while real-world problem-solving develops true mastery. Understanding sector-specific landscapes is important, as is keeping updated with emerging technologies through conferences, online courses, and independent research. Striking a balance between specialization and adaptability is key.

Expanding into adjacencies provides well-roundedness, and obtaining domain certifications demonstrates focused expertise.

### *Example pathways*

The advanced phase in a cybersecurity career is marked by the deepening of expertise, focusing on specialized domains, and understanding how to merge this expertise with the broader objectives of an organization. These professionals have already built a strong foundational and mid-level knowledge base. The journey ahead requires them to strike a balance between specialization, adaptability, and leadership.

Here are some detailed example pathways to becoming a CSA, starting from an advanced cybersecurity specialist role:

- **Starting as an application security specialist**: Leverage your expertise to expand into reviewing architecture designs and providing guidance on secure **software development life cycle (SDLC)** processes. Pursue management and leadership training. After 3–4 years, attempt to move into an application security architect role. From there, gain experience with cloud and infrastructure security to round out your skills. After 8+ years total, achieve an enterprise architect job.

- An application security engineer can earn credentials such as the **Certified Secure Software Lifecycle Professional (CSSLP)** certification:

  - **Pathway**:

    - **Initial role**: Application security specialist

    - **Intermediate steps**: Master secure SDLC processes | application security architect | diversify into cloud and infrastructure security

    - **Final destination**: Enterprise CSA

  - **Certifications**: Obtain the CSSLP certification. Dedicate 15 hours weekly for focused study over 2–3 months for each certification.

- **Training**:

  - Engage in workshops on secure application development and threat modeling.

  - Collaborate with software development teams to integrate security into the development process.

  - Attend conferences focusing on application security.

- **Pitfalls**:

  - Overspecialization in only application security.

  - Lack of experience in designing and integrating comprehensive security solutions.

- **Starting as a TI analyst**: Hone skills analyzing emerging threats, gathering adversary intelligence, and mapping attack campaigns. Pursue certifications such as **GIAC Cyber Threat Intelligence (GCTI)** and **SysAdmin, Audit, Network, and Security Forensics 578 (SANS FOR578)**. Seek opportunities to train others and present findings to leadership. After 4–5 years, attempt to transition into a strategic role focused on cyber threat modeling and intelligence-driven defense. Later, lead efforts to architect TI capabilities. After 10+ years, obtain an enterprise architect position.

- Threat analysts can take cloud security courses and get hands-on by building **Amazon Web Services (AWS)** sandbox environments. Network defense specialists can cross-train in identity management. Rotate annually to avoid narrow perspectives:

  - **Pathway**:

    - **Initial role**: TI analyst

    - **Intermediate steps**: Master threat analysis | strategic role in threat modeling | lead in architecting TI capabilities

    - **Final destination**: Enterprise CSA

  - **Certifications**: Pursue certifications such as CTI and SANS FOR578

  - **Training**:

    - Attend courses that offer hands-on experience in cloud security. Building sandbox environments in platforms such as AWS can be highly beneficial.

    - Engage in cross-training activities, perhaps diving into identity management or application security.

    - Rotate roles annually to get a well-rounded perspective on different facets of cybersecurity.

- **Pitfalls**:

  - Overemphasis on intelligence gathering without utilizing the intel to bolster organizational security posture.

  - Failing to adequately engage and communicate with stakeholders.

- **Starting as an IR specialist**: Expand into leading **incident management** (**IM**) processes and mentoring junior staff. Improve technical skills across network/host forensics, reverse engineering, and cloud investigation. After 5+ years, aim for a senior IR leader role driving continuous enhancement of detection and response capabilities. Later, lead efforts to architect modern SOCs leveraging automation and **machine learning** (**ML**). After 12+ years, attain an enterprise architect job:

  - **Pathway**:

    - **Initial role**: IR specialist

    - **Intermediate steps**: Lead IM | senior IR leader | architect modern SOCs

    - **Final destination**: Enterprise CSA

  - **Training**:

    - Engage in deep dives into network/host forensics, reverse engineering, and cloud investigations.

    - Mentor junior staff, sharing experiences and insights.

    - Explore the latest advancements in automation and ML to enhance SOC capabilities.

  - **Pitfalls**:

    - Restricting oneself to purely technical IR roles without strategic involvement.

    - Having gaps in understanding the integration and application of different technologies.

Seeking leadership roles elevates strategic impact. SOC engineers can volunteer to lead updates to IR playbooks. Application penetration testers can serve as mentors to junior team members.

Developing executive engagement abilities enables influence. Principal consultants can present cyber risk overviews to the audit committee or board. Technical leads can draft proposals to leadership on security roadmap priorities.

Attending conferences such as *Black Hat Briefings* and *RSA* expands visibility. Submitting speaking proposals raises your profile as a thought leader. Publishing articles in industry journals demonstrates communication abilities.

Create 3–5-year plans to address experience and exposure gaps. Specialists wanting enterprise CSA roles can obtain risk management credentials while leading projects, mentoring others, and presenting to executives.

The keys are diversifying technical expertise, developing leadership skills, seeking challenges outside your comfort zone, and maintaining a strong passion for continued learning before pursuing top-tier architect roles.

With a purposeful elevation of specialized skills, expanded breadth, leadership development, executive engagement, industry visibility, and multi-year roadmapping, cybersecurity professionals can avoid pitfalls and optimize preparation for top-tier enterprise architecture positions.

## Real-life example

After six years at the bank, I made a shift to a position as a city employee in California. My cybersecurity career really began in early 2006 as an information technology technician with the City of Victorville, California. In this role, I honed core infrastructure skills while pursuing foundational Microsoft certifications during off-hours.

Seeing the growing need for cybersecurity expertise even at the local government level, I completed a master's degree in information assurance that I had begun while with the bank and thereby augmented my knowledge. I also obtained pivotal certifications, including CISSP and Security+, to validate my developing capabilities.

Leveraging this education and hunger to drive progress, I spearheaded efforts to establish formal security policies, risk management processes, and technology controls for Victorville. This allowed me to gain invaluable hands-on experience while institutionalizing best practices.

By demonstrating a commitment to security and governance, I positioned myself for advancement. Soon, I attained a role as an information security manager for a **Department of Defense (DoD)** contractor. This mid-career pivot expanded my exposure to large-scale enterprise risk management, federal compliance, and leading strategic initiatives.

During this time with the DoD, I expanded my understanding of the need to adhere to standards but be flexible to support mission and capabilities. After a year, the contract ended, and it forced me to find other employment. This led me to become a contractor with a company supporting the **Department of Homeland Security (DHS)** and specifically the **Transportation Security Administration (TSA)**.

My progression into senior cybersecurity leadership roles accelerated rapidly in the late 2000s during my time as SOC oversight manager within TSA. This high-impact position provided oversight of all technical aspects of TSA's enterprise security infrastructure.

On a day-to-day basis, I led a team of federal employees and contractors handling security engineering initiatives to comply with mandates such as the **Federal Information Security Act (FISMA)** and the **National Institute of Standards and Technology's (NIST's) Risk Management Framework (RMF)**. We implemented technologies such as **Internet Security System (ISS)** network intrusion detection sensors, Sourcefire IDS, ArcSight logging, and McAfee web gateways to strengthen monitoring and threat detection capabilities.

I provided regular guidance directly to the TSA **chief information officer (CIO)**, **chief information security officer (CISO)**, and other agency executives on strategic directions for security infrastructure modernization. I also represented TSA on the DHS Senior Information Officers Council, influencing cybersecurity policy across the department.

Managing complex procurement processes, I oversaw the acquisition of all security solutions for the agency. This included directing the development of **statements of work (SOWs)**, independent cost estimates, and detailed project schedules. I also instituted robust multi-year budget planning and tracking to ensure appropriate funding of security priorities and projects.

As SCE team lead, I chaired the TSA **Systems Change Control Board (SCCB)**, providing hands-on architectural direction on integrating new technologies with minimal disruption. In addition, I was responsible for security oversight across all operational systems and infrastructure during their entire life cycle.

In addition to pivotal cybersecurity leadership roles within government and commercial sectors, I maintained an active presence as a writer, speaker, and lecturer to contribute thought leadership to the industry.

Even during demanding executive positions, I frequently published articles on popular infosec blogs and sites to share insights on topics such as cloud security, governance frameworks, and compliance.

In 2009 and 2014, I co-authored chapters in the 5th and 6th editions of the highly regarded *Computer Security Handbook*, published by Wiley Press, focusing on secure coding practices. Collaborating with established authors such as M.E. Kabay expanded my writing skills.

I ensured my perspectives directly reached both technical and leadership audiences by consistently presenting at major conferences. For example, in 2014, I delivered a well-received lecture on *Lessons Learned From Growing Web Security in a Federal Agency* at Intel's *FOCUS* conference.

On the stage, I was able to distill years of experience securing complex government sites into actionable guidance valuable for attendees driving security in public and private sector organizations.

Through writing, speaking, lecturing, and participating in conferences consistently during my career, I remained connected with the broader cybersecurity community. I exchanged perspectives with peers and helped promote best practices while further developing my own leadership presence and communication abilities.

This active industry involvement demonstrates my continual drive for learning and passion for advancing the cybersecurity field through multifaceted thought leadership.

This unique role allowed me to synthesize technical security capabilities with organizational objectives and compliance needs. These skills quickly led to a director of cybersecurity position within a rapidly growing commercial firm.

Here, I developed full-spectrum security solutions for clients in industries such as finance, energy, and healthcare. With strong stakeholder engagement skills honed in government, I worked routinely with the executive leadership of companies to craft strategies addressing their complex security and compliance needs.

I also instituted robust security governance frameworks using **Information Technology Infrastructure Library (ITIL)** and NIST standards. Expanding on my people-management experience, I built high-performance teams, mentored junior cybersecurity professionals, and instituted knowledge-sharing programs.

My next pivotal move was becoming director of infrastructure and security for a **Software as a Service (SaaS)** provider. I led teams securing **Federal Risk and Authorization Management Program (FedRAMP)**, **Health Information Trust Alliance (HITRUST)**, and commercial data center environments. We enhanced cloud security on AWS, Azure, and Oracle Cloud.

I provided guidance to the C-suite on security strategies, budget planning, and regulatory compliance. I also instituted vendor management practices and mentored junior engineers and analysts.

In retrospect, the foundation built through foundational IT roles, graduate degree attainment, and coveted certifications enabled my rapid trajectory into cybersecurity leadership. I realized progress requires proactive learning, seeking challenges beyond the day-to-day, and proving value delivery. These lessons served me well as I charted an unconventional course to meaningful security impacts.

Through these progressive steps, I mastered the fusion of technical knowledge, communication fluency, and visionary leadership required in executive cybersecurity roles. Each opportunity allowed me to synthesize security capabilities with business objectives and develop trusted-advisor relationships with stakeholders. This comprehensive experience prepared me for CISO positions overseeing cyber risk management for large-scale enterprises.

## Senior level – becoming a CSA

The culmination of the cybersecurity career journey is the architect role, integrating business objectives, technical security capabilities, and policy governance. Years of prior experience should equip architects with strategizing, designing, and communication skills to bridge gaps between stakeholders, executives, and technical teams. Ongoing technical education remains critical to assess and incorporate new tools and frameworks. Mentoring junior team members also enhances leadership abilities.

For seasoned architects seeking to reach the pinnacle of the profession, expanding scope, cultivating business acumen, and developing leadership versatility are key.

Obtaining credentials such as **CISSP-Information Systems Security Architecture Professional (CISSP-ISSAP)** demonstrates architect-level expertise. Provide executive coaching to junior employees to develop management skills. Publish thought leadership articles in industry journals to build visibility.

Pursuing an executive MBA or corporate board training develops business alignment. Take coursework in strategy, risk management, finance, and communication tailored to security leadership roles. Study 10 hours weekly for 2 years, balancing it with work.

Leading strategic cross-functional initiatives provides diverse exposure. Architects can spearhead technology modernization projects coordinating with IT, vendors, and **lines of business (LOBs)**. Or, they can drive security automation efforts leveraging **robotic process automation (RPA)**, **security orchestration, automation, and response (SOAR)**, and ML technologies.

## *Example pathways*

Attaining the esteemed position of CSA or even transitioning to a CISO role marks a zenith in the cybersecurity career ladder. This stage is characterized by the amalgamation of technical acumen, business strategy, and leadership finesse. Professionals here are tasked with bridging technical teams, executives, and other stakeholders, ensuring security aligns with broader business objectives.

Here are some detailed example pathways for career growth starting from a senior CSA role:

- **Starting as an application security architect**: Look for opportunities to broaden your scope such as leading enterprise API and microservices strategies. Pursue executive leadership training and consider an MBA to strengthen business acumen. After 4–5 years, strive for a chief architect job responsible for firm-wide cybersecurity blueprinting:

  - **Pathway**:

    - **Initial role**: Application security architect

    - **Intermediate steps**: Master enterprise API and microservices strategies | executive leadership training | MBA for business acumen enhancement

    - **Final destination**: Chief architect overseeing the comprehensive cybersecurity landscape

  - **Education**:

    - Pursue the CISSP-ISSAP certification, which showcases architect-level expertise.

    - Consider an executive MBA program, focusing on subjects such as strategy, risk management, finance, and cybersecurity-centric communication.

    - Dedicate approximately 10 hours weekly for about 2 years, ensuring a balance with work responsibilities.

  - **Training**:

    - Engage in workshops or courses on advanced application security topics, especially as they pertain to emerging technologies.

    - Attend seminars on executive communication and leadership to effectively brief top-tier stakeholders.

- **Pitfalls:**

  - Becoming overly engrossed in application security, neglecting broader enterprise security challenges.

  - Not adequately engaging with or understanding the needs of stakeholders.

- **Starting as an infrastructure security architect:** Leverage your skills to lead technology modernization initiatives, aligning security with IT objectives. Collaborate cross-functionally to understand diverse business needs. Consider pursuing board and committee leadership roles externally. After 6+ years, attain a chief architect job guiding organizational security vision:

  - **Pathway:**

    - **Initial role:** Infrastructure security architect

    - **Intermediate steps:** Lead tech modernization projects | strengthen the alignment of security with IT objectives | engage in cross-functional business collaboration

    - **Final destination:** Chief architect, shaping and guiding the entire organizational security paradigm

  - **Training:**

    - Engage in advanced courses focusing on modern infrastructure technologies and their associated security implications.

    - Develop skills related to cloud security, hybrid environments, and emerging tech trends.

    - Seek out board and committee leadership roles externally, providing a platform to influence larger industry decisions.

  - **Pitfalls:**

    - Being too entrenched in technical specifics at the expense of broader strategic considerations.

    - Not fully grasping the unique security needs of the specific industry vertical.

- **Starting as an identity and access architect:** Expand your role by leading teams focused on security governance, policy, and compliance. Pursue opportunities to brief executives and boards on security issues. Maintain technical skills, but equally develop leadership presence. After 8–10 years, achieve chief architect job overseeing infosec vision and governance:

  - **Pathway:**

    - **Initial role:** Identity and access architect

    - **Intermediate steps:** Lead teams on security governance | engage in policy and compliance | conduct frequent briefings to executives on pertinent security matters

    - **Final destination:** Chief architect overseeing information security vision, governance, and integration with enterprise objectives

- **Training**:

    - Continuously update knowledge on evolving **identity and access management (IAM)** technologies.

    - Attend workshops on governance, risk, and compliance to ensure a holistic approach to security.

    - Cultivate leadership presence through coaching, mentoring, and executive engagement programs.

- **Pitfalls**:

    - Becoming too narrowly focused on access controls without considering a holistic security strategy.

    - Not sufficiently engaging with or understanding the overarching business strategy and objectives.

At the senior level, briefing executives and boards becomes a pivotal aspect of the role. Actively providing insights on cyber risks, budgetary considerations, emerging threats, and regulatory dynamics is essential. Volunteering for industry association groups, especially those centered on standards, policy, or technology, can significantly broaden leadership horizons.

Mentorship plays a dual role—it invests in budding talent and hones leadership capabilities. Setting up regular mentorship sessions, sharing experiences, and offering opportunities for growth can be invaluable.

Transitioning from a senior CSA to the very pinnacle of the profession, be it a chief architect or even a CISO, necessitates a blend of technical mastery, strategic foresight, and leadership dexterity. It's about seeing the big picture, influencing at the highest levels, and ensuring cybersecurity strategies align seamlessly with business objectives. This journey demands continuous learning, adaptability, and a fervent commitment to the ever-evolving world of cybersecurity.

## Real-life example

After over a decade of honing my skills in various architecture and leadership positions, I attained my current role as VP of cybersecurity operations and director of security architecture for a major financial services company.

This represents the pinnacle of my career journey thus far—leveraging decades of experience to provide executive-level guidance balancing business needs and cyber risks.

On a daily basis, I lead senior architects to design innovative solutions that enable business growth while managing risk. I routinely brief the C-suite and board on cyber threats, regulatory compliance, budget planning, and strategies for resilience.

In this influential role, I lead a team of senior security architects collaborating with **business units (BUs)** and technology groups to architect solutions balancing business needs and risk management.

Drawing from decades of experience, I provide authoritative guidance on integrating security into business objectives for areas such as cloud adoption, digital transformation, mergers and acquisitions, and global expansion.

I continue mentoring junior security team members, imparting knowledge and leadership techniques honed over my career. I also institute vendor management programs to deliver cost-optimized security capabilities aligned with organizational priorities.

In addition, I lead regulatory compliance efforts for frameworks such as the **General Data Protection Regulation (GDPR)**, SOX, and the **Payment Card Industry Data Security Standard (PCI DSS)**, leveraging robust policies and controls.

This pinnacle role allows me to blend technical expertise, communication fluency, and strategic vision to architect innovative cybersecurity solutions enabling business growth. I enjoy the complexity of balancing risk management with a competitive advantage.

My progression from hands-on engineer to government security leader then industry executive has provided perspective into the multifaceted capabilities modern CISOs require. I aim to pay forward these learnings by developing security talent and guiding organizations to succeed through disruptive change.

I spearhead initiatives such as adopting **zero-trust architecture (ZTA)** and leveraging automation, allowing me to synthesize security with IT and LOB objectives. This has provided diverse exposure beyond technology alone.

My foundation gained from past technical and leadership roles prepared me for the multifaceted capabilities required as an enterprise CSA. I've been able to successfully bridge IT, security, and executives to enact strategies securing the organization amid constant change.

The cyber risk landscape continues evolving rapidly, but by applying the lessons from my comprehensive career journey, I am confident in steering institutions to maintain resilience. There are always new threats to observe, insights to orient from, decisions to make, and actions to take. I remain committed to operating inside these **Observe, Orient, Decide, and Act (OODA)** loops to secure our digital future.

For cybersecurity professionals with sights set on reaching the architect tier, perseverance and dedication to cultivating well-rounded skills are crucial. While challenging, integrating security architecture seamlessly across the entirety of complex global businesses represents the pinnacle of impact for cyber leaders.

## The big picture

This comprehensive account charts an exemplary cybersecurity career journey from humble beginnings to the highest echelons of the field. It underscores how a nonlinear path guided by continuous learning, seizing impactful opportunities, and delivering value can lead to security leadership roles securing organizations and nations.

The passage highlights how foundational technology expertise from initial IT roles laid critical groundwork for a segue into cybersecurity. Hands-on infrastructure, networking, and programming experience provided familiarity with systems essential for security analysis.

Mid-career milestones centered on deepening security knowledge through pivotal certifications, specialized roles, and progressively higher responsibilities. Compliance pressure at a bank motivated self-driven solutions showcasing security capabilities. Federal government and commercial sector stints expanded enterprise architecture skills securing critical systems.

In parallel, a commitment to well-rounded abilities manifested through diverse writing, speaking, and community engagements. These built thought leadership and communication fluency, synthesizing technical depth with strategic insight.

This storied career exemplifies how obsessive curiosity and persistence in upskilling, delivering high-impact security solutions, and nurturing talent can overcome unconventional beginnings. It serves both as a vicarious experience for aspirants and as a model for achieving the pinnacle of the cybersecurity field through comprehensive capabilities cultivated over time.

While specialized technical expertise is vital, professionals must also focus on soft skills, risk management, and effective communication. The cybersecurity career path rewards those who take a long-term, strategic approach to their professional growth. Laying a foundation across technical domains, progressively gaining specialized expertise, and proactively avoiding common pitfalls can help aspiring cybersecurity leaders reach the architect level and beyond. The journey requires persistence, but each phase brings its own rewards.

## Where to start

The parallels between aerial combat maneuvering and navigating a cybersecurity career are more than metaphorical. Both require operating within intense OODA loops—continuously observing, orienting, deciding, and acting.

Like fighter pilots, cyber professionals must voraciously absorb intelligence on the latest threats, innovations, and industry movements. This *radar sweep* of the environment equates to OODA's observation phase.

They must then orient themselves by analyzing observations and synthesizing context on where they stand relative to the frontier. What skills, certs, or experience will differentiate them from the competition?

Informed orientation enables decisive career maneuvering. Should they specialize further or expand their breadth? Pursue management or technical mastery? Switch industries or domains? The optimal decision stems from timely orientation.

Finally, prompt action is imperative—upskilling rapidly, seizing opportunities, publishing, and networking. Each action changes the landscape, producing new observations to re-orient.

The tempo of business accelerates yearly. Cybersecurity professionals must fly ever tighter OODA loops to evolve. Maintaining the cycle speed to orient, decide, and act before competitors is the key to reaching the upper tiers.

Like Boyd's ace pilots—who I will talk about in a bit—aspiring cyber experts must operate within the adversary's OODA Loop by shortening their own. They must out-learn, out-certify, out-author, and out-connect at each career stage to gain advantage. This framework, applied relentlessly, allows cyber professionals to successfully chart the course to the top.

## A bit of history

Colonel John Boyd (1927–1997) was an acclaimed American fighter pilot and military strategist renowned for revolutionizing maneuver warfare and fighter aircraft tactics. He is best known for developing the decision-making framework called the **OODA Loop**.

Boyd flew F-86 Sabre jets during the Korean War, where he analyzed why American pilots consistently performed better against rival MiG fighters despite comparable aircraft capabilities. Boyd attributed success to American pilots quickly transitioning between observing, orienting, deciding, and acting – quicker than their opponents.

This breakthrough became Boyd's OODA Loop concept. As he described it: "*Operating inside adversary's OODA loops or get inside their mind-time-space-as a basis to penetrate the moral-mental-physical being of one's adversaries in order to pull them apart and bring about their collapse.*" (Source: *Patterns of Conflict* presentation by John Boyd, December 1986)

After retiring from the US Air Force as a colonel, Boyd consulted with the Pentagon and continued expanding his theories on military strategy. His OODA Loop framework emphasized that the key was to cycle through observation-orientation-decision-action faster than one's opponent, disrupting their tempo and gaining advantage.

Boyd hypothesized the OODA Loop had applications far beyond aerial combat and could be applied to business, politics, and other arenas of competition and conflict. As he stated: "*Machines don't fight wars. Terrain doesn't fight wars. Humans fight wars. You must get into the mind of humans. That's where the battles are won.*" (Source: Interview in *John Boyd: The Fighter Pilot Who Changed the Art of War* by Robert Coram, November 2002.)

Though initially controversial, Boyd's concepts became highly influential on modern military thinking. The OODA Loop remains an essential framework taught in war colleges and applied across air, ground, and naval warfare doctrine. It has also found extensive adoption in competitive business strategy and decision-making processes.

## The OODA Loop

The four stages of the loop are the following:

- **Observe**: Gather information and inputs from your environment through your senses, tools, and systems. In cybersecurity, this could mean monitoring TI, observing new hacker techniques, or watching industry trends. In life, it could mean observing your daily experiences, challenges, and relationships.

- **Orient**: Analyze and synthesize the information observed to build mental models, context, and understanding. In cybersecurity, this means connecting threat data to determine campaign patterns. In life, it's making sense of observations to gain insights about yourself and your situation.

- **Decide**: Make a judgment about which action to take based on the enhanced orientation. In cybersecurity, this could mean deciding on a strategy to strengthen defenses based on observed threats. In life, it's making choices aligned with your goals and values based on your assessment of observations.

- **Act**: Execute decisions affecting the environment and generate new observations. In cybersecurity, this means implementing new security controls. In life, it's taking action and noticing the results to inform future observations.

The OODA Loop is depicted in the following diagram:

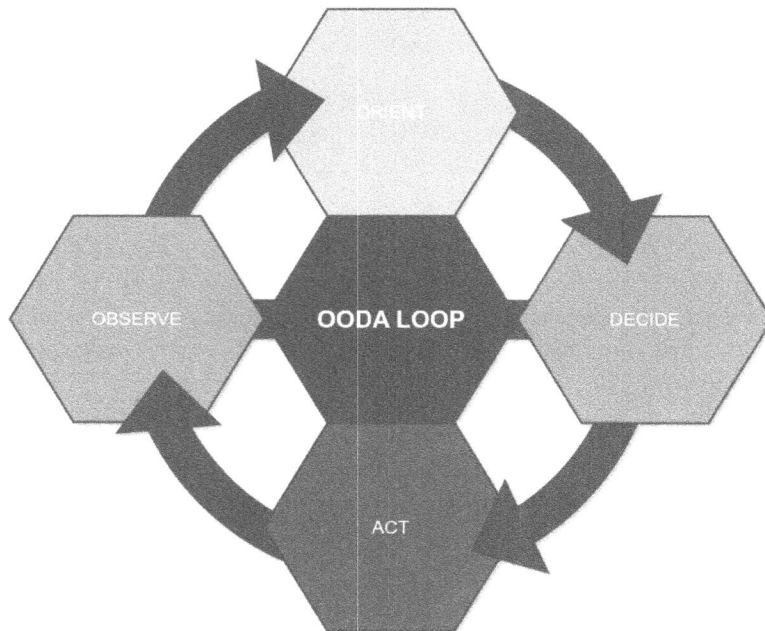

Figure 7.1 – OODA Loop flow diagram

The framework then repeats, using the effects of actions as new observations to re-orient, re-decide, and re-act. This creates a continual process of observation-guided adaptation.

By quickly cycling through the OODA Loop stages, you gain an advantage in any endeavor. You observe and orient faster to make quicker, more informed decisions. Then, you act rapidly based on those decisions, disrupting the opponent's tempo.

This could allow a cybersecurity analyst to identify and mitigate threats before major damage or help someone orient themselves amid life complexities, gain insights, and take purposeful actions. The OODA Loop is a powerful framework for out-maneuvering adversity and competition.

## Applying lessons learned

Today's business cycles in cybersecurity have accelerated, and aspiring professionals must match the pace to succeed.

To advance in their careers, cybersecurity practitioners should operate inside the OODA Loop. This means continuously absorbing new developments in the field (*observe*), contextualizing their knowledge and experience (*orient*), making strategic career moves (*decide*), and executing practical steps to attain new skills and roles (*act*).

## Entry level – analysts

At the entry level, technology or cyber analysts must voraciously observe the latest threats and technologies. They should orient themselves by connecting new learning to their existing skill set and interests. Decisions at this stage include pursuing certifications, training, and projects to fill knowledge gaps. Actions revolve around hands-on practice and experimentation:

- **Observe**: Absorb and monitor recent cyber threats, technological advancements, and emerging tools designed to counteract these challenges.
- **Orient**: Relate new knowledge to existing expertise, identifying areas of interest and potential gaps in understanding.
- **Decide**: Choose relevant certifications, courses, and hands-on projects. This might mean delving into a CEH program or enrolling in a practical cybersecurity bootcamp.
- **Act**: Engage in practical exercises, perhaps setting up personal labs, participating in CTF challenges, or contributing to cybersecurity forums and platforms.

## Mid-level – security engineers

At the mid-level, security engineers must observe industry trends, innovations, and role models to emulate. Orientation means understanding how their specialized expertise fits into the big picture.

Decisions include pivoting into an adjacent domain or diving deeper into a niche. Action steps involve structured preparation and strategic networking:

- **Observe**: Stay updated with industry advancements, technological innovations, and the career trajectories of role models and thought leaders in the field
- **Orient**: Reflect on one's unique expertise, evaluating how it complements broader industry objectives and dynamics
- **Decide**: Consider either branching out to an adjacent specialization or further deepening knowledge within a particular niche
- **Act**: Invest time in targeted training programs, seek mentorship, and expand professional networks through conferences and seminars

## Advanced level – principal consultants

At the advanced level, principal consultants observe business objectives, risk environments, and architecture best practices. Orientation requires analyzing how security capabilities align with business goals. Key decisions include embracing leadership roles and new technologies. Actions mean guiding teams through complex projects and earning executive trust:

- **Observe**: Maintain a keen understanding of evolving business goals, changing risk paradigms, and the latest in cybersecurity architectural practices
- **Orient**: Analyze and map the intersection of security capabilities with overarching business objectives
- **Decide**: Opt to step into more prominent leadership roles, spearhead the adoption of emerging technologies, or lead transformative initiatives
- **Act**: Lead and guide project teams, provide expert consultation, and build trust at executive and board levels

## CSA-to-CISO level

Those at the CSA-to-CISO level broadly observe the competitive landscape, threat horizon, and internal culture. Orientation involves connecting insights to strengthen enterprise resilience. Decisions require balancing security priorities and business outcomes. Actions revolve around fostering talent, clearly communicating risk, and enacting forward-looking strategies:

- **Observe**: Keep a bird's-eye view on global threat landscapes, industry competition, cultural shifts within the organization, and the impact of regulatory changes
- **Orient**: Synthesize insights to fortify organizational resilience and adaptability, ensuring both immediate security and long-term viability

- **Decide**: Strategize on balancing between immediate security concerns and long-term business outcomes, potentially integrating **artificial intelligence** (**AI**)-driven security solutions or embracing zero-trust models

- **Act**: Develop and nurture the next generation of cybersecurity talent, communicate complex risk scenarios to stakeholders effectively, and spearhead visionary security strategies that anticipate future challenges

Much like Boyd's fighter pilots, who thrived in high-stakes scenarios by staying agile and proactive, today's cybersecurity professionals must employ a similar strategy. The OODA Loop offers a structured yet adaptable approach, allowing professionals to not only react to the industry's ever-accelerating pace but to lead it. By embracing this philosophy and tailoring its stages to their career progression, cybersecurity experts can solidify their trajectory toward the pinnacle of their profession.

# The cold open

For those looking to pivot into a cybersecurity career from a non-technical background, the path to becoming a CSA may seem daunting. However, with proper planning and focus, it is certainly achievable. The key is to take incremental steps to methodically build both technical expertise and business acumen. While the core competency stage may rely more on self-study, later milestones benefit from structured learning.

## Taking inventory of your skills

The first stage is gaining core competencies. For those outside technology, this means learning networking basics, operating systems, and scripting. Certifications such as CompTIA IT Fundamentals, Network+, and Security+ provide initial credibility. Hands-on projects, online courses, and volunteering for tech roles during your current job can accelerate learning:

- Research job roles and skill requirements for entry-level IT and cybersecurity roles to understand expected qualifications. Identify knowledge gaps.

- Identify transferable skills from your background—communication, analytical thinking, project coordination, and so on.

## Building hands-on skills

Enroll in fundamental courses through platforms such as Coursera, edX, Udemy, or community colleges to start building core IT knowledge. Earn introductory certifications such as CompTIA IT Fundamentals:

- Follow beginner computing skills tutorials on platforms such as Codecademy (https://www.codecademy.com/catalog/subject/all) and Udemy (https://www.udemy.com/courses/it-and-software/?price=price-free&sort=popularity) to gain practical experience.

- Get books to use as training and references from resources such as `https://www.packtpub.com/`.

- Develop a learning schedule.

- Create a routine of nightly 1–2-hour study sessions. Take practice tests frequently. Adjust as needed based on your pace.

## Preparing for interviews

Update your resume to highlight any prior experience that shows analytical, problem-solving, or communication skills. Emphasize eagerness to learn. Tailor your resume to highlight transferable skills, motivation to start an IT career, and foundational knowledge/certifications.

Research the company's tech stack and products. Review common IT support and help desk interview questions. Emphasize your desire to learn.

Adapting to a steep learning curve requires perseverance. Leverage free resources, create a study routine that fits your life, and stay motivated by focusing on career goals. Entry-level tech roles provide the launchpad.

With determination and consistent effort to close knowledge gaps, those outside technology can successfully pivot into the industry through persistence and purposeful positioning of transferable abilities.

Apply to entry-level roles such as IT support technician, help desk analyst, or IT coordinator. Stress passion to start an IT career even without direct experience.

Offer to volunteer for IT-related initiatives at your current organization, even if unofficial. Seek opportunities to gain visibility and demonstrate motivation.

Next is obtaining an entry-level position to get real-world experience. This may be help desk, systems administration, or security operations. Soft skills from your previous field will be a strength. Immerse yourself in the new environment, demonstrate eagerness to learn, take on increasing responsibility, and fill knowledge gaps through continued self-study.

Once in an entry-level role, devote time outside work to continue gaining certifications, skills, and experience. Immerse yourself in the field.

## Continuing to upskill

The middle stage involves pivoting into a cybersecurity specialist role. Earn intermediate certs such as CISSP and pursue specialized training. Leverage soft skills from past roles while honing technical aptitude. Seek mentorship from senior CSAs, shadow them, and volunteer for security-focused projects:

- Attend local tech meetups or conferences to make connections in the industry. Ask about mentorship opportunities or informational interviews.

- Consider transition roles such as sales engineering or business analysis at a tech company. These build valuable knowledge before attempting a direct IT shift.

- Be prepared to speak intelligently about your existing skill set and how it applies to technology roles during interviews. Show ambition and willingness to reskill.

- Upon entering the role, maintain learning momentum. Study for the next certification such as Network+ while gaining on-the-job experience.

The later stages are about demonstrating architectural vision. Study for advanced certifications and diversify your hands-on experience. Seek opportunities to design solutions, guide projects, and brief leadership. Develop skills holistically as a communicator, leader, strategist, and technologist.

The end goal is to attain a senior architect position. This requires both technical breadth to connect diverse solutions and business acumen to align security with organizational needs. Patience and persistence are vital; expect the journey to take 5+ years. But for those with the drive to continuously learn, upskill, and deliver value, the cybersecurity field offers immense opportunity regardless of background.

# The transfer

For technology professionals seeking to advance their careers toward a CSA role, the journey requires building diverse hands-on experience and demonstrating architectural vision. While foundational technology skills provide a strong starting point, progressing through increasing levels of responsibility and capability is key.

The first milestone after gaining core competencies is obtaining an intermediate cybersecurity practitioner role, such as security analyst, network security engineer, or penetration tester. Certifications such as Security+, CISSP, and CEH validate capabilities. Immerse yourself in specific security domains while strengthening soft skills such as communication, collaboration, and project management.

The next stage involves demonstrating leadership and versatility as a security specialist or consultant. Expand the depth of skills in your chosen specialty while broadening knowledge across other areas. Pursue advanced certifications and lead complex security initiatives. Gain visibility for your accomplishments.

Later milestones center on exhibiting a strategic vision by evolving into an enterprise CSA or CISO. Broaden your perspective through cross-functional initiatives with IT, finance, risk management, and other groups. Pursue executive leadership training and bridge gaps between the technical and business sides.

Ultimately, aspiring to the top tier requires accentuating strengths while overcoming weaknesses. For technical experts, building business acumen is key. For security generalists, developing specialized expertise is crucial. Architects must synthesize technical details with organizational objectives and communicate effectively across stakeholders.

The foundation of technology skills gives those on the inside track an advantage, but resting on your laurels is a pitfall. Viewing progress as a continuum of learning and experience rather than a ladder is essential to reaching the top levels of cybersecurity leadership.

# How to expand

Launching a cybersecurity career on strong technical foundations is crucial. Common starting points are degrees in computer science or information technology, which provide fundamental knowledge of systems, networking, and programming. Hands-on roles such as systems administrator or network engineer allow burgeoning professionals to hone real-world skills in managing systems, servers, and infrastructure. During 2–4 years in these positions, continuous learning is imperative. Pursuing entry-level certifications such as CompTIA's Network+ and Security+ validates core competencies and shows commitment to growth.

## Pivoting to cybersecurity

Armed with well-rounded technical abilities, the next phase involves transitioning into cybersecurity-focused functions. Roles such as security analyst, ethical hacker, and vulnerability assessor provide a specific understanding of cyber risks, compliance standards, TI, and security testing. Immersion in these roles allows professionals to discern security vulnerabilities from an attacker mindset. After 3–5 years, intermediate certifications such as CISSP and CEH confirm progressed capabilities. Ongoing education in new attack techniques, tools, and mitigation approaches is essential to keep pace with the evolving threat landscape.

Here's an example. As a penetration tester, one could simulate cyber attacks on systems, understanding vulnerabilities from an attacker's viewpoint. This experience can be instrumental later as a CSA when designing robust systems.

Here's the timeline. Typically, after another 3–5 years in roles such as security analyst, penetration tester, or threat hunter, individuals should consider advanced cybersecurity certifications. Certifications such as CISSP or CEH can be invaluable.

## Cultivating specialized expertise

To ascend to senior positions, cultivating expertise in specific cybersecurity domains becomes advantageous. Professionals can carve out specialties aligning with their interests and organizational needs, such as application security, cloud infrastructure security, or network defense. Becoming a **subject-matter expert** (SME) enables greater impact and thought leadership. Complementary advanced certifications, such as the CCSP credential for cloud specialists, demonstrate focused knowledge. However, well-rounded skills remain important, as enterprise security requires holistic strategies across environments. Rotational programs across security functions help broaden perspectives.

Here's an example. Someone specializing in cloud security might work as a cloud security engineer, focusing exclusively on best practices and security architectures for platforms such as AWS, Azure, or **Google Cloud Platform** (**GCP**). This deep dive can later inform more holistic security strategies when architecting solutions that incorporate various platforms.

Here's the timeline. Over another 3–4 years, a specialist could also consider acquiring certifications tied to their specialization, such as AWS Certified Security—Specialty or CCSP.

## Ascending to CSA

After a decade or longer honing both specialized and cross-disciplinary security skills, professionals may ascend to the pinnacle architect role. This requires synthesizing technical expertise with business objectives, risk management principles, and communication fluency. Architects act as visionaries, designing comprehensive security blueprints spanning technologies, policies, awareness programs, and integration with business processes. They head enterprise security strategy, lead large teams, and interface with executives as trusted advisors. Maintaining a technology edge and a continuous improvement mindset is still imperative due to the ever-evolving threat landscape. For those with patience, persistence, and lifelong dedication to their craft, the architect role represents the apex of cybersecurity career achievement.

Here's an example. A CSA at a multinational corporation might be responsible for creating a unified security strategy that encompasses local office networks, cloud services, mobile devices, and remote work solutions, ensuring data integrity and security across all touchpoints.

Here's the timeline. After 10–15 years in the field, and with advanced certifications such as CISSP-ISSAP, one would be well positioned to step into the role of CSA. However, continuous learning remains key, with emerging threats and technologies always on the horizon.

The journey toward becoming a CSA is both challenging and rewarding. It requires a blend of continuous education, hands-on experiences, and strategic foresight. By systematically progressing through the stages outlined previously, punctuated with relevant certifications and specializations, aspiring professionals can chart a successful career path toward the pivotal role of CSA.

## Summary

This chapter outlined a framework for progressing through a cybersecurity career, using the journey from entry-level to architect roles as an example. It emphasized that while cybersecurity foundations seem basic, combining them creatively like musical notes into elegant solutions requires finesse gained over time.

It examined milestones at each level. Early roles focus on building diverse technical competencies and foundational certs while avoiding overspecialization. Mid-level pivots into hands-on security functions, pursuing intermediate certifications and specializing while networking to enable advancement. At the advanced stage, cultivating specialized expertise in a domain while demonstrating leadership versatility is key.

Reaching the pinnacle of the CSA role requires synthesizing technical and business capabilities. Personal examples illustrated potential pathways, such as progressing from network engineering to infrastructure security to enterprise architecture.

The chapter emphasized applying Colonel John Boyd's OODA Loop concept of continuously observing, orienting, deciding, and acting faster than opponents. Examples were provided for each career level, from entry-level analysts rapidly absorbing threats to senior architects observing competitive landscapes and strengthening resilience.

For non-technical backgrounds, the chapter outlined methodically acquiring expertise through certifications, hands-on roles, and business acumen. For technology professionals, it focused on diversifying experience, honing specialized skills, developing leadership vision, and playing to strengths.

In summary, the chapter provided a comprehensive overview of strategically advancing through cybersecurity by setting milestones, maneuvering through OODA loops, and avoiding pitfalls. It aims to help driven professionals chart a course for top-tier cybersecurity leadership roles.

In the next chapter, we will discuss a number of certifications for security architecture as well as others to help differentiate oneself from others competing for the same position, as well as the good, bad, and ugly of the certification process and how to make choices that will best match readers' overall career plan and direction.

# The Certification Dilemma

*"Foreknowledge cannot be gotten from ghosts and spirits, cannot be had by analogy, cannot be found out by calculation. It must be obtained from people, people who know the conditions of the enemy."*

*– Sun Tzu*

*"Thus we may know that there are five essentials for victory: (1) He will win who knows when to fight and when not to fight; (2) he will win who knows how to handle both superior and inferior forces; (3) he will win whose army is animated by the same spirit throughout all its ranks; (4) he will win who, prepared himself, waits to take the enemy unprepared; (5) he will win who has military capacity and is not interfered with by the sovereign."*

*– Sun Tzu*

The previous chapter provided a comprehensive roadmap for advancing through the cybersecurity field, from entry-level positions to the esteemed cybersecurity architect role.

It stressed that while foundational security concepts seem basic initially, integrating them into adaptable solutions requires extensive creativity and finesse accrued over time through practice.

Sun Tzu emphasized foreknowledge as essential for victory. In cybersecurity, certifications provide a foreknowledge of the threats, tools, and best practices needed to succeed. Like military ranks designating capabilities, certifications signal expertise. They reward those who have prepared themselves through study.

Yet as Tzu noted, true knowledge comes from people of experience. Certifications validate skills, but real mastery requires creatively combining concepts through practice. Their value depends on selection should be aligned to one's goals and continuous learning beyond paper credentials.

This chapter will explore certification pathways that can help technologists at all career levels "know when and how to fight" in the cybersecurity field. But just as an army requires spirit throughout its ranks, cybersecurity necessitates dedicating yourself to lifelong betterment, not chasing certificates alone. Use certifications to obtain new knowledge then actively apply it. With proper inclusion in a career strategy, they can provide the foreknowledge needed to be a part of the emerging professionals or cybersecurity architects alike on their security journeys.

The chapter covers the following topics:

- Certifications landscape
- Why get certified?
- Certification considerations

# Certifications landscape

Certifications have become ubiquitous across the cybersecurity industry, with hundreds of options at varying levels catering to diverse specialties. For those aspiring to become cybersecurity architects, navigating this crowded certification landscape requires a strategic approach.

While mandatory credentialing requirements are still relatively rare for cybersecurity architecture roles, obtaining respected certifications can provide several advantages. The right certifications validate a mastery of foundational knowledge needed for cybersecurity architect solutions. They demonstrate commitment to continuous learning and signal technical capabilities to employers.

However, certifications should complement rather than replace hands-on experience. Cybersecurity architects rely heavily on real-world expertise to craft innovative designs tailored to their organization's environment and objectives. The most impactful cybersecurity architects back paper credentials with the wisdom gained from practical application over many years.

Foundational and intermediate certifications can provide the essential literacy needed to speak the language of cybersecurity fluently. As they advance, cybersecurity architecture-focused certifications such as the **Certified Information Systems Security Professional – Information Systems Security Architecture Professional** (**CISSP-ISSAP**) distinguish senior-level technical prowess and strategic vision. Yet even at the pinnacle, certifications support but do not substitute for multifaceted capabilities cultivated over a career. With a judicious and proactive approach, certifications remain valuable way-points on the journey to cybersecurity architect.

In this section, we will look at certifications mentioned in previous chapters in detail, to help you understand the scope and rationale for obtaining the certification.

## CompTIA

**CompTIA**, short for the **Computing Technology Industry Association**, is a globally recognized nonprofit trade association that plays a pivotal role in the IT industry. Established in 1982, CompTIA serves as a hub for IT professionals, businesses, and educational institutions, facilitating collaboration and providing a range of vendor-neutral certifications. These certifications cover various aspects of IT, enabling individuals to validate their skills and knowledge, and helping organizations identify qualified IT professionals. CompTIA certifications have become a standard measure of IT competency, influencing career development and guiding businesses in their technology investments. The CompTIA website can be found at the following URL: `https://www.comptia.org/`. The organization's commitment to education, advocacy, and industry collaboration makes it a significant contributor to the IT community. CompTIA currently has nine certifications. The following are the certifications to look at to cement your cybersecurity architecture career:

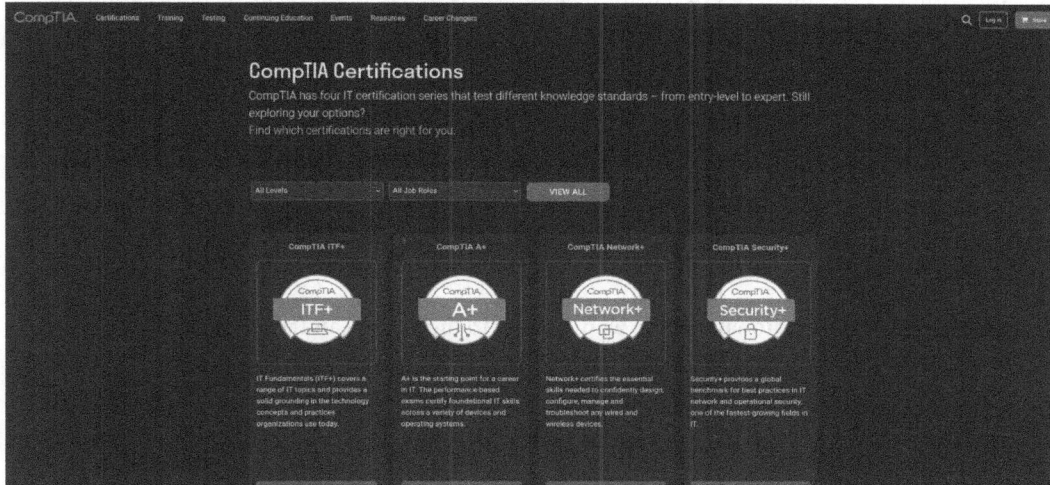

Figure 8.1 – CompTIA certifications

Let us explore them in detail.

### A+ certification

CompTIA A+ is one of the most recognized and widely respected certifications for entry-level IT professionals. It serves as the foundational stepping stone for individuals looking to establish a career in IT support and operations. The certification is designed to validate the essential skills and knowledge required to succeed in roles related to computer and network support.

The CompTIA A+ certification program covers a comprehensive range of IT topics, focusing on practical skills, troubleshooting, and problem-solving in various IT scenarios. This includes hardware, software, operating systems, basic networking, and security. A+ certified professionals are expected to demonstrate competency in tasks such as hardware maintenance, software installation, and system troubleshooting.

Key components covered in the A+ certification include the following:

- **Hardware:** Understanding computer components, peripherals, and mobile devices. Configuring and troubleshooting hardware components such as CPUs, motherboards, memory, and storage devices.

- **Software**: Installing and configuring various operating systems, including Windows, macOS, Linux, and mobile OS. Troubleshooting software issues and ensuring system functionality.

- **Networking**: Basic networking concepts, protocols, and troubleshooting. Configuring and managing wired and wireless networks.

- **Security**: Fundamentals of IT security, including best practices for securing devices and data. Identifying and mitigating security risks.

- **Troubleshooting**: Developing problem-solving skills to identify and resolve hardware and software issues. Diagnosing and resolving common IT problems.

The CompTIA A+ certification is primarily aimed at entry-level IT professionals, help desk technicians, and support specialists. The certification is ideal for individuals who are just starting their careers in IT or transitioning to roles that require foundational IT knowledge and skills. The typical audience includes the following:

- **Entry-level IT professionals**: Those who are new to the IT field and want to establish a strong foundation in IT support and operations

- **Help desk technicians**: Individuals working in help desk or technical support roles, providing assistance to end users with hardware and software issues

- **Support specialists**: IT support professionals responsible for maintaining and troubleshooting computer systems, peripherals, and software

- **Career changers**: People from non-IT backgrounds looking to enter the IT industry and build a career in IT support

- **Students and recent graduates**: Those pursuing IT-related degrees or completing IT training programs who want to gain a recognized certification

The A+ certification is vendor-neutral, meaning it does not focus on a specific technology or platform, making it relevant for a broad range of IT environments.

The benefits of CompTIA A+ certification are as follows:

- **Industry standard**: A+ is considered the industry standard for entry-level IT operational roles. It is recognized by employers globally

- **Career entry**: A+ provides a strong entry point for individuals looking to start their IT careers

- **Versatility**: The knowledge gained is applicable to various IT roles, from support specialists to IT technicians

- **Validation**: Certification validates foundational IT skills and knowledge

- **Career advancement**: A+ can serve as a stepping stone for more advanced certifications and career progression

Earning the CompTIA A+ certification is a solid investment in an IT career, as it lays the groundwork for future success in the field. It is also a requirement for many IT support and technician positions and provides a competitive edge in the job market.

### Network+ certification

The CompTIA Network+ certification is a globally recognized credential that validates the skills and knowledge necessary for a career in network administration and IT infrastructure. This certification program is designed to equip IT professionals with the essential networking expertise needed to manage, maintain, troubleshoot, and secure networks.

The Network+ certification curriculum covers a wide range of networking concepts, from the basics of network technologies to more advanced topics related to network security and cloud computing. The program ensures that certified professionals are well-versed in various networking protocols, topologies, and technologies. The key areas covered in the Network+ certification include the following:

- **Networking concepts**: Understanding of networking models, protocols, and network services
- **Infrastructure**: Knowledge of cabling, network devices (routers, switches, and access points), and virtualization technologies
- **Network operations**: Configuration, management, and monitoring of networks
- **Network security**: Identification of security threats, implementation of security solutions, and best practices in securing networks
- **Network troubleshooting and tools**: Proficiency in identifying and resolving network issues using various tools and troubleshooting methodologies

The CompTIA Network+ certification is aimed at a diverse range of IT professionals, including network administrators, network technicians, IT support specialists, and individuals aspiring to pursue careers in network administration and IT infrastructure. The typical audience includes the following:

- **Network administrators**: Those responsible for the configuration, management, and maintenance of network infrastructures
- **Network technicians**: IT professionals dealing with network troubleshooting, installations, and maintenance
- **Help desk and support specialists**: Individuals providing support for network-related issues in organizations
- **IT professionals transitioning to networking**: People with general IT knowledge looking to specialize in network administration
- **Entry-level network professionals**: Those starting their careers in the field of networking

The benefits of CompTIA Network+ certification are as follows:

- **Industry recognition**: The Network+ certification is recognized globally as a benchmark of network administration skills

- **Career advancement**: Earning this certification can lead to better job prospects, career growth, and increased earning potential

- **Vendor-neutral knowledge**: The certification focuses on vendor-neutral networking concepts, making it applicable in a variety of IT environments

- **Solid foundation**: Network+ provides a strong foundation for more advanced certifications in networking and security

- **Enhanced competence**: Certified professionals are equipped to manage and secure complex network infrastructures, contributing to an organization's efficiency and security

The CompTIA Network+ certification is especially relevant in today's interconnected world, where networks are the backbone of modern businesses. It equips IT professionals with the skills to design, manage, and secure networks, ensuring the availability and integrity of critical data and services. Earning this certification is a valuable achievement for those pursuing careers in network administration and IT infrastructure.

## Security+ certification

The CompTIA Security+ certification is a globally recognized and vendor-neutral credential that validates an individual's expertise in information security and cybersecurity. It is designed to equip IT professionals with the essential knowledge and skills required to secure and protect networks, systems, and data from various security threats and vulnerabilities.

The Security+ certification program covers a comprehensive range of security topics, including network security, compliance and operational security, threats, vulnerabilities, and risk management. It is a valuable certification for professionals seeking to establish a career in cybersecurity or enhance their existing security expertise. Key areas covered in the Security+ certification include the following:

- **Threats, attacks, and vulnerabilities**: Understanding various types of security threats, attacks, and vulnerabilities that can impact an organization's information security

- **Technologies and tools**: Knowledge of security technologies, tools, and best practices for securing data and systems

- **Architecture and design**: Designing secure network architecture, systems, and applications while considering security principles

- **Identity and access management (IAM)**: Implementing IAM solutions to control and monitor user access to systems and data

- **Risk management**: Identifying and mitigating security risks, including risk assessment, analysis, and management

- **Cryptography and public key infrastructure (PKI)**: Knowledge of cryptographic techniques and the use of PKI to secure communications and data

- **Secure communication and networks**: Ensuring the confidentiality, integrity, and availability of data during transmission over networks

The CompTIA Security+ certification is intended for a wide range of IT professionals, including security professionals, network administrators, system administrators, and individuals looking to specialize in cybersecurity. The typical audience includes the following:

- **Security professionals**: Those pursuing or advancing careers in cybersecurity and information security

- **Network administrators**: IT professionals responsible for securing network infrastructures

- **System administrators**: Individuals managing and maintaining computer systems and servers, focusing on security

- **Security analysts**: Professionals analyzing and responding to security incidents and threats

- **Security consultants**: Experts providing security consulting and advisory services to organizations

- **IT professionals transitioning to cybersecurity**: Individuals with a general IT background seeking to transition to cybersecurity roles

The benefits of the CompTIA Security+ certification are as follows:

- **Global recognition**: Security+ is widely recognized as a reputable cybersecurity certification globally

- **Versatility**: The certification is vendor-neutral and applicable across various IT environments and industries

- **Cybersecurity career path**: It serves as a foundational step for those pursuing cybersecurity careers

- **Security expertise**: Certified professionals are equipped with the skills to identify and mitigate security risks and protect critical data and systems

- **Compliance and best practices**: Knowledge of security compliance and best practices in information security management

Earning the CompTIA Security+ certification is a significant achievement for professionals aiming to excel in the field of cybersecurity. It not only validates their expertise but also provides them with the knowledge and skills to protect organizations from the increasing number of cyber threats and attacks. This certification is instrumental in building a strong foundation for a successful career in cybersecurity.

## CySA+ certification

The CompTIA **Cybersecurity Analyst (CySA+)** certification is a renowned credential designed for IT professionals who specialize in cybersecurity analysis and threat detection. CySA+ focuses on equipping professionals with the skills and knowledge required to identify, respond to, and mitigate security threats and vulnerabilities.

The CySA+ certification curriculum delves into various aspects of cybersecurity, including threat detection, analysis, and incident response. It emphasizes the ability to proactively monitor and protect an organization's systems and networks against cyber threats. Key areas covered in the CySA+ certification include the following:

- **Threat detection**: Recognizing and analyzing various types of cybersecurity threats, such as malware, attacks, and vulnerabilities

- **Data analysis and interpretation**: Analyzing and interpreting data to identify **indicators of compromise (IoCs)** and potential security incidents

- **Security technologies and tools**: Knowledge of security technologies, tools, and best practices for proactive threat detection and mitigation

- **Incident response**: Developing and implementing effective incident response plans and strategies to mitigate security incidents

- **Compliance and security frameworks**: Understanding security compliance requirements and industry-standard security frameworks

- **Network traffic analysis**: Analyzing network traffic to detect and respond to security threats and anomalies

- **Security data visualization**: Presenting security data and findings in a comprehensible manner for decision-makers

The CompTIA CySA+ certification is intended for IT professionals who aspire to work as cybersecurity analysts, threat hunters, and **security operations center (SOC)** professionals. The typical audience includes the following:

- **Cybersecurity analysts**: Those responsible for monitoring and analyzing security threats and incidents in an organization

- **SOC analysts**: Professionals working in SOC environments to identify and respond to security incidents

- **Security engineers**: Individuals involved in designing and implementing security measures and monitoring security controls

- **IT professionals transitioning to cybersecurity**: Those with general IT experience looking to specialize in cybersecurity analysis

- **Security consultants**: Experts providing security consulting services, including threat detection and incident response

The benefits of CompTIA CySA+ certification are as follows:

- **Validation of threat detection skills**: CySA+ certifies the ability to detect, analyze, and respond to security threats effectively

- **Employability**: The certification enhances job prospects and employability for cybersecurity analyst roles

- **Industry recognized**: CySA+ is recognized by organizations and government agencies as a credible certification for security analysis

- **Incident response expertise**: Professionals are equipped with the skills to develop and execute incident response plans and strategies

- **Proactive threat mitigation**: Certified individuals can proactively monitor and protect organizations from emerging cybersecurity threats

The CompTIA CySA+ certification is instrumental in preparing cybersecurity professionals for a career focused on threat detection and incident response. It equips them with the skills and knowledge necessary to safeguard organizations from a wide range of security threats. With the increasing sophistication of cyber attacks, the CySA+ certification plays a critical role in building a workforce that can effectively identify and mitigate security risks.

## PenTest+ certification

CompTIA PenTest+ is a certification that focuses on penetration testing and ethical hacking. It's designed for cybersecurity professionals who want to specialize in identifying vulnerabilities, exploiting security weaknesses, and assessing network security.

The CompTIA PenTest+ certification covers several domains, and each domain has a specific percentage of questions in the exam. The primary domains are as follows:

- **Planning and scoping**: This involves the planning of penetration tests, defining the scope, and identifying targets

- **Information gathering and vulnerability identification**: This covers gathering information about the target systems, identifying vulnerabilities, and assessing potential risks

- **Attacks and exploits**: This focuses on conducting various penetration tests, including vulnerability exploitation, post-exploitation techniques, and social engineering

- **Penetration testing tools**: This includes the use of various penetration testing tools and their practical application

- **Reporting and communication**: This encompasses the reporting of findings, communication with stakeholders, and the ethical and legal aspects of penetration testing

The CompTIA PenTest+ certification assesses the following skills and knowledge:

- Conducting penetration tests and vulnerability assessments
- Identifying security weaknesses and vulnerabilities
- Exploiting vulnerabilities and conducting ethical hacking
- Analyzing and interpreting penetration test results
- Utilizing penetration testing tools and frameworks
- Reporting and communicating security findings effectively

Why should you choose CompTIA PenTest+? Let's take a look at the benefits:

- **Vendor neutrality**: CompTIA certifications are vendor-neutral, meaning they don't focus on a specific technology or platform, making them versatile and applicable in various environments
- **Industry recognition**: CompTIA is a well-recognized certification body in the IT and cybersecurity industry
- **Career advancement**: The PenTest+ certification is an excellent choice for professionals looking to specialize in penetration testing and ethical hacking, as it validates their skills and knowledge in this crucial field
- **Cybersecurity expertise**: Penetration testing is a critical component of cybersecurity, and this certification equips professionals with the expertise to identify and mitigate security vulnerabilities
- **Comprehensive knowledge**: The exam covers a broad range of penetration testing topics, ensuring that certified professionals are well-prepared to tackle real-world security challenges
- **Continuing education**: CompTIA certifications require continuing education to stay current, which is essential in the ever-evolving field of cybersecurity

CompTIA PenTest+ is a valuable certification for professionals seeking to specialize in penetration testing and ethical hacking. It covers a wide range of topics and is recognized in the cybersecurity industry, making it an excellent choice for those looking to advance their careers in this field.

CompTIA is a leading force in the IT industry, offering a comprehensive suite of certifications that cater to professionals at all career stages. These certifications span topics such as IT fundamentals, hardware, software, security, and networking, allowing individuals to specialize in their areas of interest. CompTIA certifications are widely recognized by employers and serve as a valuable benchmark of IT competency. Whether you are an entry-level IT enthusiast looking to launch your career, a mid-level professional aiming to sharpen your skills, or an experienced IT specialist seeking to validate your expertise, CompTIA provides a certification tailored to your needs. These certifications have proven instrumental in advancing careers, enhancing job prospects, and promoting excellence in the IT field. Furthermore, CompTIA's commitment to fostering collaboration and advocating for industry best practices underscores its crucial role in the ever-evolving landscape of IT.

# EC-Council

The **International Council of E-Commerce Consultants**, commonly known as **EC-Council**, is a globally recognized leader in cybersecurity education and certification. Established in 2001, EC-Council has been instrumental in setting industry standards and equipping cybersecurity professionals with the knowledge and skills needed to combat cyber threats. The EC-Council certification website can be found at the following URL: `https://www.eccouncil.org/`. EC-Council currently has multiple certifications. With a focus on ethical hacking, penetration testing, and secure information practices, EC-Council has become a trusted authority in the field of cybersecurity, offering a wide range of certifications and training programs:

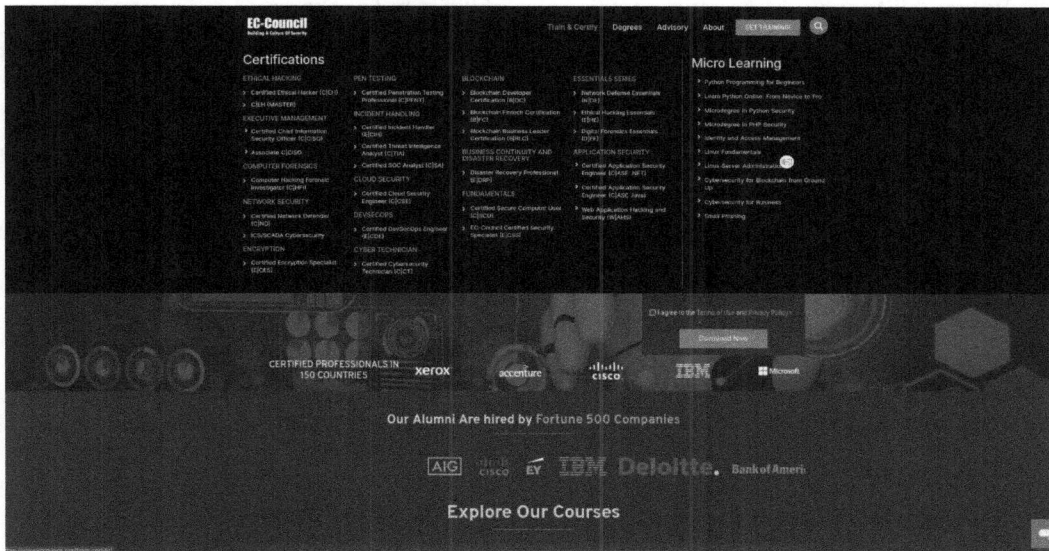

Figure 8.2 – EC-Council certifications

Let us explore them in detail.

## *Certified Ethical Hacker (C|EH)*

The EC-Council **C|EH** certification is one of the most recognized and respected credentials in the field of ethical hacking and cybersecurity. C|EH is designed for professionals who want to gain the skills and knowledge required to assess the security of computer systems and networks, identify vulnerabilities, and apply ethical hacking techniques to secure them. This certification equips individuals with the tools and techniques used by malicious hackers, allowing them to proactively defend against cyber threats.

The C|EH certification exam assesses candidates in various domains, including but not limited to the following:

- **Introduction to ethical hacking**: The basics of ethical hacking, hacking phases, and different types of hackers

- **Footprinting and reconnaissance**: Information gathering and footprinting techniques

- **Scanning networks**: Network discovery and scanning methods

- **Enumeration**: Gathering information about network services and vulnerabilities

- **Vulnerability analysis**: Assessing system vulnerabilities and weaknesses

- **System hacking**: Techniques for gaining access to systems

- **Malware threats**: Understanding malware and its countermeasures

- **Sniffing**: Packet capturing and analysis

- **Social engineering**: Techniques to manipulate individuals and gather information

- **Denial-of-service (DoS)**: DoS and **distributed denial of service (DDoS)** attacks and countermeasures

- **Session hijacking**: Techniques for hijacking network sessions

- **Hacking web servers**: Web server vulnerabilities and attacks

- **Hacking web applications**: Vulnerabilities in web applications and their exploitation

- **SQL injection**: Exploiting SQL database vulnerabilities

- **Hacking wireless networks**: Wireless network vulnerabilities and attacks

- **Evading intrusion detection systems (IDS), firewalls, and honeypots**: Techniques to bypass security measures

- **Buffer overflow**: Exploiting buffer overflow vulnerabilities

- **Cryptography**: Understanding encryption and cryptographic attacks

- **Penetration testing**: Conducting penetration tests and reporting

The C|EH certification is one of the most recognized and respected credentials in the field of ethical hacking and cybersecurity. C|EH is designed for professionals who want to gain the skills and knowledge required to assess the security of computer systems and networks, identify vulnerabilities, and apply ethical hacking techniques to secure them. This certification equips individuals with the tools and techniques used by malicious hackers, allowing them to proactively defend against cyber threats.

EC-Council stands as a formidable force in the realm of cybersecurity education and certification. Through a diverse portfolio of certifications such as C|EH, **Certified Chief Information Security Officer (C|CISO)**, and other certs, EC-Council addresses the ever-evolving challenges in the cybersecurity landscape. Its emphasis on ethical hacking and penetration testing has led to a global community of cybersecurity professionals well-versed in identifying vulnerabilities and safeguarding digital assets. EC-Council's impact extends beyond certification, as it actively contributes to the development of cybersecurity standards and practices. As the cybersecurity landscape continues to evolve, EC-Council remains at the forefront, equipping professionals with the skills and knowledge to defend against cyber threats effectively.

## Information Systems Audit and Control Association (ISACA)

**ISACA** is a globally recognized professional association dedicated to the fields of information systems governance, risk management, and cybersecurity. Established in 1969, ISACA has played a pivotal role in shaping industry standards and best practices, providing guidance to professionals in managing and securing IT systems. The ISACA website can be found at the following URL: `https://www.isaca.org/`. ISACA currently has 8 certifications. With a mission to advance the profession and help individuals and organizations navigate the complex world of technology, ISACA offers a range of certifications, resources, and a supportive community for information systems and cybersecurity professionals:

Figure 8.3 – ISACA certifications

Let us explore them in detail.

## *Certified Information Security Manager (CISM) certification*

The **CISM** certification, offered by ISACA, is a globally recognized credential tailored for professionals who design and manage an enterprise's information security program. CISM focuses on the management and governance aspects of information security, making it ideal for individuals in leadership roles. It validates expertise in information risk management, governance, incident response, and strategic alignment of security with organizational goals.

The CISM certification exam covers the following domains:

- **Information security governance**: Establishing and maintaining an information security governance framework and supporting processes

- **Information risk management**: Identifying and managing information security risks to achieve business objectives

- **Information security program development and management**: Establishing and managing the information security program

- **Information security incident management**: Planning, establishing, and managing the capability to respond to and recover from information security incidents

CISM certifies the following skills and knowledge:

- Information security governance and risk management

- Development and management of an information security program

- Information security incident management and response

- Strategic alignment of security with organizational goals

- Compliance with regulatory requirements

- Effective management and governance of information security

Let us look at the benefits of CISM:

- **Industry recognition**: CISM is widely recognized and respected globally, often sought after for leadership roles in information security

- **Management focus**: The certification emphasizes the management and governance aspects of security, making it suitable for professionals in leadership positions

- **Career advancement**: CISM is a significant asset for career advancement, particularly for those aspiring to become CISOs or senior security leaders

- **Compliance expertise**: CISM holders excel in compliance and risk management, which are vital in today's regulatory landscape
- **Global network**: ISACA's global community provides resources, networking opportunities, and continuous professional development

The CISM certification is a highly esteemed credential for professionals in information security management. It verifies the knowledge and skills required to establish and manage an organization's information security program. CISM-certified professionals are equipped to address the strategic aspects of information security, making them valuable assets for organizations seeking effective leadership in information risk management and governance.

ISACA stands as a leading authority in the domains of information systems, audit, control, and cybersecurity. The association's certifications, including **Certified Information Systems Auditor (CISA)** and CISM, have become industry benchmarks, reflecting the expertise of professionals in information assurance and management. ISACA also provides valuable resources, research, and guidelines to navigate the evolving landscape of technology and security. With a global presence and a commitment to professional growth and ethical practices, ISACA continues to be a vital resource for individuals and organizations seeking to excel in information systems governance, risk management, and cybersecurity.

## The International Information System Security Certification Consortium (ISC2)

**ISC2** is a globally recognized and esteemed organization dedicated to advancing the field of information security. Founded in 1989, ISC2 has played a pivotal role in shaping the cybersecurity landscape, fostering a community of certified professionals committed to protecting critical information systems and data. The organization is renowned for its rigorous certifications, including the **Certified Information Systems Security Professional (CISSP)**, which is considered a benchmark for security professionals worldwide. The ISC2 website can be found at the following URL: `https://www.isc2.org/`. ISC2's mission is to empower and certify individuals who are on the front lines of the cybersecurity battle, ensuring the integrity, confidentiality, and availability of critical information. ISC2 currently has 15 certifications. The following certifications are the ones to look at to cement your cybersecurity architecture career:

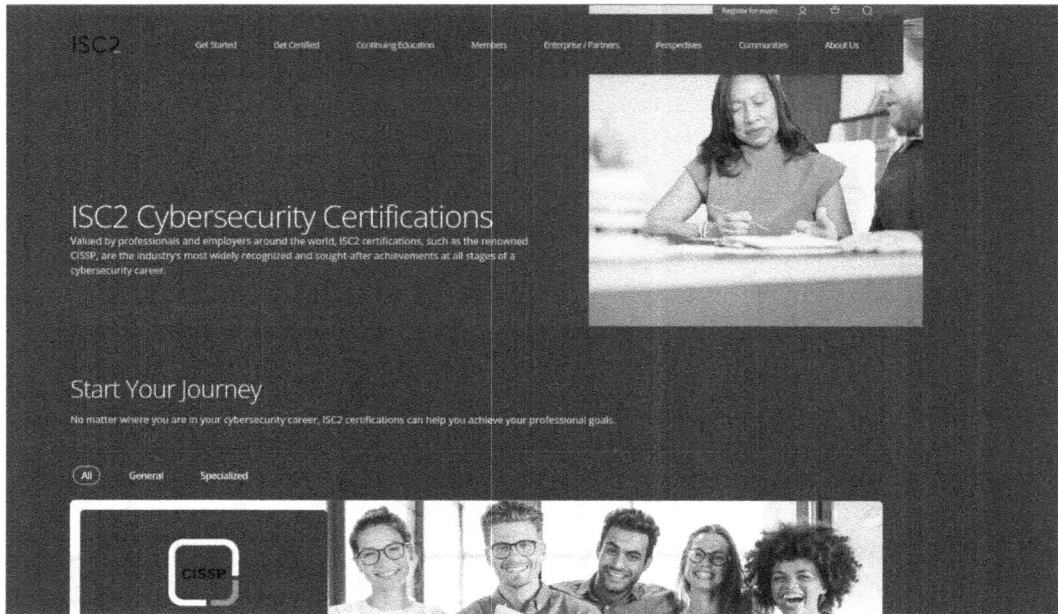

Figure 8.4 – ISC2 certifications

Let us explore them in detail.

## *CISSP certification*

The CISSP certification, offered by ISC2, is a globally recognized credential for information security professionals. It validates expertise in designing, implementing, and managing a comprehensive security program. CISSP is well regarded in the field of cybersecurity and is ideal for individuals who aim to excel in roles related to security architecture, engineering, and management.

The CISSP certification exam assesses knowledge in eight key domains:

- **Security and risk management**: Principles of governance, compliance, ethics, and security policies
- **Asset security**: Protecting the confidentiality, integrity, and availability of information assets
- **Security architecture and engineering**: Building and maintaining secure systems and environments
- **Communication and network security**: Protecting the transmission and storage of information
- **IAM**: Controlling access and managing identity
- **Security assessment and testing**: Designing and validating security measures
- **Security operations**: Foundational concepts of security operations
- **Software development security**: Integrating security into the software development process

CISSP certifies the following skills and knowledge:

- Security program design and management
- Security architecture, engineering, and models
- Risk management and governance
- Security policy development and implementation
- IAM
- Cryptography and network security
- Security assessment and testing
- Incident response and recovery

The CISSP certification is of paramount importance for security architecture due to the following reasons:

- **Comprehensive knowledge**: CISSP provides a broad understanding of security principles and best practices across various domains, including security architecture.
- **Security design expertise**: CISSP-certified professionals have the knowledge to design secure systems, applications, and networks.
- **Risk management**: Security architecture is closely tied to risk management. CISSP equips individuals with risk assessment and management skills, critical in architectural decisions.
- **Security models**: CISSP covers security models and frameworks, helping professionals choose the most suitable model for their architecture.
- **Compliance and policy**: CISSP addresses security policies and legal and regulatory compliance, which are integral to security architecture.
- **Networking and cryptography**: Security architecture often involves secure communication and data protection, areas covered by CISSP.
- **Business continuity**: CISSP addresses incident response and recovery, essential for ensuring the resilience of security architecture.

The CISSP certification is a prestigious credential that holds immense value for professionals in security architecture. It signifies a comprehensive understanding of information security concepts and is highly regarded in the industry. For security architects, CISSP provides the foundational knowledge and skills needed to design and implement secure and resilient systems, making it an essential certification for those aiming to excel in this field.

## CISSP-ISSAP certification

The CISSP-ISSAP is a specialized certification offered by ISC2. It is designed for professionals who possess the CISSP certification and have expertise in designing and implementing security solutions and systems. The CISSP-ISSAP focuses specifically on security architecture, making it an ideal choice for individuals who want to excel in this critical aspect of information security.

> **Prerequisite**
>
> To pursue the CISSP-ISSAP certification, candidates must hold the CISSP certification, showcasing their foundational knowledge in information security.

The CISSP-ISSAP certification exam assesses knowledge in four key domains related to security architecture:

- **Access control systems and methodology**: Designing access control systems and implementing access controls to protect resources

- **Communications and network security**: Designing secure communication and network systems, including secure protocols and technologies

- **Cryptography**: Implementing cryptographic solutions for securing data and communication

- **Security architecture analysis**: Analyzing security architectures and identifying vulnerabilities and risks, as well as proposing solutions

CISSP-ISSAP certifies the following skills and knowledge:

- Security architecture design and analysis

- Access control systems and methodologies

- Secure communication and network design

- Cryptographic solutions and their implementation

- Identifying vulnerabilities and proposing solutions

The CISSP-ISSAP certification is crucial for security professionals focusing on information systems security architecture for several reasons:

- **Specialization**: CISSP-ISSAP is a specialized certification that demonstrates expertise in security architecture, providing recognition of advanced knowledge

- **Career advancement**: For individuals aiming for roles as senior security architects or consultants, CISSP-ISSAP is a valuable credential

- **Security architecture expertise**: CISSP-ISSAP holders have in-depth knowledge of designing and analyzing security architectures, making them indispensable for organizations

- **Risk management**: A strong emphasis is placed on analyzing security architecture to identify vulnerabilities and risks, crucial for managing information security risk

- **Secure communication**: CISSP-ISSAP covers secure communication design, which is essential for safeguarding data in transit

The CISSP-ISSAP certification is a highly specialized credential for professionals who want to excel in information systems security architecture. It provides the recognition of advanced skills in designing, analyzing, and proposing secure architectures, making it essential for individuals pursuing senior security architecture roles and offering valuable expertise in the field of information security.

ISC2 stands as a bastion of excellence in the realm of information security, offering a portfolio of highly respected certifications that are synonymous with expertise in the field. The organization's commitment to education, certification, and the development of best practices has made it a leader in the global cybersecurity community. Through its rigorous certification programs, ISC2 not only recognizes the proficiency of security professionals but also serves as a driving force behind the advancement of cybersecurity practices. As the world becomes increasingly interconnected and digital threats continue to evolve, ISC2 remains dedicated to preparing professionals to defend against emerging challenges, ultimately safeguarding the digital realm.

## Global Information Assurance Certification (GIAC)

**Global Information Assurance Certification**, widely known as **GIAC**, is a prestigious and globally recognized entity that specializes in information security education and certification and be found at the following URL: https://www.sans.org/cyber-security-certifications/. Established in 1999, GIAC has earned a stellar reputation for its rigorous and practical certification programs, each meticulously designed to equip cybersecurity professionals with the skills and knowledge required to defend against the ever-evolving landscape of cyber threats. With a mission to provide cutting-edge information security certifications, GIAC has become a cornerstone in the field, fostering a community of certified experts who are at the forefront of safeguarding data and information systems. GIAC offers over 85+ classes that align to over 40 certifications. There are far too many certifications, but the ones listed here are a starting point to advance your cybersecurity architecture career:

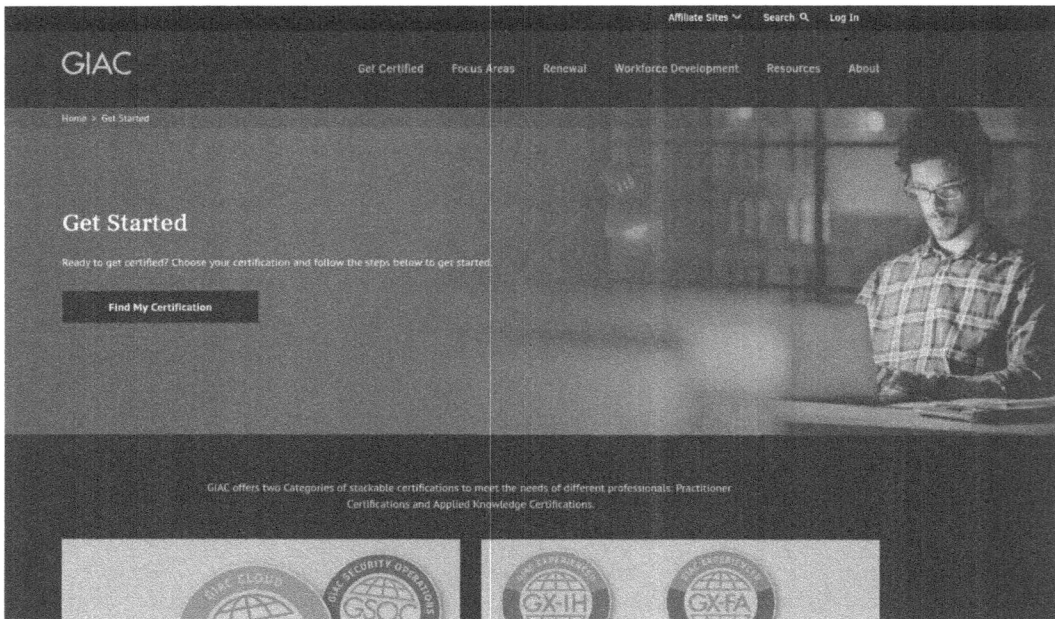

Figure 8.5 – GIAC certifications

Let us explore them in detail.

## *GIAC Certified Enterprise Defender (GCED) certification*

**GCED** is a certification offered by the GIAC organization. It is designed for professionals responsible for defending their organization's systems and networks. The GCED certification focuses on the practical aspects of enterprise defense, making it a valuable credential for individuals in roles related to security operations and incident response.

> **Note**
> While there are no specific prerequisites for taking the GCED certification exam, GIAC recommends that candidates have a strong foundation in information security concepts.

The GCED certification exam assesses knowledge in the following key domains:

- **Security essentials**: Foundational security concepts and principles
- **Network and host-based security**: Understanding network and host security and implementing measures

- **Enterprise defensive solutions**: Implementing defensive security measures, including firewalls and intrusion detection/prevention systems
- **Incident handling and threat detection**: Detecting and responding to security incidents and threats
- **Vulnerability assessment and penetration testing**: Assessing vulnerabilities and conducting penetration testing
- **Security operations and monitoring**: Effective security operations and monitoring strategies
- **Security policy and program management**: Developing and managing security policies and programs

GCED certifies the following skills and knowledge:

- Network and host security
- Incident detection and response
- Vulnerability assessment and penetration testing
- Security policy development and management
- Security operations and monitoring

The GCED certification is important for professionals in the field of enterprise defense for several reasons:

- **Practical expertise**: GCED focuses on real-world skills and knowledge relevant to securing enterprise environments, making it a practical and valuable certification
- **Incident response**: With an emphasis on incident detection and response, GCED equips professionals with the skills needed to effectively respond to security incidents
- **Defensive solutions**: GCED covers the implementation of defensive security solutions, which is essential for preventing and mitigating security threats
- **Vulnerability assessment**: Knowledge of vulnerability assessment and penetration testing is critical for identifying and addressing weaknesses in an organization's systems
- **Policy and program management**: The certification also addresses security policy development and program management, key elements in establishing a robust security posture

The GCED certification is a valuable credential for professionals responsible for securing their organization's systems and networks. It focuses on practical skills, including incident detection and response, vulnerability assessment, and security policy management. As such, it equips individuals with the knowledge and expertise needed to effectively defend enterprises against modern security threats.

## *GIAC Certified Web Application Defender (GWEB) certification*

**GWEB** is a certification offered by the GIAC organization. It is designed for professionals who work with web applications, such as developers, security analysts, and penetration testers. The GWEB certification focuses on web application security, providing the knowledge and skills required to identify and mitigate vulnerabilities in web applications.

> **Note**
> While there are no specific prerequisites for taking the GWEB certification exam, GIAC recommends that candidates have a strong foundation in web application security concepts.

The GWEB certification exam assesses knowledge in the following key domains:

- **Web application technologies**: Understanding web application architecture, components, and technologies

- **Web application security fundamentals**: Fundamentals of web application security, including common vulnerabilities and threats

- **Web application attacks and defenses**: In-depth knowledge of various web application attacks and how to defend against them

- **Secure software development and Software Development Life Cycle (SDLC)**: Secure software development practices and integrating security into the SDLC

- **Web application security assessment**: Conducting security assessments of web applications, including vulnerability scanning and penetration testing

- **Security policies and procedures**: Development and implementation of web application security policies and procedures

GWEB certifies the following skills and knowledge:

- Web application architecture and technologies

- Understanding common web application vulnerabilities and threats

- In-depth knowledge of web application attacks and how to defend against them

- Secure software development practices

- Conducting security assessments of web applications

- Developing and implementing web application security policies and procedures

The GWEB certification is important for professionals working with web applications for several reasons:

- **Web application security expertise**: GWEB focuses specifically on web application security, providing professionals with in-depth knowledge and skills in this critical area

- **Vulnerability identification**: Professionals holding the GWEB certification can effectively identify vulnerabilities in web applications, which is crucial for securing them

- **Secure development**: The certification covers secure software development practices and integrating security into the SDLC, fostering a proactive approach to security

- **Comprehensive coverage**: GWEB addresses various aspects of web application security, from fundamentals to security assessments, making it a well-rounded certification

The GWEB certification is a valuable credential for professionals working with web applications. It equips individuals with the knowledge and skills needed to understand web application security, identify vulnerabilities, and implement secure software development practices. In a digital landscape where web applications are prevalent, GWEB-certified professionals play a crucial role in securing these applications against potential threats and attacks.

## GIAC Defensible Security Architecture (GDSA) certification

**GDSA** is a certification offered by the GIAC organization. This certification is designed for professionals involved in designing and implementing secure and defensible network architectures. GDSA focuses on developing expertise in creating robust security architectures that can withstand and mitigate modern cybersecurity threats.

> **Note**
> While there are no specific prerequisites for taking the GDSA certification exam, candidates are expected to have a solid understanding of network security concepts and experience in designing and implementing network security solutions.

The GDSA certification exam assesses knowledge in the following key domains:

- **Security architecture fundamentals**: Understanding the core principles and concepts of security architecture

- **Network security technologies**: In-depth knowledge of network security technologies, including firewalls, intrusion detection systems, and VPNs

- **Secure network design**: Designing secure and defensible network architectures that protect against cyber threats

- **Security operations and monitoring**: Implementing effective security operations and monitoring strategies to ensure the ongoing security of the network

- **Security policy and program development**: Developing and managing security policies and programs that align with the security architecture

GDSA certifies the following skills and knowledge:

- Understanding security architecture fundamentals

- In-depth knowledge of network security technologies

- Designing secure and defensible network architectures

- Implementing effective security operations and monitoring

- Developing and managing security policies and programs

The GDSA certification is important for professionals involved in security architecture for several reasons:

- **Defensible security architecture**: GDSA focuses specifically on creating security architectures that are robust and defensible, which is crucial in the modern threat landscape

- **In-depth knowledge**: The certification covers a wide range of security architecture topics, providing professionals with comprehensive knowledge

- **Effective design**: GDSA equips professionals with the skills needed to design secure networks that can withstand cyber threats and attacks

- **Security policy development**: GDSA also covers the development of security policies and programs, ensuring that security is an integral part of an organization's strategy

The GDSA certification is a valuable credential for professionals involved in designing and implementing secure network architectures. It focuses on core security architecture principles, network security technologies, secure design, and effective security operations. In an era where cyber threats are continuously evolving, GDSA-certified professionals play a critical role in creating and maintaining security architectures that can protect organizations from these threats.

GIAC serves as a formidable force in the realm of cybersecurity education and certification, offering a diverse array of certifications that are highly respected across industries. The organization's commitment to practical, hands-on assessments ensures that certified professionals possess the practical skills and knowledge required to tackle real-world cybersecurity challenges. GIAC-certified individuals are recognized for their proficiency in identifying vulnerabilities, implementing security measures, and safeguarding critical information. In an era where cyber threats continue to escalate in sophistication, GIAC plays a crucial role in preparing professionals to protect organizations and individuals from an array of digital risks, securing the digital future.

# Cloud Vendor – Amazon Web Services/Azure/Google Cloud Platform

Cloud computing has become the backbone of modern IT infrastructure, and cloud vendors such as **Amazon Web Services (AWS)**, Microsoft Azure, and **Google Cloud Platform (GCP)** are pivotal players in this ecosystem. Each of these cloud giants offers a comprehensive range of cloud services and solutions to meet the diverse needs of businesses and organizations. To ensure that professionals are equipped to harness the full potential of these platforms, AWS, Azure, and GCP provide certification programs that validate individuals' expertise in cloud services, architecture, and best practices. These certifications are highly regarded in the IT industry, serving as a testament to a professional's ability to design, manage, and secure cloud-based solutions.

## *AWS certifications*

AWS offers a comprehensive range of certifications that cater to different aspects of cloud computing, including security. For professionals looking to specialize in security careers within the AWS ecosystem, there are several certifications to consider. These certifications are essential in demonstrating expertise in securing AWS environments, protecting data, and mitigating security threats. The AWS website can be found at the following URL: `https://aws.amazon.com/certification/exams/`. AWS currently has 13 certifications. Let's break down the AWS certifications that are relevant to a security-focused career:

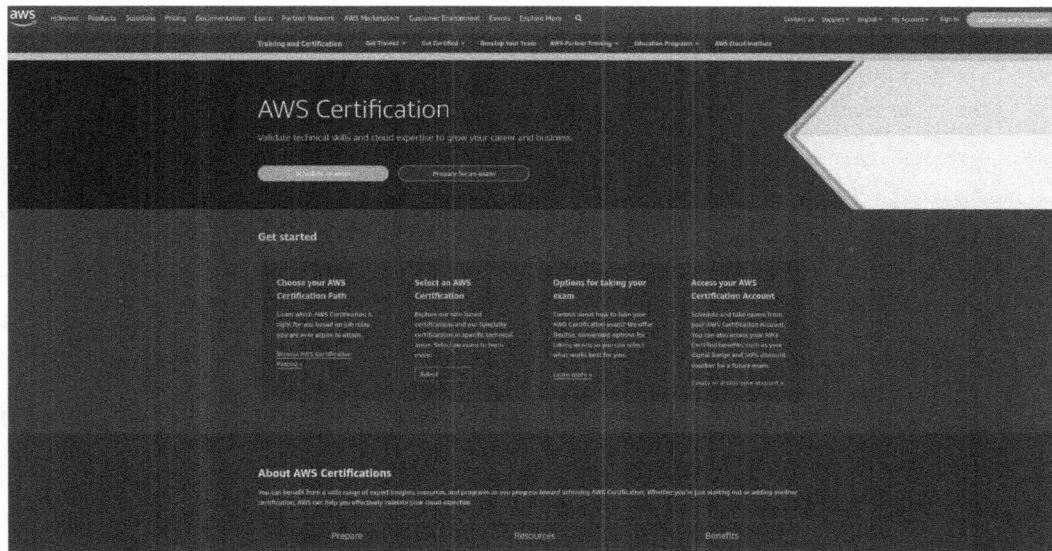

Figure 8.6 – AWS certifications

Let us explore them in detail.

## AWS Certified Security – Specialty

This certification is specifically designed for security professionals. It covers a wide array of security topics related to AWS, including IAM, encryption, incident response, and compliance:

- **Audience**: It is ideal for security professionals who want to demonstrate their expertise in securing AWS environments and services
- **Key topics**: These include security and compliance, IAM, encryption, and incident response

## AWS Certified Solutions Architect – Professional

While not solely focused on security, the AWS Certified Solutions Architect – Professional certification covers security extensively in the context of architecture design and best practices. Security is a vital component of this certification:

- **Audience**: It is recommended for solutions architects, security architects, and professionals responsible for designing and implementing secure AWS architectures
- **Key topics**: These include security best practices, IAM, encryption, and compliance

## AWS Certified DevOps Engineer – Professional

This certification focuses on automating security practices within a DevOps pipeline, emphasizing the integration of security into the development and deployment process:

- **Audience**: It is suitable for DevOps engineers, security professionals, and anyone involved in automating security in a DevOps environment
- **Key topics**: These include security automation, continuous security monitoring, and compliance

## AWS Certified SysOps Administrator – Associate

While not purely a security certification, it includes essential security topics. It covers system administration within AWS, including security tasks such as access control, data protection, and compliance:

- **Audience**: It is designed for system administrators, but it's valuable for those responsible for managing security within AWS
- **Key topics**: These include IAM, data protection, compliance, and monitoring

## AWS Certified Cloud Practitioner

This is a foundational certification providing an overview of AWS services and best practices, including basic security concepts and compliance:

- **Audience:** It is suitable for individuals new to AWS and those looking to understand the foundational security aspects of AWS
- **Key topics:** These include security best practices, compliance, and AWS service overview

In summary, AWS certifications offer a well-rounded approach to security within the AWS ecosystem. AWS Certified Security – Specialty is the primary choice for professionals looking to specialize in AWS security, but other certifications, such as the Solutions Architect – Professional and DevOps Engineer – Professional, include significant security components. These certifications are invaluable for security professionals looking to advance their careers in AWS cloud security.

## Microsoft Azure certifications

Microsoft Azure, one of the leading cloud providers, offers a variety of certifications that cater to different aspects of cloud computing, including security. For professionals looking to specialize in security careers within the Azure ecosystem, there are several certifications to consider. These certifications are essential in demonstrating expertise in securing Azure environments, protecting data, and mitigating security threats. The Azure certification website can be found at the following URL: `https://learn.microsoft.com/en-us/credentials/`. Microsoft has over 100 certifications. Here's a detailed analysis and breakdown of Azure certifications relevant to a security-focused career:

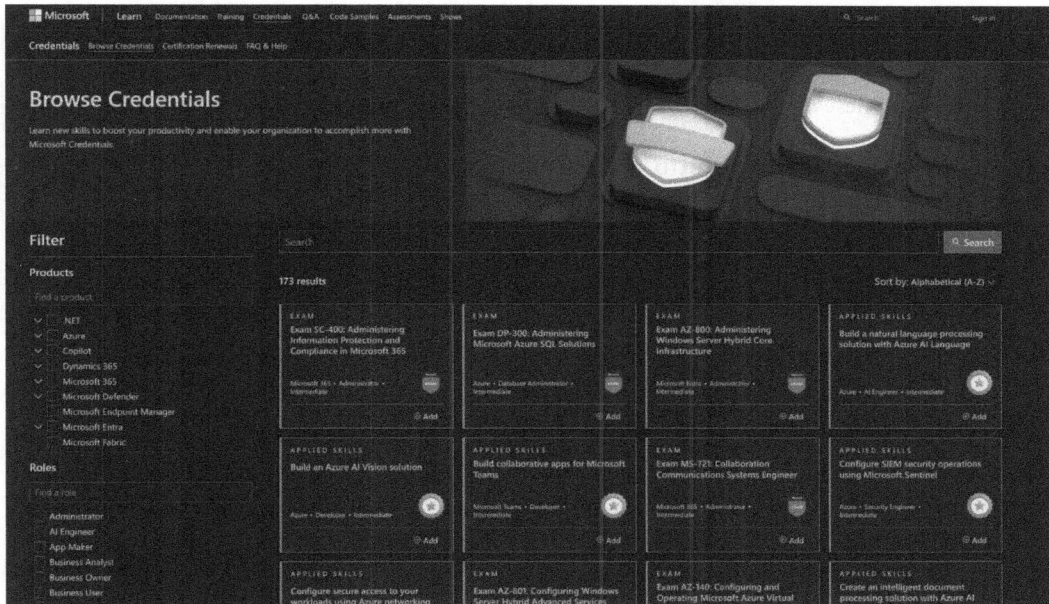

Figure 8.7 – Microsoft Azure certifications

Let us explore them in detail.

## Microsoft Certified: Azure Security Engineer Associate

This certification is tailor-made for security professionals who want to demonstrate their expertise in implementing and managing security controls on the Azure platform. It covers various security aspects, including IAM, data protection, and threat protection:

- **Audience**: This is ideal for security professionals responsible for implementing security controls and managing security within Azure environments

- **Key topics**: These include IAM, data protection, threat protection, security monitoring, and governance

## Microsoft Certified: Azure Administrator Associate

While not solely focused on security, this certification covers security extensively within the context of Azure administration. It includes key security topics such as IAM, data protection, and security monitoring:

- **Audience**: This is suitable for Azure administrators, including those responsible for configuring and managing security settings

- **Key topics**: These include IAM, data protection, security monitoring, and governance

## Microsoft Certified: Azure Solutions Architect Expert

This certification is focused on architecting solutions on the Azure platform, including security aspects. It's essential for solutions architects who need to design secure and compliant Azure environments:

- **Audience**: It is designed for solutions architects responsible for designing secure Azure solutions

- **Key topics**: These include security best practices, IAM, encryption, and compliance

## Microsoft Certified: Azure DevOps Engineer Expert

While primarily focused on DevOps and automation, this certification emphasizes integrating security practices into the DevOps pipeline. It covers topics related to automating security and compliance checks within Azure:

- **Audience**: It is suitable for DevOps engineers, security professionals, and anyone involved in integrating security into DevOps processes

- **Key topics**: These include security automation, continuous security monitoring, and compliance

In summary, Microsoft Azure certifications provide a comprehensive approach to security within the Azure cloud ecosystem. The Microsoft Certified: Azure Security Engineer Associate certification is the primary choice for professionals specializing in Azure security. Other certifications, such as Azure Administrator Associate and Azure Solutions Architect Expert, also include significant security components and are valuable for security professionals looking to advance their careers in Azure cloud security. These certifications are essential for professionals seeking to secure Azure environments effectively.

## GCP certifications

GCP offers a range of certifications that cover various aspects of cloud computing, including security. For professionals aiming to specialize in security careers within the GCP ecosystem, there are several certifications to consider. These certifications are essential for demonstrating expertise in securing GCP environments, protecting data, and mitigating security threats. The *Google Cloud Certification* website can be found at the following URL: `https://cloud.google.com/learn/certification`. GCP currently has 11 certifications. Here's a detailed analysis and breakdown of GCP certifications relevant to a security-focused career:

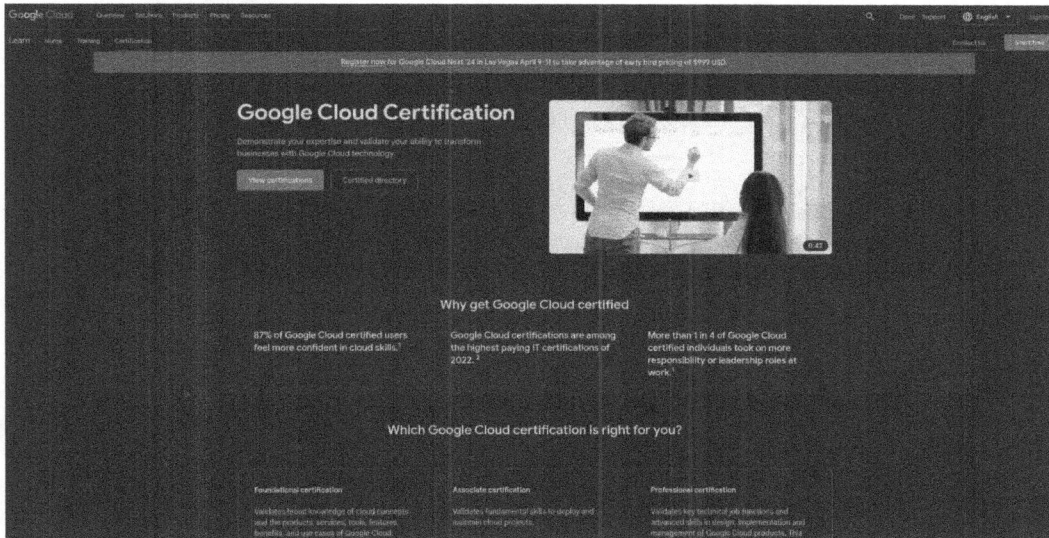

Figure 8.8 – Google Cloud certifications

Let us explore them in detail.

### Google Cloud Certified – Professional Cloud Security Engineer

This certification is designed for security professionals who want to demonstrate their expertise in designing and implementing security solutions on the GCP platform. It covers a wide array of security topics, including IAM, data protection, and threat detection:

- **Audience**: It is ideal for security professionals responsible for implementing security controls and managing security within GCP environments
- **Key topics**: These include IAM, data protection, threat detection, security monitoring, and compliance

## Google Cloud Certified – Professional Cloud Architect

While not solely focused on security, this certification covers security extensively within the context of architecture design and best practices. Security is a critical component of this certification, as architects need to design secure and compliant solutions:

- **Audience**: This is recommended for cloud architects, security architects, and professionals responsible for designing and implementing secure GCP architectures
- **Key topics**: These include security best practices, IAM, encryption, and compliance

## Google Cloud Certified – Professional DevOps Engineer

This certification focuses on automation and DevOps practices but emphasizes integrating security practices into the DevOps pipeline. It covers topics related to automating security and compliance checks within GCP:

- **Audience**: This is suitable for DevOps engineers, security professionals, and anyone involved in integrating security into DevOps processes
- **Key topics**: These include security automation, continuous security monitoring, and compliance

In summary, GCP certifications offer a well-rounded approach to security within the GCP ecosystem. The Google Cloud Certified – Professional Cloud Security Engineer certification is the primary choice for professionals specializing in GCP security. Other certifications, such as Professional Cloud Architect and Professional DevOps Engineer, also include significant security components and are valuable for security professionals looking to advance their careers in GCP cloud security. These certifications are essential for professionals seeking to secure GCP environments effectively.

AWS, Azure, and GCP certifications are emblematic of a professional's mastery of cloud technologies, offering an industry-recognized path to validate their cloud expertise. These certifications span a spectrum of roles and expertise levels, ranging from foundational to specialized, and cover various aspects of cloud computing, including cloud architecture, security, and data management. Earning these certifications not only enhances one's career prospects but also demonstrates their ability to leverage the full potential of cloud platforms for building, deploying, and managing scalable and secure solutions. As businesses increasingly rely on cloud services for innovation and agility, AWS, Azure, and GCP certifications have become essential in guiding professionals toward mastering the cloud and driving digital transformation.

This overview explores the key security-focused certifications offered by the leading cloud providers – AWS, Microsoft Azure, and GCP.

It begins by introducing AWS certifications relevant to security careers, such as the AWS Certified Security – Specialty certification designed specifically for security professionals. Other highlighted AWS certifications such as Solutions Architect – Professional and SysOps Administrator – Associate also incorporate security domains.

It then shifts to analyzing Microsoft Azure certifications, with a focus on the Azure Security Engineer Associate credential tailored for security implementation in Azure environments. Certifications such as Azure Administrator Associate and Azure Solutions Architect Expert are also noted for their security components.

Finally, it examines GCP certifications, emphasizing the Professional Cloud Security Engineer certification for those securing GCP platforms. Additional certifications such as Professional Cloud Architect and Professional DevOps Engineer are also called out for their security coverage.

### The Sherwood Applied Business Security Architecture (SABSA) certification

The **SABSA** certification provides training on a framework and methodology for developing enterprise security architecture aligned with business needs. SABSA was created by David Lynas and John Sherwood in the 1990s on the principle of building customizable security solutions tailored to an organization's requirements.

Its core components are as follows:

- **Six-layer model**: SABSA has six layers (contextual, conceptual, logical, physical, component, operational) spanning from high-level goals to technical specifics to integrate security with business objectives
- **Attribute profiling**: A key aspect is attribute profiling, which is used to define security characteristics such as confidentiality, availability, and compliance that guide strategy
- **Matrix approach**: The framework maps business needs to security controls using a matrix of architecture layers and operational aspects such as governance and risk management
- **Adaptive methodology**: The methodology is adaptable across organizations of differing sizes and scalable to changing business and technology landscapes
- **Life cycle mindset**: SABSA employs a life cycle approach for continuous improvement as risks and requirements evolve

SABSA certification develops expertise in applying the framework across multiple levels:

- **Foundation**: This covers fundamentals for individuals looking to gain a basic understanding of SABSA
- **Practitioner**: This takes a deep dive into applying SABSA principles and methodology to real-world scenarios and use cases
- **Advanced Practitioner**: This focuses on equipping individuals to provide SABSA training and advice to enterprises
- **Instructor**: This develops capabilities to teach others and certify companies on the framework
- **Consultant**: This enables strategic consulting on implementing the SABSA methodology

Here is the SABSA website:

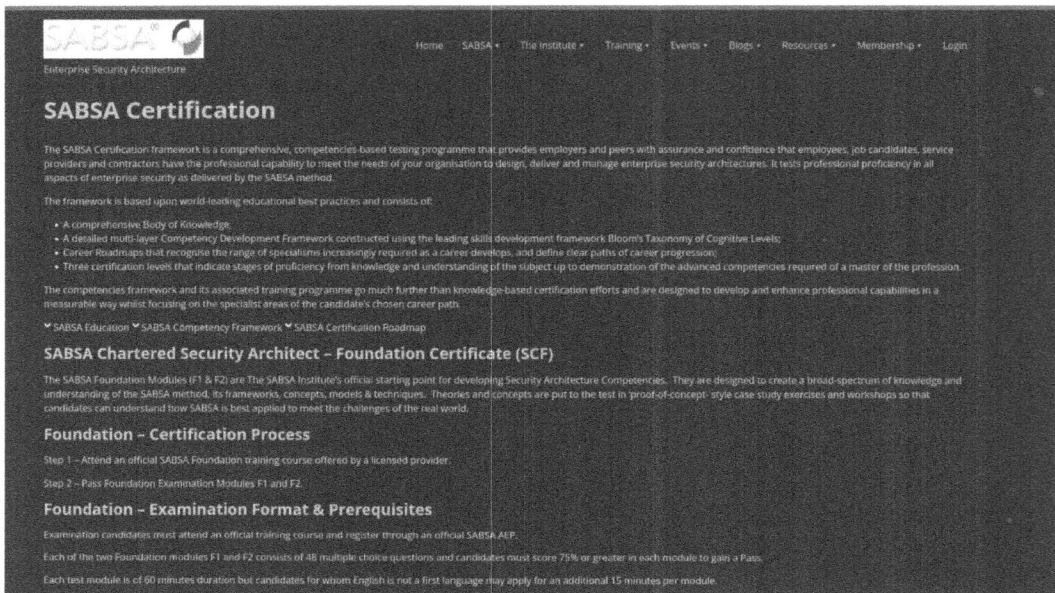

Figure 8.9 – SABSA certifications

The SABSA certification offers a layered, attribute-driven approach to developing flexible and adaptive security architecture aligned with evolving business needs. As threats and technologies continue to progress rapidly, frameworks such as SABSA provide organizations with the methodologies to embed security strategies into their DNA. The certification pathways allow professionals to gain expertise from foundation to strategic consulting levels, enabling widespread adoption across all industry verticals. For enterprises looking to build robust cybersecurity postures, having personnel trained on the nuances of mapping organizational imperatives to security capabilities can make a significant difference. As SABSA gains more prominence globally, professionals with this certification stand to be leaders steering their organizations securely into the digital age.

In summary, the overview provides a concise yet comprehensive analysis of key credentials offered by the top cloud providers for validating expertise in cloud security architecture, engineering, and operations. It serves as a valuable guide for security professionals seeking industry-recognized certifications to advance cloud security careers.

## Why get certified?

As you can see from the previous section, the certification landscape can be daunting, and the list of certifications referenced and highlighted only scratches the surface of what is potentially an option. As a hiring manager for several companies and agencies, while I have hired personnel with certifications, there have been times that I chose someone without. At times, it was experience or capabilities that trumped the certifications.

Certifications can provide valuable benefits but should not be the sole focus for cybersecurity architects and aspiring professionals. Here are some key points on the merits and limitations of certifications.

The benefits of certifications are as follows:

- **Validation of knowledge**: Certifications test and validate comprehension of key concepts, tools, and best practices. They provide *foreknowledge*, as Sun Tzu emphasized.

- **Career advancement**: Certifications can improve prospects for promotions, leadership roles, and higher salaries. They signal expertise to employers.

- **Industry recognition**: Certifications from respected bodies such as GIAC, ISC2, and CompTIA are globally recognized as benchmarks of competency.

- **Specialization**: Advanced certifications allow concentration in specific domains such as cloud security, ethical hacking, or risk management.

- **Continuous learning**: Renewal requirements ensure certified professionals stay updated on evolving threats and technologies.

The limitations of certifications are as follows:

- **No substitute for experience**: While certifications validate knowledge, real-world expertise is irreplaceable. Seasoned architects rely more on hard-won experience than credentials alone.

- **Over-focusing**: Collecting certificates as checkboxes without retaining knowledge is ineffective. Integrating and practicing concepts learned is key.

- **Rapidly evolving field**: New attack vectors and tools emerge constantly. Certifications cannot fully keep pace with the state-of-the-art.

- **Business alignment**: Certifications emphasize technical security but cannot substitute for understanding an organization's business needs and objectives.

- **Creativity not tested**: Designing adaptable solutions requires ingenuity beyond proven textbook concepts. Certification exams cannot assess creativity.

Certifications serve as milestones and guideposts but not destinations themselves. Cybersecurity architecture requires fusing broad technical capabilities with business acumen and creative problem-solving ability. The most effective architects view certifications as knowledge springboards to launch their practical experience and career growth, rather than chasing credentials alone. With the right expectations and career integration, certifications remain useful waypoints on the journey to becoming an accomplished cybersecurity architect.

# Certification considerations

Certification considerations can vary substantially for cybersecurity architects across different industries, government entities, and individual career goals. It's important to carefully evaluate potential certifications based on your specific circumstances. Here are some examples to illustrate the factors that can influence certification decisions.

## Industry variations

When evaluating which certifications to pursue, cybersecurity architects must carefully consider variations across industries. Different business sectors often have specific certification needs and preferences based on their operating environments and requirements:

- **In cloud computing**: AWS certifications such as Cloud Security – Specialty and Azure certifications such as Security Engineer Associate are highly sought after. These validate expertise in proprietary cloud security tools.

- **In banking/finance**: Certifications such as CISA, CISM, and CISSP that cover compliance, auditing, and risk management are preferential. These meet regulatory requirements such as the **Sarbanes-Oxley Act (SOX)**, the **Gramm-Leach-Bliley Act (GLBA)**, and the **Federal Financial Institutions Examination Council (FFIEC)**.

- **In software**: Application security certifications such as the CSSLP and GWAPT, which focus on secure coding, testing, and vulnerability management, are desirable. These align with the emphasis on software assurance.

- **As a consultant**: Platform-agnostic certifications such as CompTIA CySA+ and C|EH can demonstrate versatility across diverse client environments using varied tools/frameworks.

## Government requirements

When pursuing a cybersecurity architecture career working with government entities, specialized certification and clearance considerations come into play. The public sector often has stringent requirements when it comes to information security credentials and access:

- **For federal positions**: Certifications help but clearances such as **Top Secret/Special Compartmentalized Information (TS/SCI)** with polygraphs are often mandatory, especially in defense/intelligence agencies where classified access is required.

- **As a government contractor**: Certain approved credentials such as Security+ or CISSP may be explicitly required to bid on contracts, especially to provide IT services to federal agencies.

- **For state governments**: Baseline certifications such as Network+ or Security+ may be stipulated for IT security personnel. California requires CASPs for IT management roles.

## Cost considerations

The costs associated with cybersecurity certifications can heavily influence the decisions of aspiring architects. Top credentials require significant financial investment and the costs only increase as certifications get more advanced:

- Top certifications such as CISSP cost $699 just for the exam. Add $1,000+ for training materials and classes, and renewals add $125+ yearly costs.

- If paid by employers, certification costs are covered but time away from work is limited. Self-funded efforts allow more prep time but at significant out-of-pocket costs.

- The earnings increase from a certification needs to justify the investment. Entry-level certifications can pay for themselves quickly. Advanced ones require calculating **return on investment (ROI)**.

In summary, industry, government, and personal factors should all be carefully weighed when evaluating certifications. The prudent cybersecurity architect selects certifications tailored to their unique professional situation and goals. One-size-fits-all rarely applies for certifications in this field.

# Summary

In this chapter, we learned that certifications can serve as valuable milestones for cybersecurity professionals when approached strategically. They validate knowledge and signal expertise, but real-world experience remains irreplaceable. Certification value depends on prudent selection aligned with industry, government, and personal career factors. Costs should be weighed against potential benefits.

While certifications have limitations and should not be the sole focus, they can augment practical skills when used judiciously. Renewal requirements promote continuous learning about evolving threats. With the right expectations, certifications provide useful guideposts, not destinations themselves, on the journey to becoming an accomplished cybersecurity architect.

The most effective architects view certifications as knowledge springboards to launch their hands-on expertise. They understand one size does not fit all roles or careers. Savvy professionals evaluate certifications based on their industry's specialized needs, applicable government mandates, and personal growth goals. With careful consideration of their unique circumstances, certifications remain valuable waypoints. However, integrating new concepts via practice is the path to mastery.

Sun Tzu emphasized that true foreknowledge comes from experienced people, not just paper credentials. While certifications validate skills, real-world experience is irreplaceable for senior architects. The prudent cybersecurity professional sees certifications as milestones to augment hard-won knowledge, not to replace it. With the right strategic alignment, certifications grant useful foreknowledge to combine with practice for victory. But credentials alone cannot substitute for creativity, business acumen, and versatility, which distinguish cybersecurity architecture masters.

The next chapter cuts through the tool overload facing modern architects by providing clarity on tool categories, selection principles, and business factors.

# Part 3: Advancements

The concluding part of the book focuses on elevating cybersecurity architecture expertise to greater heights. It explores both technical and personal advancements enabling architects to implement evolving protections with increasing mastery.

*Chapter 9* and *Chapter 10* provide strategies for rationalizing and future-proofing complex security technology toolsets tailored to organizations' unique needs. *Chapter 11* revisits cybersecurity best practices, providing guidance on customizing gold-standard controls for maximum effectiveness per environment.

Recognizing adaptability is imperative in a field of constant change, *Chapter 12* discusses cultivating versatile mindsets and skills without sacrificing secure foundations. *Chapter 13* synthesizes lessons into architectural considerations for holistic security strategies.

Finally, *Chapter 14* continues from the previous chapter and also summarizes key insights, and looks ahead to future learning. It aims to inspire you to continue advancing architecture through a lifelong dedication to this essential and ever-evolving profession.

Together, these chapters detail avenues for cybersecurity architects to build upon foundations and career pathways. They provide perspectives to reach the pinnacle of architecture excellence, securely empowering organizational success amid emerging opportunities and obstacles.

This part has the following chapters:

- *Chapter 9, Decluttering the Toolset – Part 1*
- *Chapter 10, Decluttering the Toolset – Part 2*
- *Chapter 11, Best Practices*

- *Chapter 12, Being Adaptable as a Cybersecurity Architect*
- *Chapter 13, Architecture Considerations – Design, Development, and Other Security Strategies – Part 1*
- *Chapter 14, Architecture Considerations – Design, Development, and Other Security Strategies – Part 2*

# 9
# Decluttering the Toolset – Part 1

*"Water shapes its course according to the nature of the ground over which it flows; the soldier works out his victory in relation to the foe whom he is facing."*

*– Sun Tzu*

*"By method and discipline are to be understood the marshaling of the army in its proper subdivisions, the graduations of rank among the officers, the maintenance of roads by which supplies may reach the army, and the control of military expenditure."*

*– Sun Tzu*

*"Hence the skillful fighter puts himself into a position which makes defeat impossible, and does not miss the moment for defeating the enemy."*

*– Sun Tzu*

*"Thus it is that in war the victorious strategist only seeks battle after the victory has been won, whereas he who is destined to defeat first fights and afterwards looks for victory."*

*– Sun Tzu*

In the previous chapter, we discussed trying to make sense of the certification landscape. In this chapter, we'll look at the landscape of tools available to cybersecurity architects.

As Sun Tzu emphasized, the shrewd combatant tailors their approach to the unique circumstances at hand, rather than relying on prescribed formulas alone. This adaptable mindset is equally crucial for cybersecurity architects to select tools to secure their organization's digital assets and data.

With hundreds of products flooding the market, it is easy to get overwhelmed by the hype of the latest offerings. However, the most effective cybersecurity architects thoughtfully curate their security toolkit based on their unique threat landscape, business drivers, and operational constraints – just as a river shapes its course according to the ground it flows over.

Rather than blindly adopting every new tool, the discerning cybersecurity architect focuses on developing a tailored toolkit to address their specific challenges. They marshal resources judiciously, striking the right balance between capabilities and costs. Through rigorous methods and discipline, they invest wisely to get the optimal return on security.

This chapter will explore how to thoughtfully assemble your cybersecurity architecture toolkit by filtering through solutions to find the right fit. It emphasizes understanding your distinct vulnerabilities and risk profile first, then matching appropriate defenses accordingly. With the proper toolkit in hand, you can nimbly respond to any adversary that appears, seizing opportunities to strengthen protections before trouble arises. By preparing the perfect set of tools in advance, victory is assured.

The chapter covers the following topics:

- What's in the toolbox?

## Technical requirements

This chapter was originally going to be a lab/exercise-heavy chapter that allowed you to explore various tools that can be used and deployed within the enterprise but still be accessible in the home lab. Well, the labs ended up creating a chapter that was over 200 pages in length and consisted of pictures and step-by-step instructions. This caused the editors to have a mild heart attack. All kidding aside, these labs are still a critical part of this chapter. With that in mind and understanding the need to have them available, Packt has made them available at the following GitHub link: `https://github.com/PacktPublishing/Cybersecurity-Architects-Handbook`.

The labs associated with this chapter include the following labs and exercises:

- **Lab 1**: Microsoft Threat Modeling Tool
- **Lab 2**: OWASP Threat Dragon
- **Lab 3**: Intrusion detection/prevention systems using Snort
- **Lab 4**: Firewall configuration using OPNsense
- **Lab 5**: SIEM solution using Graylog
- **Lab 6**: Antivirus software implementation using ClamAV
- **Lab 7**: Endpoint detection and response using Wazuh
- **Exercise**: Exercise – setting up and configuring Keycloak for IAM
- **Lab 8**: Data encryption with VeraCrypt

- **Lab 9**: Vulnerability scanning with OpenVAS
- **Lab 10**: Security configuration management using Ansible
- **Exercise**: Patch management with WSUS
- **Lab 11**: Digital forensics with The Sleuth Kit and Autopsy
- **Lab 12**: Incident response with Security Onion
- **Exercise**: Static application security testing with SonarQube
- **Lab 13**: Dynamic application security testing with OWASP ZAP
- **Lab 14**: Setting up and securing an AWS environment
- **Lab 15**: Implementing and configuring a GRC tool
- **Lab 16**: Penetration testing with Kali Linux and Metasploit
- **Lab 17**: Security automation with StackStorm

# What's in the toolbox?

Selecting the right tools is fundamental to building an effective cybersecurity architecture. With the overwhelming array of solutions on the market, architects must thoughtfully curate a toolkit tailored to their organization's specific risks, constraints, and use cases.

Rather than reactively adopting every new technology, discerning professionals take a systematic approach based on established frameworks such as NIST or MITRE ATT&CK. This provides a stable taxonomy for evaluating tools by common categories and security functions.

The following sections will explore major classes of security tools, providing examples and analyzing their purpose within a defense-in-depth toolkit. While not exhaustive, these categories encompass core solutions for threat detection, prevention, and response. In addition, the various labs and exercises associated with each tool set vary in complexity, from basic to more advanced, but all of them should be accessible to you.

By filtering through the hype and noise, architects can assemble a lean but potent arsenal of mutually reinforcing safeguards. Just as military leaders must select equipment suited for the battlefield, cybersecurity tool selection demands matching defenses to your terrain. With a precise understanding of needs and options, victory becomes simply a matter of preparation and execution.

## Threat modeling and risk assessment tools

Threat modeling and risk assessment tools are essential for cybersecurity architects to employ early in the system design process. These tools provide a systematic methodology for proactively identifying and analyzing potential threats, vulnerabilities, and risks before production deployment. Using threat modeling helps security architects preemptively address risks when mitigation costs are lower.

The key benefits of using threat modeling and risk assessment tools are as follows:

- **Visualize the system architecture**: Create detailed diagrams of components, data flows, trust boundaries, and interactions. This provides visibility into the *attack surface* that's been exposed.

- **Methodically identify risks**: Leverage libraries of known threats and automated analysis rules to reveal potential issues such as data leaks, injection flaws, or insufficient authentication.

- **Prioritize mitigation**: Quantify and rank risks to focus efforts on addressing high-probability and high-impact threats first.

- **Guide secure design**: Threat modeling results steer architectural decisions, security control selection, and requirements for secure coding practices.

- **Cost optimization**: Fixing issues earlier in the **software development life cycle (SDLC)** is exponentially cheaper than later remediation. Threat modeling saves resources.

- **Meet compliance requirements**: Demonstrating systematic threat analysis helps satisfy industry standards such as PCI DSS, ISO 27001, and NIST.

- **Continuous improvement**: Updating the threat model throughout the SDLC adapts defenses as risks evolve.

- **Collaboration**: Threat modeling reports and diagrams provide objective artifacts to communicate risks with stakeholders and drive mitigation.

By facilitating continuous, comprehensive analysis of risks from design through deployment, integrating threat modeling practices lays a foundation for building secure systems resistant to real-world attacks. Threat modeling and risk assessment tools empower architects to orchestrate a proactive defense, rather than reacting to threats. Threat modeling and risk assessment tools play a critical role in proactively identifying vulnerabilities early in the system development life cycle. By enabling architects to systematically analyze potential threat vectors and exposures, these tools help address risks before production deployment. Two prominent examples are **Microsoft Threat Modeling Tool** and **OWASP Threat Dragon**.

## Network defense and monitoring tools

Examples of such tools are **intrusion detection systems (IDSs)** such as Snort, **intrusion prevention systems (IPSs)**, network firewalls, and **security information and event management (SIEM)** systems such as Splunk or IBM QRadar.

Their purpose is to monitor network traffic for suspicious activity, prevent unauthorized access, and log security events for further analysis.

Architects can systematically uncover risks and guide threat mitigation in a structured, repeatable manner. This provides crucial visibility into attack surfaces early when issues are easiest to address. Network defense and monitoring tools are essential components of a robust cybersecurity architecture. They provide continuous visibility into traffic, activities, and events across the environment, enabling threat prevention, timely detection, centralized monitoring, and coordinated response. Here are the key benefits:

- **Access control**: Firewalls and IPS solutions actively block malicious traffic and enforce security policies based on traffic inspection. This prevents threats from penetrating protected assets.

- **Threat detection**: IDS and SIEM solutions passively analyze traffic and event logs to identify indicators of compromise, suspicious activities, and potential exploits for investigation.

- **Centralized monitoring**: Security operations teams use SIEM dashboards for holistic monitoring rather than reviewing individual tool alerts. This improves efficiency.

- **Threat hunting**: Rich network telemetry empowers analysts to proactively hunt for sophisticated threats that evade preventative controls.

- **Incident response**: Comprehensive event data enables rapid tracing of the timeline, cause, and impact of security incidents.

- **Forensics and correlation**: Centralized data lakes facilitate forensic analysis and correlation of events across tools for root cause identification.

- **Regulatory compliance**: Logging and reporting support adherence to legal/regulatory mandates around event auditing and retention.

- **Baseline visibility**: Network behavior analytics establish a baseline to identify anomalies indicative of emerging threats, misconfigurations, or malicious activities.

By implementing layered network monitoring and control tools, security architects can gain broad situational awareness and the ability to dynamically adapt defenses against continuously evolving threats. Integrating these capabilities is foundational for rapid detection, coordinated response, and continuous security improvement. Network monitoring, access control, and event logging tools provide visibility into traffic and activities across the environment to detect and prevent threats.

## Endpoint protection tools

Endpoint protection tools safeguard devices such as laptops, desktops, and servers from compromise, and facilitate swift response to contain threats. Core examples are antivirus software and **endpoint detection and response (EDR)** solutions.

Let's elaborate on the specific functionality provided by leading endpoint protection platforms. Reviewing these details through hands-on evaluation enables organizations to validate effectiveness and select solutions tailored to their specific protection needs and budgets:

- Block known malware before it can execute using signature-based detection

- Machine learning models enable the detection of new malware variants

- Prevent script-based, fileless attacks residing solely in memory

- Gain visibility into suspicious processes and registry and file activity on endpoints

- Identify compromised endpoints attempting to move laterally

- Rapidly isolate infected endpoints to prevent threat spread

- Accelerate incident response with centralized visibility and alerts

- Contain threats by terminating processes or quarantining files

- Remediate endpoints remotely without physical access

- Free tools allow protection even on tight budgets

- Hands-on experience highlights effectiveness and usability firsthand

- Fine-tune configurations and analytics to minimize false positives

- Ensure optimal performance impact via controlled testing

- Updated signatures and detection rules counter emerging threats

In summary, the multilayered prevention, detection, visibility, and response capabilities of modern antivirus and EDR solutions deliver comprehensive endpoint protection. Evaluating specific solutions hands-on enables organizations to validate functionality, optimize configurations, and select tools that deliver the best protection for their needs and budget.

Creating labs for the implementation and evaluation of open source or free antivirus software and EDR solutions necessitates a practical, hands-on approach that mirrors real-world applications. The labs can be found at the aforementioned URL in the *Technical requirements* section.

## Identity and access management (IAM) tools

IAM tools control who can access systems and data by managing user identities, authentication, and authorization. Example technologies include **multi-factor authentication** (**MFA**), **single sign-on** (**SSO**) solutions, **identity providers** (**IdPs**), and **privileged access management** (**PAM**) solutions:

- MFA requires users to provide an additional factor such as a **one-time-password** (**OTP**), along with their username/password

- SSO streamlines authentication by enabling a single login to access multiple applications

- PAM secures privileged account access with enhanced controls and monitoring

IAM tools play an indispensable role in cybersecurity by enabling architects to centrally control and secure access to applications, systems, and data. Robust IAM provides several key advantages:

- **Improved security posture**: IAM limits access to authorized users only via centralized authentication, authorization, and auditing. This reduces the attack surface.

- **Enforce least privilege**: Grant users just the minimum access required to perform their role, limiting the exposure of sensitive systems and data.

- **Better visibility**: Centralized identity stores and access policies provide visibility into who has access and under what conditions.

- **Regulatory compliance**: Comprehensive access controls, auditing, and logging help adhere to regulations related to privacy, data security, and segregation of duties.

- **Efficiency**: SSO enables single login access to many applications, reducing redundancy. Automated provisioning streamlines onboarding.

- **User convenience**: Frictionless authentication options such as SSO and biometric MFA improve experience while balancing security.

- **Cost savings**: Automating manual provisioning and deprovisioning processes reduces administrative overhead for access management.

- **Threat mitigation**: Advanced capabilities such as session recording, privileged access management, and behavioral anomaly detection counter sophisticated identity threats.

Given that compromised credentials are a leading attack vector, robust IAM serves as a critical line of defense. Integrating IAM tools provides the identity control plane required to implement least privilege and zero trust architectures.

## Data protection tools

Data protection tools encrypt data at rest, in transit, and in use to prevent unauthorized access or leaks. Key technologies include encryption, **data loss prevention** (DLP), and database security solutions:

- Encryption encodes data to render it unreadable without the cryptographic key

- DLP tools detect and block potential data leaks

- Database security solutions provide user access controls, auditing, encryption, and masking

Safeguarding sensitive data via robust data protection tools and technologies is imperative for security architects to incorporate into their cybersecurity strategies. Here are some key reasons why comprehensive data protection is vital:

- **Prevents data theft**: Encryption renders data unreadable to unauthorized parties, thwarting common attack vectors such as credential compromise or network snooping.

- **Maintains compliance**: Data security mandates such as HIPAA, PCI-DSS, and GDPR require protection for sensitive data through encryption, access controls, and activity logging.

- **Reduces data leak risks**: DLP blocks overt data exfiltration channels. Rights management limits exposure through need-to-know access.

- **Protects integrity**: Encryption and hashing ensure data has not been manipulated in storage or transit, maintaining integrity.

- **Cost avoidance**: A single data breach can cost millions in recovery, fines, and reputational damage. Robust data protection helps avoid these costs.

- **Preserves privacy**: Controls around access, transmission, and logging curtail unnecessary exposure of personal information.

- **Fosters trust**: Demonstrating *security by design* builds confidence in customers, partners, and stakeholders that their data is protected.

- **Enhanced visibility**: Classifying and tracking sensitive data facilitates risk analysis, compliance reporting, and targeted safeguards.

Layered data protection safeguards sensitive information throughout its life cycle and across infrastructure. In the context of cybersecurity, data protection tools are utilized to safeguard data from unauthorized access, disclosure, alteration, and destruction. Implementing these tools involves understanding encryption, data masking, and rights management, among others.

## Vulnerability management tools

Vulnerability management tools play an indispensable role in cybersecurity strategies by empowering organizations to continuously identify and remediate security weaknesses before they can be exploited by attackers. Here are some of the key benefits:

- **Vulnerability visibility**: Systematic scanning detects vulnerabilities across the environment including assets that may get overlooked

- **Risk prioritization**: Severity ratings based on **Common Vulnerability Scoring System** (CVSS) scores and threat intelligence enable a focus on fixing high-risk flaws first

- **Continuous monitoring**: Repeated scans catch new vulnerabilities that emerge as assets change over time, tracking status

- **Improved patching**: Scan results guide smarter patch management by correlating vulnerabilities to relevant patches

- **Regulatory compliance**: Auditors often require proof of routine vulnerability scanning to validate security hygiene

- **Attack surface reduction**: Eliminating vulnerabilities proactively reduces the entry points malicious actors can abuse to infiltrate networks

- **Incident prevention**: Many breaches exploit known unpatched vulnerabilities that could have been detected and remediated beforehand

- **Resource optimization**: By eliminating false positives and highlighting the most severe risks first, scanning helps focus strained security resources on issues that matter most

- **Post-remediation validation**: Rescanning assets verifies that vulnerabilities have been effectively mitigated after remediation efforts

By institutionalizing continuous vulnerability discovery and remediation powered by specialized scanning tools, organizations can achieve tighter security and compliance while optimizing the use of scarce security resources. Vulnerability management tools such as Nessus, Qualys, and OpenVAS systematically scan for security misconfigurations and software vulnerabilities across an environment. This allows for the prioritization and remediation of risks.

## Security configuration and patch management tools

Security configuration and patch management tools serve crucial functions in cybersecurity architectures by enabling standardized, secure system configurations and timely patching. Here are some of the key benefits:

- **Prevent vulnerabilities**: Hardened operating system configurations and prompt patching block the exploitation of known weaknesses

- **Improve resilience**: Patches fix bugs that can cause crashes, instability, and disruptions when attacked

- **Compliance**: Tools validate and document compliance with configuration benchmarks such as the **Center for Internet Security (CIS)**, **Defense Information Systems Agency (DISA)**, and **Security Technical Implementation Guides (STIGs)**

- **Automation efficiency**: Automated configuration and patch deployment frees IT teams from tedious and error-prone manual work

- **Consistency at scale**: Centrally defining and enforcing desired configurations provides consistency across large environments

- **Change control**: Change monitoring on configuration state and patch levels provides visibility into drift and unapproved changes

- **Attack surface reduction**: Security configuration hardening and vulnerability patching greatly reduce the attack surface targeted by threat actors

- **Cost optimization**: Standard tools such as Configuration Management and **Windows Server Update Services (WSUS)** minimize licensing costs for capabilities available in most operating systems

- **Risk mitigation**: Rapidly closing vulnerabilities through patching eliminates exposure that attackers actively exploit in the wild

With cyber attacks constantly evolving, tools that bring consistency, automation, and current security best practices to system configuration and patching provide a critical element of defense in depth for security architects. Configuration and patch management tools such as Ansible, Chef, Puppet, and WSUS ensure environments remain securely configured and up to date.

## Incident response and forensics tools

Incident response and forensic tools empower security teams to thoroughly investigate, understand the root cause, quantify the impact, and recover rapidly when security events occur. Here are some of the key benefits:

- **Incident analysis**: Collecting and examining forensic artifacts reconstructs details of security events to determine how attackers breached defenses

- **Maintain operations**: Orchestration and automation enable business operations to continue while an incident is being handled

- **Compliance**: Forensic data provides audit trails and the evidence needed to comply with breach disclosure laws

- **Eliminate backdoors**: Identifying and eliminating all remnants of an attacker's presence is crucial to prevent continued access after an incident

- **Network defense improvement**: Deep investigation highlights gaps in visibility, tools, or processes so that defenses can be bolstered

- **Shortened dwell time**: Expert use of tools and platforms accelerates incident response, allowing for the rapid eviction of threats, thus limiting damage

- **Enhanced situational awareness**: Centralized incident management provides visibility into all ongoing security events across the organization

- **Legal evidence**: Forensic techniques adhere to the evidentiary standards that are needed for civil or criminal proceedings against attackers

- **Metrics and reporting**: Platforms capture response performance metrics, thus helping measure and improve programs

With threats inside networks, assuming compromise will occur is prudent. Preparing skilled responders armed with advanced tools enables resilient organizations to swiftly neutralize threats and restore operations. Incident response and forensics tools support threat investigation and recovery after security events. Core examples include forensics software such as EnCase or Autopsy and incident response platforms such as TheHive.

## Application security tools

Application security testing tools provide vital capabilities for architects to incorporate into their software assurance strategies. Robust application testing delivers several key benefits:

- **Identify vulnerabilities early**: Detecting flaws during development enables remediation before deployment when fixes are cheaper

- **Reduce exploited weaknesses**: Applications often contain vulnerabilities that allow threats such as injection attacks or unauthorized access without proper testing

- **Enforce secure coding practices**: Application testing reinforces secure coding standards, encouraging developers to eliminate common weaknesses

- **Meet compliance requirements**: Tools support application security mandates in regulations such as HIPAA and PCI-DSS

- **Improve quality**: Applications developed using security testing tools tend to be higher quality and more stable since vulnerabilities are removed

- **Facilitate security monitoring**: IAST tools provide visibility into application attacks and anomalies in production

- **Enable automation**: Application testing tools integrate into CI/CD pipelines allowing automation of security checks

- **Attack surface reduction**: Eliminating weaknesses minimizes the attack surface malicious actors can exploit to compromise applications

- **Quantify risk**: Tools provide metrics to measure residual application risk over time as vulnerabilities are addressed

Given the prevalence of application vulnerabilities, integrating rigorous security testing practices using automated tools is essential for architects seeking to develop more secure and resilient software applications cost-effectively. Application security testing tools analyze application code, configurations, and runtime behavior to identify vulnerabilities developed applications may contain. Here are some core examples:

- **Static application security testing (SAST)** analyzes application source code for vulnerabilities

- **Dynamic application security testing (DAST)** tests applications while running by attacking the surface

- **Interactive application security testing (IAST)** instruments applications to analyze attacks on running software

## Cloud security tools

As cloud adoption accelerates, employing dedicated cloud security tools is imperative for architects to extend their security strategies to the cloud. Here are the key benefits of robust cloud security tools:

- **Visibility**: Tools such as **Cloud Access Security Brokers (CASBs)** provide centralized visibility into cloud usage, data patterns, and threats given limited native controls

- **Data protection**: Cloud encryption, tokenization, and rights management safeguard sensitive data and prevent loss

- **Threat prevention: Cloud Workload Protection Platforms (CWPPs)** block identified threats targeting cloud workloads and infrastructure

- **Compliance enablement:** Tools facilitate compliance with regulations around data security, residency, and chain of custody in the cloud

- **Misconfiguration remediation: Cloud Security Posture Management (CSPM)** services detect over-privileged or non-compliant configurations in cloud infrastructure that increase risk

- **Consistent security:** Apply existing security tools and policies consistently across on-premises and cloud environments

- **Automated assessment:** Continuously assess cloud environments for new risks and deviations from best practices as they evolve

- **Attack surface reduction:** Eliminating cloud vulnerabilities and misconfigurations removes potential attack vectors exploiting the cloud

- **Cost optimization:** Reducing cloud data exposures and threats lowers the potential costs of cloud data loss and breaches

As organizations shift from data centers to the cloud, purpose-built tools enable security architects to successfully secure cloud migrations and new environments cost-effectively. Cloud security tools extend data and threat protection policies to cloud environments. Core examples include CASB, CWPP, and CSPM solutions:

- CASB monitors cloud access and data use

- CWPP secures cloud workloads

- CSPM monitors cloud resource configurations

## Cybersecurity governance and compliance tools

Cybersecurity governance and compliance tools are essential for managing policies, demonstrating compliance, and providing visibility into the overall security program. Here are the key benefits:

- **Centralized policy library:** Provides a single source of truth for controls such as standards and procedures

- **Maintains compliance:** Automates mapping controls to regulations and validates adherence

- **Reporting:** Produces audit-ready reports to demonstrate compliance with mandates

- **Reduces risk:** Identifies control gaps and ensures the execution of critical policies that prevent threats

- **Enhances visibility:** Dashboards give leadership visibility into security program effectiveness and risk

- **Workflow automation**: Automates processes such as policy review/attestation and control assessment

- **Consistency**: Ensures technical controls remain consistent and aligned across the organization

- **Accountability**: Associates policy responsibility and acceptance by role across the workforce

- **Collaboration**: Provides a central repository for cross-team collaboration on control development

- **Cost efficiency**: Reduces the overhead required for manual policy/compliance management

**Governance, risk, and compliance (GRC)** tools enable architects to effectively govern security programs at scale, ensuring provable adherence to expanding regulations – a critical concern for modern organizations. GRC tools centralize the management of security policies, controls, assessments, and regulatory mandates. Example solutions include dedicated GRC platforms and policy/document management systems.

## Penetration testing and red team tools

Penetration testing and red team tools provide a crucial capability for cybersecurity professionals by enabling simulated adversarial attacks against production infrastructure to proactively identify vulnerabilities and security gaps. Here are some of the key benefits:

- **Find unknown risks**: Mimics threat behaviors to uncover weaknesses defenders are blind to

- **Test protections**: Validates that controls such as firewall policies isolate critical assets as intended

- **Improve prevention**: Identified weaknesses inform enhancements to logging, detection, and prevention systems

- **Justify investments**: Quantifies defensive gaps to justify the budget for security tools and resources

- **Verify skills**: Tests and improves incident response team readiness by exercising workflows with realistic drills

- **Inform the design**: Security architecture and engineering deficiencies are revealed through testing feed improved designs

- **Cost-efficiency**: Internal red teams amplify a limited penetration testing budget for continuous assessments

- **Attack surface reduction**: Eliminating uncovered vulnerabilities decreases the attack surface for malicious intruders

- **Build confidence**: Demonstrating the ability to find and fix gaps internally reassures customers and executives

While requiring careful scoping and authorization, seasoned penetration testers who apply appropriate tools safely can identify weaknesses and meaningfully improve resilience. Penetration testing and red team tools simulate real-world attacks to proactively identify security gaps before adversaries abuse them. Here are some examples:

- Metasploit is an open source penetration testing framework

- Kali Linux provides a full security toolkit for testing

- Cobalt Strike is a commercial penetration testing and red team platform

## Automation and orchestration tools

**Security automation, orchestration, and response** (SOAR) platforms provide indispensable capabilities to security architects by enabling workflow standardization, security orchestration, and automation. Let's look at some of the benefits:

- **Increased efficiency**: Automating manual repetitive tasks allows security staff to focus on high-value efforts

- **Improved response times**: Security playbooks enact end-to-end incident response processes faster than manual approaches

- **Consistency at scale**: Playbooks enforce consistent workflows across the organization

- **Reduced errors**: Automation eliminates human errors that often occur in manual processes

- **Flexibility**: Modular playbooks allow continuous customization as needs evolve

- **Process visualization**: Workflow modeling provides visibility into security processes that require improvement

- **Simplified integrations**: SOAR platforms integrate disjointed products via APIs into unified workflows

- **Institutional knowledge capture**: Playbooks codify tribunal knowledge, making it accessible across the team

- **Improved metrics**: Dashboards provide data to measure and optimize security operations

As threats increase in sophistication, leveraging SOAR tools mitigates manual inefficiencies and human limitations through workflow automation, which is essential for modern security programs.

SOAR platforms optimize security operations by stitching together disparate tools, accelerating workflows, and enabling automation. Here are some examples:

- Demisto SOAR integrates security technologies through its automated playbooks

- Swimlane helps manage security workflows

# Summary

This chapter explored strategies for thoughtfully assembling a cybersecurity architecture toolkit by evaluating solutions to find the optimal fit. It emphasized understanding unique organizational vulnerabilities and risks first, then matching appropriate defenses accordingly.

This chapter covered several major classes of security tools:

- Threat modeling tools such as Microsoft TMT systematically uncover risks and guide mitigation early in system design
- Network monitoring, firewalls, and SIEM solutions provide visibility into activities across environments to detect and prevent threats
- Endpoint protection platforms use layered antivirus, EDR, and advanced analytics for device security
- IAM tools manage access to resources by enforcing least privilege authorization
- Data protection technologies such as encryption and rights management safeguard sensitive information
- Vulnerability management scanners continuously assess weaknesses across attack surfaces
- Configuration and patch tools automate consistent hardening and update processes
- Incident response platforms accelerate threat investigation, impact analysis, and recovery coordination
- Cloud security services extend visibility, data, and threat controls to cloud environments
- GRC tools centralize policy and compliance artifact management for consistency
- Penetration testing toolsets emulate adversary behaviors to uncover security gaps proactively
- SOAR solutions optimize workflows, orchestration, and automation for improved efficiency

This chapter emphasized matching defenses to unique organizational terrain, risks, and constraints – customizing toolkits versus one-size-fits-all approaches. By preparing specialized security tools in advance tailored to their environment, architects can achieve victory against attackers.

The upcoming chapter is part two of a two-part discussion that distills lessons from renowned military strategist Sun Tzu to guide security architects in mastering strategy and execution when designing, building, and managing enterprise cybersecurity solutions. Core focus areas include tailoring robust technical architectures to address unique organizational risks and adopting security best practices throughout solutions' life cycles. This finale synthesizes the key principles that have been covered in this book into strategic blueprints for architects seeking victory over cyber adversaries.

# 10
# Decluttering the Toolset – Part 2

*"Water shapes its course according to the nature of the ground over which it flows; the soldier works out his victory in relation to the foe whom he is facing."*

*– Sun Tzu*

*"By method and discipline are to be understood the marshaling of the army in its proper subdivisions, the graduations of rank among the officers, the maintenance of roads by which supplies may reach the army, and the control of military expenditure."*

*– Sun Tzu*

*"Hence the skillful fighter puts himself into a position which makes defeat impossible, and does not miss the moment for defeating the enemy."*

*– Sun Tzu*

*"Thus it is that in war the victorious strategist only seeks battle after the victory has been won, whereas he who is destined to defeat first fights and afterwards looks for victory."*

*– Sun Tzu*

In the previous chapter, we discussed how to make sense of the certification landscape. In this chapter, we will now look at the landscape of tools available to the cybersecurity architect.

As Sun Tzu emphasized, the shrewd combatant tailors their approach to the unique circumstances at hand, rather than relying on prescribed formulas alone. This adaptable mindset is equally crucial for cybersecurity architects selecting tools to secure their organization's digital assets and data.

With hundreds of products flooding the market, it is easy to get overwhelmed by the hype of the latest offerings. However, the most effective cybersecurity architects thoughtfully curate their security toolkit based on their unique threat landscape, business drivers, and operational constraints – just as a river shapes its course according to the ground it flows over.

Rather than blindly adopting every new tool, the discerning cybersecurity architect focuses on developing a tailored toolkit to address their specific challenges. They marshal resources judiciously, striking the right balance between capabilities and costs. Through rigorous methods and discipline, they invest wisely to get the optimal return on security.

This chapter will explore how to thoughtfully assemble your cybersecurity architecture toolkit by filtering through solutions to find the right fit. It emphasizes understanding your distinct vulnerabilities and risk profile first, then matching appropriate defenses accordingly. With the proper toolkit in hand, you can nimbly respond to any adversary that appears, seizing opportunities to strengthen protections before trouble arises. By preparing the perfect set of tools in advance, victory is assured.

The chapter covers the following topics:

- What tool to use?
- Business considerations

# What tool to use?

Selecting the optimal set of cybersecurity tools from the multitude of options available can seem daunting. Just look at the previous section to see that only the surface was scratched regarding the potential tools. However, by methodically aligning tools to organizational needs and infrastructure, architects can assemble the ideal toolkit. Cybersecurity tool selection is a critical strategic decision that impacts the overall security posture of an organization. To navigate the complex landscape of available options, a structured approach aligning tools with the organization's unique requirements and risk profile is essential. This section delves into the methodology for selecting the optimal set of tools. Here are some key considerations when deciding which tools to implement.

## Clearly define requirements

Start by identifying your specific use cases and requirements. Determine where you have gaps in visibility, protection, or response capabilities. Define technical and business requirements such as scalability needs, budget constraints, and compliance mandates. This focuses tool selection on fulfilling unmet needs. Understanding the specific security needs of an organization is the first step in the tool selection process. A clear set of requirements must be articulated, which encompasses the following:

- Gaps in the current security posture
- Protection, detection, and response capabilities needed
- Scalability to support growth and adapt to dynamic workloads

- Compliance with industry and government regulations

- Alignment with the organization's budget and resources

By clearly defining these parameters, organizations can narrow down their tool selection to those that directly address defined needs.

## Assess organizational risk profile

Factor in your organization's risk profile, including critical assets, known vulnerabilities, and prior incidents. For example, regular IP theft incidents warrant more emphasis on data protection and rights management tools, and highly regulated data calls for access controls and activity monitoring tools. The risk profile of an organization dictates the prioritization of tool selection. Critical considerations include the following:

- Identifying and categorizing assets based on their criticality

- Historical analysis of security incidents and their impact

- Current vulnerabilities and potential attack vectors

- The nature of the data handled, including personal, financial, and intellectual property

For instance, organizations with high-value intellectual property may prioritize DLP and encryption tools over others.

## Map to core security frameworks

You should align your tools to widely adopted frameworks such as NIST CSF, mapping capabilities to core functions such as **Identify**, **Protect**, **Detect**, **Respond**, and **Recover**. This structures tool selection by high-level security objectives. Adherence to established cybersecurity frameworks, such as the NIST **Cybersecurity Framework** (CSF), provides a structured approach to selecting tools. By mapping tools to the five core functions—Identify, Protect, Detect, Respond, and Recover—organizations can ensure comprehensive coverage across all facets of their cybersecurity program.

### Layer complementary safeguards

You should assemble complementary tools that provide defense in depth. Choose preventive controls such as firewalls and access managers, coupled with detective controls such as SIEMs and IDS to surface threats that bypass the first line. A multi-layered, or defense-in-depth, approach is fundamental. This involves the integration of the following:

- Preventative tools, such as firewalls, antimalware, and access controls

- Detective tools, including IDS and SIEM systems

- Corrective tools such as patch management and incident response platforms

Ensuring these tools work in concert can provide a robust security posture capable of responding to a multitude of threat scenarios.

## Right-size investment

Consider cost and complexity trade-offs, resisting the temptation to over-invest in the latest offerings. For example, open source tools such as Snort offer excellent detection without SIEM price tags. Focus the spending on priorities first. Fiscal responsibility should not be overlooked in the selection process. Organizations must do the following:

- Balance the cost against the expected benefit
- Avoid overspending on unnecessarily complex solutions
- Evaluate the **return on investment** (**ROI**) of security expenditures

Selecting open source solutions or less expensive alternatives for certain use cases, where appropriate, can optimize the security budget.

## Evaluate ease of use

Factor in usability, integration overhead, and required personnel expertise. More usable tools lower adoption barriers. Consider leveraging platforms or suites integrating multiple capabilities. Usability and manageability are often undervalued factors in tool selection. This includes tools with the following characteristics:

- Intuitive and easy to operate, which can increase the efficiency of security teams
- Capable of integration with existing systems, which reduces complexity
- Supported by a strong vendor or community, which ensures longevity and support

The availability of skilled personnel to operate the tools should also influence the decision-making process.

## Incorporate future plans

Anticipate how needs and infrastructure will evolve to select extensible tools. For example, today on-premises SIEMs should support cloud and container telemetry to accommodate future migration. Organizations should anticipate future changes in the threat landscape, as well as shifts in their own IT environment, by selecting tools that offer the following features:

- Flexibility to scale and adapt to new technologies and threats
- Extensibility to integrate with future platforms and services
- Forward compatibility with emerging industry standards

Considering the potential for cloud migration, the growing relevance of mobile and IoT devices, and the rise of artificial intelligence and machine learning in cybersecurity practices is crucial.

## Leverage trials and proof of concepts (POCs)

Test tools in non-production environments to validate capabilities, usability, and integration viability. Many vendors offer free trials and **POCs** for in-depth evaluation prior to purchase. Before finalizing any purchase, organizations should do the following:

- Conduct trials and POCs in controlled environments

- Validate the efficacy, compatibility, and manageability of the tools

- Engage with vendor support to understand the level and quality of service provided

This hands-on evaluation is invaluable in confirming that a tool delivers on its promises and fits seamlessly into the existing security architecture.

By meticulously aligning cybersecurity tools with organizational objectives, infrastructure requirements, and the overarching strategic framework, architects can construct an effective security toolkit. Regular reassessment of this toolkit is imperative to ensure it keeps pace with evolving business needs, technological advancements, and the dynamic threat landscape. With a judicious selection process, organizations can establish a resilient defense that not only protects but also empowers their business operations.

# Business considerations

In the realm of cybersecurity, aligning technical decisions with business considerations is paramount. The optimal toolset must not only safeguard the organization's assets but also support its strategic objectives and operational efficiencies. This section examines the business realities that cybersecurity architects must balance during the tool selection process.

## Total cost of ownership (TCO)

Look beyond upfront software/hardware costs to account for ongoing maintenance, training, integration expenses, and staffing requirements. Cloud services can reduce capital outlay but have subscription fees. Evaluating the **TCO** is vital in understanding the long-term financial impact of cybersecurity tools:

- Factor in not only initial acquisition costs but also recurring costs such as maintenance fees, subscription models for cloud-based services, and potential scaling expenses

- Assess the need for ongoing training and the potential for these costs to vary with employee turnover

- Consider integration costs, including those associated with custom development or consulting services to ensure interoperability with existing systems

- Account for staffing costs, as some tools may require additional or more specialized personnel to manage effectively

## Alignment to business initiatives

Tools should ultimately support business goals, not hinder them. For example, desktop antivirus software must avoid impeding employee productivity. You need to evaluate whether tools deliver sufficient value. Cybersecurity tools should facilitate, not impede, the achievement of business goals:

- Evaluate tools in the context of how they will support key business initiatives, ensuring they bolster rather than burden productivity
- Align tool selection with the organization's strategic direction, ensuring that security measures are enablers of business functions rather than inhibitors

## Impact on users

Assess workflow disruption and change impact on end users. Complex tools with steep learning curves face user resistance, so you should opt for intuitive solutions with minimal disruption. The influence of cybersecurity tools on user experience and workflow must be assessed:

- Consider the user interface and overall user experience of the tools, aiming to minimize the learning curve and resistance
- Evaluate how security measures will impact day-to-day operations and ensure that these tools streamline rather than complicate the workflow

## Executive mandates

Factor in executive directives, which often dictate the adoption of specific vendors or offerings. Procurement may require executive sign-off for large expenditures. Decisions around tool selection often require concurrence with executive-level directives:

- Acknowledge that certain decisions may be driven by executive strategies, including preferences for certain vendors or product categories
- Be prepared to advocate for alternatives if they offer superior alignment with business objectives, presenting a clear business case to executives

## Vendor viability and support

Consider long-term vendor viability, support capabilities, and product roadmap. Start-ups carry more risk. Established vendors offer stability but can lack innovation or flexibility. The selection of a vendor is a crucial aspect that extends beyond the technical capabilities of the tool:

- Assess the stability and market presence of vendors, considering their track record and financial health

- Look into the vendor's support infrastructure, responsiveness, and quality of service, as this will impact the effectiveness of the tool over its life cycle

- Review the vendor's product development roadmap for alignment with the organization's anticipated future needs

## Interoperability and integration

Evaluate how tools integrate with the existing environment and other solutions. Open APIs, common protocols, and pre-built connectors ease integration. The ability of cybersecurity tools to integrate within the existing business ecosystem is crucial:

- Prioritize tools that offer open APIs, adherence to common standards, and support for widely used protocols

- Look for solutions that come with pre-built connectors or those that demonstrate a track record of successful integration with a variety of systems

- Here are some examples:

  - Tenable integration with cloud infrastructure such as AWS and Azure for frictionless scanning

  - ProofPoint integration with Palo Alto Wildfire Analysis

## Scalability needs

Account for current and projected capacity in tools such as SIEMs as the organization grows. Plan for scale-up and scale-out options to support expanding needs. Tools must be able to grow with the organization, and therefore you must do the following:

- Select tools that can scale in performance and capacity to meet future business demands

- Ensure that the toolset can accommodate growth without requiring a complete overhaul or significant additional investment

## Resource constraints

Consider cybersecurity team skill sets and bandwidth. Complex tools require specialized expertise. Budget staff training time and supplement teams if needed. A realistic assessment of internal capabilities and constraints is essential:

- Match the complexity of the tools with the skill level and availability of the cybersecurity team

- Plan for training and knowledge enhancement to enable the team to effectively use more sophisticated tools

- Be prepared to hire or outsource expertise for highly specialized toolsets

Cybersecurity tool selection transcends the evaluation of technical features and capabilities. It requires a holistic approach that incorporates a multitude of business considerations. By comprehensively analyzing factors such as TCO, alignment with business initiatives, user impact, executive mandates, vendor viability, interoperability, scalability, and resource constraints, architects can select tools that not only secure the organization but also enhance its ability to achieve business objectives. Integrating this multi-faceted perspective into the selection process ensures that the cybersecurity infrastructure is robust, adaptable, and in harmony with the overarching goals of the enterprise.

## Summary

In closing, this chapter emphasized the importance of thoughtfully curating a cybersecurity toolkit tailored to an organization's unique risk profile, infrastructure, and strategic drivers. Rather than getting overwhelmed by the endless tool options and feature hype cycles, architects must take a methodical approach rooted in clearly defining security requirements and gaps. Tight alignment with security frameworks, layered defenses, future-proofing, and business considerations are all critical factors during selection as well.

The key takeaways include the following:

- Clearly identify your specific use cases, vulnerabilities, requirements, and infrastructure first before assessing tools

- Map tools to core security framework functions such as NIST CSF to ensure comprehensive coverage

- Implement complementary preventive, detective, and corrective controls for defense in depth

- Evaluate total cost, business alignment, usability, and other practical factors

- Validate effectiveness through POCs and trials before purchase

- Revisit selections continuously as threats and infrastructure evolve

By following this structured methodology centered on their unique risk profile and priorities, organizations can assemble an optimized toolkit to strengthen their security posture. Proper tools empower architects to shape robust defenses tailored to their landscape, much like rivers carving their course based on the ground traversed. With the right toolkit secured, organizations can nimbly combat adversaries and seize opportunities to harden protections well before trouble arises.

As we've explored so far, cybersecurity architecture demands navigating complex trade-offs between business objectives, technical realities, and security imperatives. Adopting best practices bolsters defenses, but rigid dogma ignoring unique organizational needs courts failure. With threats and technology continuously evolving, adaptability becomes critical.

In the next chapter, we will build on previous discussions to examine this delicate balancing act further. Cybersecurity architects must remain versatile, designing integrated solutions that empower operations rather than inhibit them. By bridging the gap between security and other organizational goals, architects enable objectives rather than obstruct them.

However, this requires mastering nuanced risk analysis and mitigation. Architects must objectively weigh dangers against benefits, steering stakeholders towards prudent trade-offs. Through immersive collaboration and clear communication, they constructively align perspectives. With patient diplomacy, architects shape win-win outcomes advancing all interests.

The chapter will provide frameworks and examples demonstrating this adaptive methodology. Success relies on more than technical prowess alone – it necessitates navigating needs diplomatically to arrive at optimal, tailored solutions. Cybersecurity architecture is an interdisciplinary practice requiring both strategic and interpersonal dexterity. We will explore the soft skills and mindsets that turn sound security into an organizational asset rather than a liability.

# 11
# Best Practices

*"The skillful tactician may be likened to the shuai-jan. Now the shuai-jan is a snake that is found in the Ch'ang mountains. Strike at its head, and you will be attacked by its tail; strike at its tail, and you will be attacked by its head; strike at its middle, and you will be attacked by head and tail both."*

*– Sun Tzu*

*"The art of war is of vital importance to the State. It is a matter of life and death, a road either to safety or to ruin. Hence it is a subject of inquiry which can on no account be neglected."*

*– Sun Tzu*

*"Bravery without forethought, causes a man to fight blindly and desperately like a mad bull. Such an opponent, must not be encountered with brute force, but may be lured into an ambush and slain."*

*– Sun Tzu*

*"Hence that general is skillful in attack whose opponent does not know what to defend; and he is skillful in defense whose opponent does not know what to attack."*
*– Sun Tzu*

In the previous chapter, we saw that selecting impactful cybersecurity tools requires harmonizing technical capabilities with business goals. As Sun Tzu emphasized, we must adapt strategies to circumstances. Setting requirements and mapping tools to frameworks grounds selection in objectives, layering aligned controls provides defense-in-depth, and right-sizing investments balances protection with fiscal prudence.

However, technical expertise alone does not guarantee adoption. Tools must empower users and processes, not hinder them. Success requires equipping defenders while avoiding operational disruption, just as Tzu warned against exhausting resources while battling strongholds directly.

By fusing security with usability, and recalibrating as conditions change, organizations can assemble resilient architectures. However, avoiding rigid dogma remains critical; as Tzu noted, strategies must adapt to unique terrain.

In cybersecurity, one-size-fits-all rarely prevails. The most effective leaders curate toolkits that have been tailored and scaled to their ground. Through disciplined selection and continuous reassessment, architects construct defenses positioned for victory.

Implementing cybersecurity using established best practices is like mastering Sun Tzu's strategies for victorious warfare. Both require skillfully evaluating conditions and adapting judiciously to unique circumstances.

Blindly attacking strong points wastes effort and reveals intentions, as Sun Tzu cautioned. Rigidly forcing security best practices regardless of impact squanders resources while empowering adversaries. However, disregarding best practices altogether courts ruin, akin to neglecting warcraft's vital importance, something that Tzu emphasized.

Navigating these challenges necessitates tapping collective wisdom while allowing for selective flexibility. As Tzu noted, the shrewd tactician strikes unpredictably, varying their point of attack. The most potent cybersecurity leverages best practices fluidly, not dogmatically, keeping opponents off balance.

By blending adherence with nuance, architects can bolster defenses while supporting operations. Like a general who masks intentions through versatile assault and defense, mature security leaders expertly adapt best practices to their terrain. Standards provide the foundation, but prudent customization for unique environments helps avoid blind, brute-force methods.

Let's face it – there are whole volumes dedicated to each individual best practice within technology and security. As such, this chapter will drill down to explore key best practices and their judicious implementation, recognizing cybersecurity demands and their practicality as much as their principles. Skilled professionals shape standards to business objectives, just as Tzu's victorious general harmonizes strategies to circumstances. For cybersecurity, success flows from melding real-world goals with hard-won wisdom – and recognizing when exceptions warrant flexibility rather than rigidity.

This chapter covers the following topics:

- Least privilege
- Patching and development
- **Multi-factor authentication (MFA)**
- Security training
- Vulnerability scanning

# Least privilege

The **principle of least privilege (PoLP)** is fundamental to constructing robust cybersecurity architectures. It asserts users should only have the bare minimum access required to perform duties. Architects must master least privilege to erect resilient protections. By restricting unnecessary access, risks are reduced, accountability is enabled, and attack surfaces shrink. In the context of zero trust, PoLP also applies to device identities, not just user identities. This is a comment to be applied throughout this chapter regarding least privilege.

Cybersecurity architects hold crucial responsibility for translating least privilege into technical implementations and governing policies. They must audit entitlements and align controls to curtail excess while empowering productivity. The most potent architects embed least privilege judiciously, not dogmatically, leveraging automation and awareness to balance security with usability.

This section will explore best practices around access reviews, role-based access, MFA, just-in-time elevation, and other aspects you must consider. It will provide examples and labs to tangibly demonstrate enforcing least privilege aligned with business needs.

For cybersecurity architects overseeing identity and access, least privilege remains essential for minimizing risk. It epitomizes the philosophy of *need to know*, crystallizing security wisdom accrued over decades. When woven comprehensively into the fabric of technologies and processes, least privilege becomes a formidable obstacle for adversaries yet invisible to users. Architects must master its application to help organizations do more with less risk.

Remember, good documentation practices contribute to an organization's success by providing a solid foundation for information management. Therefore, it's crucial to understand the importance of documentation and implement best practices in your organization.

## Understanding least privilege

Least privilege is a cybersecurity principle that requires restricting user, service, and process access rights and permissions to the smallest possible subset required so that they can perform their authorized functionality. Implementing least privilege strictly limits a user or process' access to only the computing resources, data, and capabilities essential for their specific role, denying anything further. Applying the checks of least privilege through precise identity and access configurations significantly reduces the risk surface of systems and the potential impact of breaches by bounding allowed actions. Effective implementation requires accurate role definitions, rigorous access controls, and continuous validation across technological layers, from network controls to user policies to process isolation.

The PoLP is a cornerstone of cybersecurity, asserting that users should be granted the minimum levels of access – or permissions – necessary to perform their job functions. This section outlines the best practices for implementing PoLP and provides detailed labs and examples to illustrate these practices.

# Best practices for implementing least privilege

Implementing least privilege remains fundamental to cybersecurity, yet realizing this vision demands nuance. While constraints guard against misuse, inflexibility impedes productivity. Balancing security with usability requires carefully constructed guardrails together with cultural accountability.

Through comprehensive access reviews, judicious entitlement modeling, privileged access oversight, and automation, architects can implement least privilege comprehensively. However, technical controls alone will inevitably fail without human buy-in. Fostering an understanding of risks cultivates ethical access mindsets.

Mature implementations move beyond all-or-nothing prohibitions toward flexible elevation, which is available instantly when warranted and then automatically revoked. By granting temporary privilege escalation on-demand, architects can enable business needs securely.

For least privilege to persist, rigorous auditing, logging, and access analytics provide ongoing guardrails. Combined with training that resonates, architects help the workforce become partners in upholding access integrity. With balanced technical and cultural measures, responsible privilege becomes a competitive advantage.

## *Conduct a user access review*

Implementing least privilege necessitates thoroughly auditing existing access controls to identify and remedy excessive entitlements. Comprehensive user access reviews are foundational.

Architects must catalyze and govern access reviews to map the current permission landscape across systems, accounts, and data. Rigorously auditing roles, groups, credentials, privileges, and permissions reveals cases of misalignment with least privilege standards.

The most impactful access reviews involve cross-functional collaboration between cybersecurity, IT infrastructure, identity management, and application owners. Architects must spearhead the process of examining identities, roles, and access across this diverse digital terrain.

By methodically scrutinizing who has been granted what level of access and why, security leaders can illuminate gaps between actual and ideal authorization postures. Architects must guide remediation by revoking unnecessary entitlements and continuing to monitor to detect privilege creep.

Ongoing access reviews represent the bedrock for implementing least privilege as they provide visibility to prune excess access and maintain alignment. They catalyze the continuous adaptation required in dynamic environments. For architects, instilling comprehensive access review programs is foundational to maturing cybersecurity and securing progress.

Cybersecurity architects should do the following:

- Audit current user roles and permissions to understand the existing access landscape

- Identify any instances of excessive privileges that contradict PoLP

### Establish role-based access control (RBAC)

To actualize least privilege, architects must champion RBAC methodologies that curtail implicit trust in identities. RBAC shifts access decisions to the roles required for specific duties.

Architects need to instill frameworks that grant users the roles justified by their responsibilities, and no more. Roles should be defined according to essential job functions and configured with the minimally necessary permissions to enable those functions.

However, static role definitions defy the dynamic realities of evolving business needs. Architects must foster regular reevaluation and adjustments of role permissions as workflows transform. This responsive adaptation contains permission creep.

Well-crafted RBAC also centralizes access management overhead. Modifying a role adjusts access for all occupants, unlike individual credential maintenance. Changes can be governed centrally through structured approval and re-authorization processes.

For least privilege to persist amid shifting needs, RBAC must become a living framework. Architects play a key role in right-sizing initial role design and formalizing processes for continuous role optimization based on use cases.

Cybersecurity architects should do the following:

- Define roles according to job functions and assign the minimum necessary permissions to these roles
- Regularly update roles to reflect changes in job functions or organizational policies

### Implement strong authentication mechanisms

To actualize least privilege, access to powerful accounts must be safeguarded stringently. Architects need to drive the adoption of strong MFA as a best practice, especially for privileged users.

MFA combats the risks of stolen credentials, which could unlock the *keys to the kingdom*. By requiring an additional factor such as a **one-time password** (**OTP**) token or biometrics, compromised passwords alone cannot enable access.

Cybersecurity architects need to spearhead expansive MFA adoption for not only external-facing login but also internal administration and credentialed services. Frictionless MFA options are available to balance security and usability.

For highly privileged accounts such as those owned by domain administrators, architects should mandate MFA coupled with **privileged access management** (**PAM**) tools to enforce **just-in-time** (**JIT**) access. By adding checks and balances around the most potent identities, MFA and PAM safeguard the reach of compromised credentials.

To actualize least privilege, architects must champion MFA together with layered access controls. By intelligently hardening authentication, they can curtail catastrophic access enabled by compromised passwords. With MFA as a best practice, architects can restrict adversary tradecraft while empowering users.

Cybersecurity architects should enforce MFA to ensure that enhanced security controls protect privileged accounts.

### Employ JIT access

To actualize least privilege amid dynamic business needs, architects should champion JIT access models that provision temporary privilege escalation.

JIT access ensures users can gain the necessary elevated permissions to perform certain tasks, but on a restricted time frame with automated revocation once the work is complete.

As per policy, architects can mandate that privileged activities follow a JIT workflow. This avoids standing access, which can be abused. Administrators obtain the access they need, when it's needed, without accumulating persistent God-like privileges.

JIT access workflows include layered approvals, extensive monitoring, and logging of activities. Architects must drive the adoption of PAM tools to automate policy enforcement and oversight.

To crucially empower productivity without sacrificing security, least privilege requires nuance – not just all-or-nothing prohibitions. Thoughtfully designed JIT access workflows provide this flexibility securely. Architects enable trusted privilege elevation while preventing standing abuse.

Cybersecurity architects should grant privileges on an as-needed basis, with a limited and audited time frame for access.

### Conduct regular audits and logs

To sustain least privilege protections, rigorous auditing and logging of access must become standard practice. Architects need to spearhead access analytics as of form of ongoing hygiene.

Comprehensive activity logging provides the forensic foundation to detect unauthorized use so that it can be contained and remediated. Logs allow events to be reconstructed when incidents occur.

Equally crucial are proactive access reviews to uncover subtle or emerging threats such as insider risks. Analyzing patterns of usage and changes from baselines reveals anomalies that warrant investigation.

Architects must drive the adoption of tooling for log centralization, aggregation, and retention to facilitate automated analysis. Alerting based on high-fidelity detections enables rapid response.

Access logs and audits demonstrate due diligence while enabling oversight. For least privilege to persist, architects must champion comprehensive logging and proactive access analytics. Otherwise, unnoticed privilege creep risks unraveling protections over time.

Cybersecurity architects should do the following:

- Continuously monitor and log access to sensitive systems and information
- Periodically review logs to detect any inappropriate access patterns or policy violations

### Educate employees

Technical controls alone cannot actualize least privilege without human accountability. Architects need to spearhead education that cultivates a culture upholding least privilege.

Through training, architects must convey how excessive access invites risks from abuse, misuse, and exposure. Educating users on policy, technology constraints, and responsibilities fosters mindset shifts that reinforce access ethics.

Creative messaging can help employees view access requests as a privilege rather than an entitlement. Gamification through tools such as CybSafe helps instill security mindfulness.

Cybersecurity architects also need to educate IT teams on least privilege principles to foster compliance. Checklists detailing access review and log auditing responsibilities raise accountability.

Training that resonates both informs and motivates. For least privilege to take root, architects should champion education that enlists the workforce as partners upholding access integrity.

Cybersecurity architects should train employees on the importance of PoLP and the risks associated with excessive privileges.

### Leverage automated tools

While human accountability is still essential, architects should utilize automated tools to aid least-privilege enforcement. Access governance solutions can automatically detect and revoke excessive permissions through ongoing access reviews and attestations.

By integrating access requests with workflow engines, organizations can require approvals before granting access. Automated policy and rule engines then provision access at precise privilege levels based on attributes and roles.

Machine learning algorithms can also analyze user activity patterns to identify anomalies that may indicate permission creep or redundant access. Any high-risk access can automatically trigger alerts and workflows for review.

Access modeling tools even allow architects to visualize entitlements, simulate policy changes, and analyze potential blast radius prior to deployment.

Automated reporting also promotes accountability by tracking usage, changes, attestations, and risks across all systems. Dashboards give architects continuous visibility to detect and address any privilege gaps.

Though tools handle the heavy lifting, human guidance is still key. Automation creates efficiency so architects can focus on governance strategy while enabling least privilege at scale.

Cybersecurity architects should use privilege management tools to automate the assignment and revocation of privileges.

## Exercise

Although we'll be using **Active Directory** (**AD**) in this exercise, creating an AD account is beyond the scope of this book. However, I recommend the Packt book *Mastering Windows Security and Hardening*, by Mark Dunkerley and Matt Tumbarello, if you want to learn how to do this. This exercise demonstrates how to implement PoLP within a Windows AD environment. You will create user roles, configure permissions, and review access controls.

The prerequisites are as follows:

- A virtual machine with Windows Server installed
- A Windows client that's virtual-machine-joined to the domain
- Administrative access to Windows Server

Let's look at the steps:

1.  Set up a lab environment:

    A.  Create a Windows Server virtual machine and install Windows Server.

    B.  Open Server Manager and add the **Active Directory Domain Services** role.

    C.  Configure AD Domain Services and create a new domain.

    D.  Create and join a Windows client virtual machine to the domain.

2.  Review existing access controls:

    A.  On Windows Server, open **Active Directory Users and Computers**.

    B.  Audit user accounts and group memberships.

    C.  Identify any excessive or unnecessary privileges.

3.  Implement RBAC:

    A.  Define organizational roles such as Helpdesk, Finance, and HR based on duties.

    B.  Create AD security groups for each defined role.

    C.  Assign user accounts to the appropriate role group(s).

    D.  Set permissions on resources by granting access to role groups rather than individual users.

4.  Require MFA:

    A.  Install an MFA provider such as Duo Security on Windows Server.

    B.  Enable MFA prompts for administrators and privileged users when they log in.

5.  Configure JIT privileged access:

    A.  Enable the **Privileged Access Management** feature on Windows Server.

    B.  Define privileged roles eligible for JIT access.

    C.  Users must request temporary privilege elevation through PAM when needed.

6.  Review activity logs:

    A.  Open Event Viewer and review Windows security logs.

    B.  Create alerts to detect anomalous access attempts.

    C.  Forward logs to a centralized SIEM for correlation and long-term retention.

7.  Educate end users :

    A.  Create a training module explaining least privilege principles.

    B.  Ensure all employees complete privilege awareness training.

8.  Automate privilege management:

    A.  Deploy a PAM tool such as Microsoft Identity Manager.

    B.  Integrate with AD to orchestrate access based on policies.

    C.  Automate privilege elevation and revocation.

## Example scenarios

Let's look at example 1 – restricting database access:

- **Situation**: A company's database contains sensitive customer information
- **Action**: Implement PoLP by creating specific database roles that define what each user can read, write, or modify
- **Outcome**: Users are only able to interact with the database in ways that are essential to their role, minimizing the risk of data leakage or unauthorized modifications

Now, let's look at example 2 – admin account restrictions:

- **Situation**: IT staff frequently log in with administrative privileges for tasks that do not require such high levels of access

- **Action**: Enforce JIT access for admin tasks, requiring IT staff to request temporary admin privileges

- **Outcome**: A reduction in the window of opportunity for privilege abuse or exploitation

To summarize, while technical controls provide the foundation for least privilege, realizing true least privilege requires an organizational culture that's upheld by accountability and education. Cybersecurity architects must spearhead ongoing training that conveys the security imperatives of restricted access, enlisting the workforce as partners in upholding access integrity through policy adherence. With motivated users and privilege management automation reinforcing the technological safeguards, organizations can make strides toward robust, risk-reducing least privilege.

# Patching and development

Proactively patching vulnerabilities represents fundamental cyber hygiene, yet this mundane maintenance can easily lapse amid rapid development. Vigilant patch management must permeate the entire software life cycle to preempt exploitation. Architects hold crucial responsibility for governing patch excellence despite competing priorities.

By championing continuous asset discovery, automated scanning, and policy rigor, architects can embed proactive patching into workflows. Testing and staging updates verify quality assurance before promotion to production. Regular cycles maintain currency amid a threat landscape in constant flux.

However, technical controls alone cannot sustain robust patching. Architects need to foster security-minded cultures through training that underscores patch diligence. Complacency threatens to unravel all preventative work.

With development velocity accelerating, architects must bring discipline to patch operations. Mastering patch management involves not just tools and tests but instilling vigilance and accountability at all levels. When patching is collectively owned as essential hygiene, organizations can efficiently build resilience into the very fabric of their software DNA.

## Best practices for patch management

Patch management is the practice of systematically identifying, acquiring, installing, and verifying patches and other software updates across an organization's computers, devices, and systems. Effective patch management incorporates continuous vulnerability monitoring, automated mechanisms to test and deploy patches, and rigorous change governance processes. With threats continually evolving, proactively keeping software, applications, and firmware current through patches that address flaws is imperative for managing risk. Patch management requires maintaining an accurate inventory of assets, their software versions, and update status, as well as policies that define deployment windows and patching responsibilities to ensure cyber resilience. Robust patch management practices provide a key safeguard against the exploitation of known weaknesses and misconfigurations.

Patching is an essential security practice that involves updating software to address vulnerabilities, enhance functionality, or improve performance. In the context of software development, patch management not only applies to third-party components and operating systems but also to the software being developed. This section will detail the best practices for effective patch management within a development life cycle, complemented by a lab exercise and examples.

Robust patch management must intersect with the **software development life cycle (SDLC)**. Building security into development workflows from the start enables more seamless, sustainable patching practices.

In the requirements phase, you must document expected patching needs based on plans for dependencies such as third-party libraries. Track these through design and development. Perform vulnerability static analysis on code pre-production, flagging common weakness types that often have patched frameworks.

Incorporate patch testing and validation sprints into existing QA cycles. Construct regression test suites assessing patch impacts on functionality and performance.

Train developers on secure coding to avoid common bug types, many of which have patched variants. Highlight examples from public vulnerability databases tied to poor coding practices.

Release management processes should notify IT teams of new patch requirements. Monitor issue trackers and mailing lists of incorporated open source components.

By tying patch management into the SDLC, organizations gain development team partnership in life cycle ownership. Developers gain greater motivation to build secure code that minimizes patching overhead. These include the likes of acceptable use policies, defining the contours of authorized resource usage and data classification policies, stipulating how data should be sorted based on its sensitivity level, and incident response procedures, which outline action plans for security incidents.

### *Continuous inventory of assets*

To fully secure software amid complexity, patching necessitates methodical asset management to provide total visibility. Cybersecurity architects must spearhead comprehensive inventorying as the foundation for diligent patching. While patch management focuses on software, proper inventorying is also true of assets, not just applications. It's imperative to know what's in your environment and to automate the isolation of anomalous assets. This is particularly true given environments now contain a mix of IoT, BYOD, GFE (where applicable), unmanaged devices dialing in through VDIs, and all the associated virtual compute instances that come and go in the environment.

By thoroughly cataloging each application component and dependency, from kernels to containers, cybersecurity architects gain an accurate bill of materials detailing the full attack surface. Automated discovery tools integrate these inventories into development pipelines to maintain currency.

With reliable visibility into all constituent elements and versions, cybersecurity architects can systematically cross-reference vulnerabilities to pinpoint required patches. Gaps in inventories invite unseen risks. Architects must champion asset hygiene as the bedrock that enables proactive issue resolution before trouble arises.

Establishing centralized repositories of up-to-date software bills of materials empowers response agility when new threats emerge. Cybersecurity architects use inventories to determine assets that have been affected and then expedite targeted patching. With continuous discovery as a best practice, architects gain control to preempt chaos.

Therefore, cybersecurity architects must maintain an up-to-date inventory of all software assets to ensure that no application or dependency is overlooked during the patching process.

### Automated vulnerability scanning

To sustain secure software, vigilant scanning must check for new issues continuously. Cybersecurity architects need to spearhead automation that bakes vulnerability assessments into development pipelines.

By integrating scanning into build and release processes, cybersecurity architects can institute regular checks for version-specific issues such as log4j or Heartbleed. Dashboards centralize visibility, enabling rapid response.

Automated scanning also assesses dependencies and third-party components for downstream risks. Architects must scan not just their code but ecosystems integrated into the software bill of materials.

Tools such as Snyk and Black Duck integrate natively with popular DevOps orchestrators, testing and then flagging vulnerabilities at code commit time. Cybersecurity architects leverage automation to scan faster than adversaries can build exploits.

Scanning yields actionable intelligence about the current security posture. Cybersecurity architects use automation to shift vulnerability management left into development, resolving issues decisively before reaching users.

Therefore, cybersecurity architects must implement automated tools to scan for vulnerabilities within the software and its components regularly.

### Adopt a patch management policy

To instill patch diligence, architects need to drive the adoption of formal governance procedures. Comprehensive policies encode accountability into vulnerability management.

Patch policies should delineate requirements around asset inventory, scanning cadence, patch sourcing, risk ratings, testing, staging, and release timing. Approval gates ensure quality and regulatory alignment before sensitive systems are patched.

By codifying patch procedures and owners into the policy, architects can embed security into development, release, and change management processes. Frequent iterative updates encourage continuous improvement.

Documented policies demonstrate due diligence to auditors that vulnerability management receives appropriate attention proportional to risk. Without a formal policy, patching easily falls prey to competing priorities.

Cybersecurity architects use common frameworks such as the **NIST Vulnerability Disclosure Policy** as guides to implement comprehensive patch management suited for their environment. Policies foster security hygiene resilience amid churning threats.

Therefore, cybersecurity architects must establish a formal patch management policy that defines how patches are identified, tested, approved, and deployed.

### Prioritize patches

With vast attack surfaces and limited resources, cybersecurity architects must drive vulnerability prioritization schemes to deliver maximum risk reduction.

Tools that rate vulnerabilities via frameworks such as the **Common Vulnerability Scoring System (CVSS)** empower data-driven prioritization based on exploit likelihood and potential impact. Cybersecurity architects focus resources on addressing critical patches first.

Asset criticality also factors into priority. A severe browser bug may warrant lower priority than a moderate flaw in an internet-facing banking application. Cybersecurity architects tailor ranking to their unique environment.

However, even lower-ranked patches require eventual remediation to shrink the attack surface. Cybersecurity architects balance priorities while ensuring comprehensive coverage over time.

Through continuous asset discovery, automated scanning, and calibrated ranking, architects can enable judicious patching, thereby averting crises without overtaxing staff. Prioritization brings order to the barrage of vulnerabilities to methodically shrink risk.

Therefore, cybersecurity architects must assess and prioritize patches based on the severity of the vulnerabilities they address and the criticality of the affected systems.

### Test patches before deployment

To sustain uptime amid constant patching, rigorous validation in staging safeguards against instability or incompatibility. Cybersecurity architects need to mandate layered testing environments to validate updates thoroughly.

Non-production environments mirror production configurations to assess patch impact across integrated systems without business disruption. Cybersecurity architects architect modular environments to test patches for different applications and versions in parallel.

With CI/CD automation, validation workflows deploy patched builds into staging while executing extensive regression test suites. Cybersecurity architects instrument staging to catch any flaws from patch conflicts or component incompatibilities.

Testing outputs clear go/no-go signals on patch quality. Cybersecurity architects leverage validation environments to verify patches and remedy target vulnerabilities without introducing new defects.

While impeding immediate deployment, staged testing ensures patches bolster resilience, not undermine it. Cybersecurity architects enable accelerated delivery of secure software by designing layered test environments into the patch process.

Therefore, cybersecurity architects must validate the stability and compatibility of patches in a non-production environment before rolling them out.

### Regular patch cycles and out-of-band updates

Consistent patching requires cybersecurity architects to implement regular update cycles while enabling urgent off-cycle deployment when priority threats emerge.

Standard cadences such as monthly patching weekends bring predictability for change management. Cybersecurity architects schedule cycles based on release tempo, balancing agility with stability.

For major updates or platform migrations, cybersecurity architects designate change windows that are reserved well in advance to accommodate extensive testing.

However, zero-day risks necessitate exception handling to expedite patches outside of routine maintenance. Cybersecurity architects design emergency workflows with controls to rapidly deploy tested fixes without excessive red tape.

Cybersecurity architects empower response agility by designing change mechanisms that support both routine and urgent patching. Consistent cycles sustain baseline hygiene while exception paths address critical threats that require immediate action.

With rigorous validations enabled by automation and modular cybersecurity architectures, cybersecurity architects can confidently accelerate patch deployment without compromising resilience.

Therefore, cybersecurity architects must schedule regular patch cycles and have procedures in place for urgent out-of-band updates when critical vulnerabilities are identified

### Documentation and audit trails

To demonstrate governance and enable oversight, patching processes require extensive documentation trails. Cybersecurity architects need to mandate detailed logging for audit readiness.

Comprehensive patching records validate adherence to vulnerability management policies, retention requirements, and release procedures. Thorough evidence also satisfies legal and compliance obligations.

Cybersecurity architects instrument build orchestrators, testing frameworks, and deployment pipelines to output immutable logs of all patching events, including tool outputs. Dashboards centralize patch reporting for management visibility.

Audit trails chronicle details such as patch contents, associated **common vulnerabilities and exposures (CVEs)**, validation methods and outputs, deployment dates, and patch statuses across environments.

With diligent documentation enabled by automation, architects can prove security hygiene to auditors and leadership. Evidence transforms patching from ad hoc efforts into disciplined governance-sustaining defenses over time.

Therefore, cybersecurity architects must keep detailed records of all patching activities for accountability and auditing purposes.

### User and developer training

To sustain robust patching, cybersecurity architects must champion education reinforcing human accountability at all levels of the SDLC.

User training emphasizes the risks of delayed patching and the importance of disruption tolerance during maintenance windows. Client-side patching enables endpoint resilience between release cycles.

For developers, secure coding techniques minimize patchable vulnerabilities. Training highlights risks such as memory management errors and injection flaws together with secure alternatives.

Cybersecurity architects also need to foster cross-team connections, allowing infrastructure personnel to better grasp release pacing and developers to learn deployment constraints.

Ongoing patch education combats the complacency that deprioritizes patching amid deadlines. Holistic training cultivates a culture that values patching as an organizational success metric as much as feature releases.

Technical controls enable patching at scale, but disciplined execution relies on people and process excellence. Cybersecurity architects sustain security hygiene through coalition building and training, which reinforces mutual commitment to patching rigor.

Therefore, cybersecurity architects must educate both end users and developers about the importance of patching and secure coding practices.

## Exercise

Like the Windows exercise, this exercise is beyond the initial scope of this book. The concepts and discussion are important, but we also recommend that you read the Packt book *The Software Developer's Guide to Linux*, by David Cohen and Christian Sturm, which provides a good understanding and use of Git and CI/CD. This exercise demonstrates how to implement a patch management process within a development environment. The process includes the use of a version control system, an automated build server, and a testing server.

The prerequisites are as follows:

- A version control system (for example, Git)
- A CI/CD server such as Jenkins
- A testing/staging environment
- An application with third-party component dependencies

Let's look at the steps:

1.  Maintain a software bill of materials:

    A.  Use OWASP Dependency Check to inventory all third-party libraries and versions used.

    B.  Integrate inventory checks into the CI pipeline so that they run continuously.

2.  Configure automated vulnerability scanning:

    A.  Install a scanner such as Snyk into the CI/CD pipeline.

    B.  Configure rules to flag new vulnerabilities as tickets/issues.

3.  Create a patch management policy:

    A.  Draft a policy document detailing processes for patch identification, testing, and release.

    B.  Store the policy in version control so that it can be accessed by all stakeholders.

4.  Prioritize patches by severity:

    A.  Configure scanner severity ratings to inform patch priority.

    B.  Auto-deploy non-critical patches to a testing branch.

5.  Test patches before release:

    A.  Use CI/CD automation to deploy the patched testing branch to the staging environment.

    B.  Execute regression tests to validate patch stability.

6.  Release patches on regular cycles:

    A.  Merge validated patches from the testing branch into the main branch.

    B.  Deploy the main branch to production on scheduled monthly cycles.

7.  Maintain detailed patch logs:

    A.  Instrument CI/CD tools to output immutable patch release logs.

    B.  Review the logs monthly to validate compliance.

8.  Educate personnel on patching best practices:

    A.  Develop training content tailored for developers and end users.

    B.  Track training completion to ensure all personnel stay current.

### *Example scenarios*

Let's look at example 1 – patching a web application framework:

- **Situation**: A critical vulnerability is reported in a web application framework that's used across various company projects

- **Action**: Developers follow the patch management policy to test and apply the framework update

- **Outcome**: Applications are secured with minimal disruption, demonstrating the efficacy of the patch management process

Now, let's look at example 2 – responding to a zero-day vulnerability:

- **Situation**: A zero-day vulnerability is discovered in a third-party library used in the company's payment processing system

- **Action**: The security team follows an out-of-band update

- **Outcome**: The zero-day is identified and the system is patched based on the severity of the attack and its risk

To summarize, robust patch management relies on cultivating human accountability, not just deploying technical controls. Cybersecurity architects must champion continuous education that motivates developers to minimize patchable flaws through secure coding while emphasizing users' roles in tolerating disruption from maintenance. Training that bridges teams also fosters a mutual understanding of release pacing and deployment constraints between developers and operations. Most crucially, ongoing patch education combats the dangerous complacency that deprioritizes this vital task. By reinforcing patching rigor and shared responsibility through coalition building and tailored training, architects can realize more consistent, resilient patch management.

# MFA

As threats become more sophisticated, sole reliance on passwords for authentication is no longer tenable. MFA strengthens identity verification by requiring multiple credentials representing independent factors. Architects hold responsibility for judiciously driving MFA adoption to protect against unauthorized account takeover while enabling productivity.

MFA should be mandatory for privileged accounts and highly sensitive systems, given the risks of lateral movement upon compromise. However, overzealous mandates undermine usability, prompting workarounds and resistance. Architects must create nuanced policies and choose frictionless MFA options that balance security with efficiency.

By complementing passwords with an additional factor such as biometrics or one-time codes, the attack surface is greatly reduced. Yet MFA also necessitates contingency mechanisms should factors become temporarily unavailable. With training and layered options, MFA can be strengthened without hampering operations.

This section will examine best practices for rolling out MFA aligned with access sensitivity. It will provide examples and hands-on labs demonstrating pragmatic MFA implementation while upholding usability and continuity. For authentication, MFA has become indispensable. Architects need to shepherd adoption to prudently safeguard identities without impeding progress.

## Best practices for MFA implementation

MFA is a cybersecurity access control method that necessitates at least two independent credential factors to definitively authenticate a user's identity during system login or transaction attempts. The credential factors span distinct authentication categories, including knowledge factors such as passwords, possession factors such as tokens, and inherent factors such as biometrics. Requiring multiple factors for verification creates additional layers of protection, significantly reducing the risk of compromised individual factors. MFA can halt many attacks, such as compromised password reuse and credential stuffing, in their tracks by demanding additional proof of identity. MFA represents a best practice for identity assurance and access security. This section outlines the best practices for implementing MFA, enhancing security posture, and ensuring that access to systems and data is protected appropriately.

### *Understand the types of authentication factors*

To architect robust MFA, understanding the types of authentication factors is essential. Each factor represents a separate credential category that must be compromised for impersonation.

Knowledge factors such as passwords and PINs are the most common but also the most vulnerable to theft and brute forcing. Possession factors protect against stolen passwords by requiring a physical object only the legitimate user has. Inherence factors such as biometrics uniquely authenticate individuals and cannot be transferred.

By combining multiple factors spanning categories, successful impersonation requires compromising various differing credentials. For example, an OTP token as a possession factor augmented by a fingerprint biometric inherence factor creates multi-layered identity assurance.

Cybersecurity architects should advocate MFA policies requiring at least one factor of each type from users for maximized security. This ensures that the loss of any single factor cannot enable unauthorized access. With comprehensive coverage across factor types, MFA creates formidable identity barriers.

Let's look at each of the factors:

- **Knowledge factors**: Something the user knows (password or PIN)
- **Possession factors**: Something the user has (security token or smartphone)
- **Inherence factors**: Something the user is (biometrics)

## Adopt a layered security approach

To maximize protection, MFA should augment other controls within a defense-in-depth model. MFA alone cannot compensate for weak foundational security. Cybersecurity architects need to advocate for its inclusion into a mesh of mutually reinforcing safeguards.

MFA combined with stringent password policies, endpoint protections, access controls, and monitoring provides overlapping identity protection that's difficult for attackers to overcome. No single control becomes a silver bullet or compensating control.

By itself, MFA only gates initial access – not preventing post-login malicious behavior. Other controls, such as privileged access management, remain essential for managing elevated rights. MFA offers one type of control surface hardening among the many required.

Cybersecurity architects should champion MFA as one component of a resilient identity governance framework that includes stringent authorizations, access reviews, activity monitoring, and anomaly response. MFA raises the bar but other measures fill the remaining gaps.

With MFA woven into a layered model, organizations can efficiently mitigate common attacks such as stolen credentials while empowering users, avoiding productivity disruptions, and enabling oversight.

Therefore, cybersecurity architects must implement MFA as part of a layered defense strategy, ensuring it complements other security measures.

## Mandatory for privileged accounts

To limit lateral movement and contain breaches, architects must mandate MFA for privileged accounts that are capable of catastrophic access if compromised.

MFA requirements for administrator, service, and emergency access accounts add essential checkpoints that restrict unauthorized use. Compromised credentials alone cannot grant adversaries the *keys to the kingdom*.

While MFA broadly increases the security posture, prioritizing protection for the highest-risk administrative identities eliminates the most dangerous attack pathways. Progressively expanding MFA from high to standard risk areas maximizes risk reduction per implementation effort.

For account types such as domain administrator and root, cybersecurity architects should require an additional possession factor coupled with an inherence factor such as biometrics for optimal assurance. The most potentially impactful accounts warrant the strongest MFA.

By proactively hardening authentication for administrative access rather than waiting for incidents, cybersecurity architects can vastly shrink the attack surface and limit potential breach impact. MFA for the highest privileges mitigates devastating credential theft.

Therefore, cybersecurity architects must ensure all administrative and privileged accounts are secured with MFA.

## User education and training

For MFA adoption to succeed, cybersecurity architects must prioritize user education to explain the importance of MFA and its proper usage.

Through training, cybersecurity architects can convey how sole reliance on passwords leaves accounts susceptible and demonstrate how MFA practically eliminates many threat vectors.

Education should walk through MFA registration, factor enrollment, and usage workflows. Instructions should cover scenarios such as replacing lost tokens, enrolling new mobile devices, and recovering access.

Creative training reinforces that the minor added login friction secures accounts from compromise, emphasizing benefits over minor drawbacks. Short training videos keep messaging consistent organization-wide.

User education enables self-service success with MFA while curtailing avoidance attempts by communicating the necessity. For MFA to take hold, cybersecurity architects need to proactively inform and support the user community.

Therefore, cybersecurity architects must conduct user training sessions to educate about the importance and usage of MFA.

## Policy enforcement

For consistent MFA adoption, cybersecurity architects need to champion comprehensive policies mandating MFA for appropriate use cases.

Formal policies should delineate where MFA is required, such as for privileged access and external-facing services. Policies should also outline suitable MFA options, management processes, and user responsibilities.

By codifying MFA requirements into your policy, exceptions become visible violations rather than accepted norms. Strict enforcement compels adoption while allowing tailoring for unique cases through policy exception workflows.

Cybersecurity architects need to instrument systems to automatically enforce MFA rules for covered accounts and scenarios – for example, blocking external VPN logins without a second factor via agents.

MFA policies provide the guardrails that guide consistent adoption. Paired with user education, enforcement formalizes secure access as a standard practice rather than a negotiable choice.

Therefore, cybersecurity architects must create a comprehensive MFA policy that mandates its use where necessary and enforce it strictly.

## Regularly review and update

Sustaining robust MFA requires regular reviews validating proper configuration and identifying the need for updates as user accounts and environments evolve.

Cybersecurity architects need to institute periodic audits while examining users' enrolled authentication factors for issues such as outdated mobile devices nearing end-of-life. Reviews should also confirm that recently added systems and accounts comply with MFA policies.

Updating and testing fallback authentication mechanisms for locked-out users ensures reliable continuity of access. Cybersecurity architects also need to integrate MFA management into off-boarding procedures when employees leave.

To accommodate personnel changes, user contact details that enable MFA recovery notifications should be kept current. Failure to promptly deactivate old accounts and tokens heightens the risk of lingering access.

While tedious, recurring MFA hygiene prevents degradation over time as personnel and systems change. By championing reviews and updates, cybersecurity architects reinforce MFA as living protection.

Therefore, cybersecurity architects must regularly review MFA settings, update the recovery methods, and ensure the contact information is current.

### Fallback mechanism

While strengthening authentication, MFA introduces reliance risks if factors become unavailable unexpectedly, necessitating fallback measures. Cybersecurity architects need to architect reliable alternatives to allow users to recover access.

Fallback options such as OTP codes via SMS, printed codebooks, or security questions provide contingency authentication should primary factors fail. However, fallback effectiveness relies on keeping alternative factors secure yet available.

Cybersecurity architects also need to design automated workflows for assisting locked-out users and enacting temporary emergency access when warranted by urgency. Integrations enabling self-service factor resets improve continuity.

By planning for real-world challenges such as lost devices, cybersecurity architects can ensure MFA enhances security without introducing availability risks. Holistic MFA strategies allow frictionless usage under normal conditions and reliable recovery during disruptions.

With comprehensive fallback mechanisms in place, organizations can mandate stringent MFA confidently across critical systems, improving security posture without concerns about loss of access.

Therefore, cybersecurity architects must implement fallback mechanisms for MFA, ensuring users can regain access if one factor is compromised or inaccessible.

### Compliance and standards adherence

To demonstrate due care, cybersecurity architects should align MFA implementations with relevant compliance mandates and industry best practice standards.

Regulations such as PCI DSS, HIPAA, and GDPR explicitly require MFA for secure access in numerous scenarios. Mapping policies to these frameworks validates compliance rigor.

Standards such as NIST SP 800-63B codify leading practices for digital identity assurance. Following NIST guidelines for acceptable MFA types and robust management ensures an organization implements MFA thoughtfully.

Certifications such as ISO 27001 also mandate multi-factor controls within identity frameworks. Cybersecurity architects employ standards as guardrails that guide secure implementations following vetted examples.

Adhering to applicable compliance and standards reduces audit findings, the risk of fines, and the likelihood of breaches. Architecting MFA around recognized frameworks demonstrates a duty of care.

Therefore, cybersecurity architects must align with regulatory requirements and industry standards, such as those specified by NIST or ISO.

## Exercise

In this exercise, we will set up MFA for a web application using an authenticator app as the secondary factor. This will help you understand the process of enhancing authentication mechanisms.

The prerequisites are as follows:

- A web application with user login capabilities
- Administrative access to the web application server
- An MFA plugin that's compatible with the web application platform (for example, Duo Security)
- A mobile device with an authenticator app (for example, Google Authenticator)

Let's look at the steps:

1. Assess the authentication flow:

    A.  Identify the login points in the application's user workflows to integrate MFA.

2. Install and configure the MFA plugin:

    A.  Install the MFA plugin on the web application server.

    B.  Enable and configure MFA in the plugin settings.

3. Initialize the MFA registration process:

    A.  Provide users with an enrollment option via their user profile.

    B.  Ensure the user scans the QR code with the authenticator app to register their device.

4.  Test MFA:

    A.  Log out and log back in to validate MFA prompts for a verification code.

    B.  Input the code from the authenticator app and confirm that access has been granted.

5.  Formalize MFA policies:

    A.  Mandate MFA for all users, especially administrators.

    B.  Communicate MFA requirements and provide training.

6.  Document the implementation details:

    A.  Record configurations, policies, and user workflows.

    B.  Confirm that the MFA approach complies with security standards.

7.  Set up fallback options:

    A.  Provide alternative verification methods, such as mobile SMS.

    B.  Create a self-service workflow for users to reset MFA.

By following these steps, you can deploy MFA to harden the sign-in process for a web application against threats such as stolen credentials.

## Example scenarios

Let's look at example 1 – MFA for corporate email:

- **Situation**: Corporate email accounts are only protected by passwords
- **Action**: IT implements MFA, requiring a push notification to a smartphone before access is granted
- **Outcome**: Email accounts are more secure, and unauthorized access attempts are reduced

Now, let's look at example 2 – MFA in online banking:

- **Situation**: An online banking platform requires enhanced security for customer accounts
- **Action**: The bank integrates MFA, requiring customers to use a biometric factor after entering their password
- **Outcome**: Customer accounts are significantly more secure, and confidence in the bank's security measures is increased

Through these best practices and the step-by-step lab provided, you now have a clear understanding of how to implement MFA and underscore its importance in the modern security landscape. By following these guidelines, organizations can significantly improve their security posture against the backdrop of an ever-evolving threat environment.

# Security training

In cybersecurity, humans represent both the weakest link and the strongest defense. While technical controls form the foundation, resilient protection relies on an aware, responsive workforce – a vigilant human firewall. Security architects hold crucial responsibility for instilling comprehensive training that informs, engages, and empowers employees at all levels to identify and prevent threats.

By championing strategic, personalized programs, architects can shape training into an asset that pays perpetual dividends. Immersive simulations and labs provide hands-on experience in recognizing and responding to real-world attacks. Customized training demonstrates relevance for each learner's unique role.

However, classroom instruction alone has limited impact without cultural reinforcement. Training should be sustained through continuous micro-learning that keeps security top of mind. Architects need to constantly cultivate human defenses through training as a long-term investment that ultimately determines the strength of organizational defenses.

This section will detail how to implement modern security training that sticks. It will provide examples and labs to make concepts tangible. While technology erects security guardrails, humans serve as the sentries that ultimately decide victory or defeat. With comprehensive training, architects can transform workforces into the most perceptive, agile frontline defenses.

## Best practices for effective security training

Security training is an essential element in strengthening an organization's human firewall. It equips employees with the knowledge and skills needed to recognize and prevent security threats. This section highlights the best practices for conducting effective security training while incorporating both strategic insights and practical steps and also includes labs and real-world examples.

### Tailored training programs

Impactful security training recognizes audiences have diverse needs and tailors content accordingly. Cybersecurity architects need to advocate role-based customization addressing learners' unique requirements and risks.

For software engineers, training should provide secure coding techniques preventing vulnerabilities such as injection flaws or buffer overflows. HR personnel warrant training on data privacy risks. Phishing simulation labs help strengthen human defenses against this threat vector.

Front desk staff represent the public face of the organization and require customer service training coupled with education spotting social engineering tactics. Security teams benefit most from emerging attacker tradecraft research and response drills.

While foundational concepts apply universally, architects need to champion personalized, relevant training. By mapping programs to audience needs, cybersecurity architects boost engagement, job-specific capabilities, and motivation to apply learning.

Therefore, cybersecurity architects must customize the training content to the roles and responsibilities of employees. For instance, developers should receive secure coding training, while finance staff should be educated on the risks of phishing scams related to financial transactions.

## *Engagement and interactivity*

For training to stick, cybersecurity architects need to advocate engaging modalities such as gamification, discussions, and simulations to make security concepts tangible through hands-on experience.

Well-designed gamification platforms use rewarding experiences to reinforce secure practices. Quizzes provide knowledge checks and reinforce retention. Immersive simulation labs bring threats to life in a controlled environment.

Cybersecurity architects should discourage monotonous slideware-based training that lacks meaningful interactivity. Taking the extra effort to integrate commanding examples, compelling storytelling, and opportunities for input amplifies learner receptiveness and recall.

By championing training mimicking real-world environments, architects can transform passive instruction into active skill-building where participants practice response muscle memory. Interactive training embeds security instincts.

Therefore, cybersecurity architects must use interactive content such as gamification, quizzes, and simulations to engage participants, making the training memorable and practical.

## *Relevance and realism*

For maximum resonance, training should emphasize real-world relevance through compelling use cases and illustrative incidents underscoring risks and harms. Cybersecurity architects need to champion the integration of contextual examples to demonstrate why security matters.

Recent breaches provide sobering case studies on tangible damage from security failings. Realistic scenarios such as ransomware incidents make consequences visceral. Examples specific to the organization further reinforce stakes, such as past phishing emails that evaded users.

With tangible, credible examples, architects can transform abstract theory into actionable understanding. Employees recognize that poor security invites real harm to themselves, their colleagues, and the organization's mission. This galvanizes retention and culture change.

Through urgency and context, architects shape training that persuades rather than just prescribes. Security fundamentals take on new meaning when tied to relatable implications.

Therefore, cybersecurity architects must integrate real-life examples and recent security incidents into the curriculum to highlight the actual risks and consequences of security lapses.

### Continuous learning

Sustaining strong human defenses requires continuous learning to keep security top of mind. Cybersecurity architects need to champion recurring training addressing evolving threats through micro-learning, refreshers, and updated materials. Ongoing education combats complacency and strengthens institutional memory.

Annual training struggles to impart durable skills given workforce turnover and rapidly advancing attacks. Cybersecurity architects need to advocate regular micro-learning modules, online refreshers, and lunch-and-learn sessions updating learners on emerging risks such as new phishing tactics.

Continuous training also repeatedly stresses fundamentals such as MFA, password management, and social engineering identification. Architects need to foster gamified experiences that frequently reinforce concepts through repetition.

With continuous learning, cybersecurity architects help the workforce internalize lifelong security habits. Well-designed micro-learning curricula transform episodic training into an embedded cultural fixture that strengthens defenses over time.

Therefore, cybersecurity architects must implement a continuous learning approach with periodic refreshers and updates to the training content to address new threats and reinforce previous lessons.

### Measurable outcomes

To demonstrate training impact, cybersecurity architects should institute quantitative metrics such as improved phishing detection rates or faster incident reporting. Measurements validate program efficacy, guiding continual improvement. Clear metrics also help justify further investment in leadership.

Cybersecurity architects need to define **key performance indicators** (**KPIs**) tailored to training objectives, such as users clicking simulated phishing emails or successful malware quarantine rates. Surveys should quantitatively capture comprehension gains.

Effective metrics require baseline measurements establishing starting levels for comparison. For example, sending phishing templates pre-training reveals susceptibility rates to inform curriculum priorities and measure progress.

Post-training, cybersecurity architects need to analyze KPI trends to identify strengths and weaknesses. For example, plateauing metrics may signal the saturation of certain content areas, which results in them requiring refreshers.

With quantitative insights established by metrics, architects can empirically calibrate training for optimal return on investment. Measurements make the case for adequate training resourcing.

Therefore, cybersecurity architects must establish clear metrics to evaluate the effectiveness of the training, such as reduced phishing susceptibility or increased reporting of security incidents.

### Management buy-in

Without executive sponsorship, training struggles for priority. Architects need to obtain managerial buy-in while emphasizing how workforce education protects the organization and enables objectives. Leadership backing is essential for security training to receive adequate focus and resourcing.

By conveying metrics demonstrating training return on investment, cybersecurity architects can build credible business cases warranting adequate budgets. Leadership signoff also enables training to be incorporated into employee performance frameworks to underscore its significance.

Visible executive participation, such as introducing training sessions or participating in simulations, signals priority. This example sets the tone organization-wide.

Cybersecurity architects need to regularly inform leadership of training participation rates and KPI impacts. Consistently positive reporting cements training as a boardroom priority yielding multifaceted dividends.

With a clear managerial mandate, cybersecurity architects can implement continuous training on the scale needed to harden defenses across the entire workforce. Executive-level sponsorship is foundational for success.

Therefore, cybersecurity architects must gain executive support to underscore the importance of security training, ensuring it's perceived as a priority throughout the organization.

### Clear communication

Training relevance relies on framing employee duties and organizational objectives. Cybersecurity architects need to advocate mapping program messaging to each learner's context, explaining how applied learning secures their unique environment. This instills personal investment in the training process.

By tailoring messaging to role-specific risks such as phishing or social engineering, cybersecurity architects can convey why training matters to each group. Breakout exercises encourage small group discussion reinforcing relevance.

Training should provide employees with clear guidance in applying concepts through their daily tasks, such as securely handling sensitive data or identifying suspicious emails. Real-world habit-building requires translation to individual workflows.

With a context promoting personal relevance, architects enable *what's in it for me* moments that drive home the importance of training for learners' unique needs. Tangible connections between tasks and training cement retention and application.

Therefore, cybersecurity architects must communicate the purpose and benefits of the training to participants, explaining how it applies to their day-to-day tasks.

### Regular assessment

Ongoing assessments reinforce retention while uncovering knowledge gaps to refine training. Cybersecurity architects should champion instruments such as pre-post surveys, embedded quizzes, and periodic simulations to gauge comprehension, identify areas for improvement, and confirm concepts stick.

Pre-training assessments establish baseline analytics to inform curriculum priorities and enable objective measurement of progress. Quizzes during training validate learning in real time.

Post-training, assessments confirm retention and long-term application. Follow-up phishing simulations and on-the-job observation provide empirical insight into knowledge durability and practical integration.

By continually measuring outputs through layered assessments, cybersecurity architects can fine-tune training plans and quantify durable workforce security competence. Assessments transform subjective training perceptions into empirical insights.

With comprehensive pre-post assessments enabled by tools, cybersecurity architects can confidently confirm training outcomes while guiding strategic improvements year-over-year. Measurement fuels excellence.

Therefore, cybersecurity architects must conduct pre- and post-training assessments to gauge knowledge gaps and learning progress.

## Exercise

This exercise provides a hands-on experience in simulating a phishing attack, aiming to train employees on how to identify and respond to such threats.

The prerequisites are as follows:

- A group of employees willing to participate in the training simulation
- A controlled training environment where simulated phishing emails can be safely sent and received
- Training material on phishing identification techniques
- Assessment tools to evaluate participant responses

Let's look at the steps:

1. Pre-assessment:

    A. Assess participants' existing knowledge of phishing threats via a questionnaire.

    B. Identify common misconceptions or knowledge gaps to tailor the training.

2. Training setup:

   A. Set up a controlled environment where simulated phishing emails can be sent without actual risk.

   B. Ensure monitoring tools are in place to track participant interactions with the emails.

3. Interactive learning session:

   A. Conduct an interactive session explaining the indicators of phishing emails.

   B. Utilize real-world examples to demonstrate various phishing techniques.

4. Phishing simulation:

   A. Send out the simulated phishing emails to the participants.

   B. Monitor how many participants interact with the email and in what way.

5. Post-interaction debrief:

   A. Gather participants for a debriefing session.

   B. Discuss the simulation results, highlighting successful detections and areas for improvement.

6. Follow-up training:

   A. Based on the simulation results, provide additional targeted training to address specific weaknesses.

   B. Repeat the simulation at a later date to measure improvement.

## Example scenarios

Let's look at example 1 – a phishing awareness campaign:

- **Situation**: An organization has experienced a rise in phishing incidents
- **Action**: A phishing awareness campaign is launched, including a lab-based simulation, to train staff on recognizing suspicious emails
- **Outcome**: Post-training assessments show a 40% improvement in phishing email identification among participants

Now, let's look at example 2 – social engineering defense training:

- **Situation**: Customer service representatives frequently handle sensitive information and are targets for social engineering

- **Action**: A security training lab is conducted, simulating social engineering attempts via phone and email
- **Outcome**: Employees are better prepared to handle such attempts, and there's a notable decrease in information leaks

Security training, through both structured programs and practical simulations, is vital for maintaining an aware and responsive workforce capable of defending against evolving cyber threats. Integrating these best practices and step-by-step labs into security training initiatives can significantly bolster an organization's human defense mechanism against cybersecurity threats.

# Vulnerability scanning

In today's complex and ever-evolving cybersecurity landscape, a team approach to documentation is essential for achieving comprehensive and up-to-date governance. By strategically dividing responsibilities and employing a diverse set of specialized and general-purpose tools, teams can collaborate effectively throughout the documentation life cycle. Utilizing centralized platforms for collaboration further enhances the efficiency and accuracy of the documentation process.

## Best practices for conducting vulnerability scanning

Vulnerability scanning is a critical cybersecurity practice that involves the use of automated tools to identify security vulnerabilities in network devices, systems, and applications. This proactive measure enables organizations to detect and mitigate vulnerabilities before they can be exploited by attackers. This section delves into the best practices for conducting effective vulnerability scans and includes a detailed lab exercise to illustrate the process.

### Comprehensive coverage

For vulnerability management to meaningfully shrink risk, scans must provide total coverage across the entire attack surface. Cybersecurity architects need to champion exhaustive assessment encompassing all assets, on-premises and in the cloud.

Scans should include not just servers and network infrastructure but also user endpoints, mobile devices, OT environments, and APIs. New cloud assets require immediate visibility. Un-assessed assets invite unseen flaws.

Certain domains, such as compliance, require evidence of full-fleet scanning. To holistically reduce weaknesses, cybersecurity architects need to meticulously eliminate coverage gaps through tool integration and unified reporting.

Comprehensive assessment provides security teams and leadership visibility to make data-driven decisions on risks and priorities. Without complete scoping, vulnerability management delivers limited, fragmented value.

Cybersecurity architects must architect scans as ongoing full-spectrum health checks assessing the entirety of hybrid environments. Only comprehensive coverage empowers comprehensive risk reduction.

Therefore, cybersecurity architects must ensure that all assets are covered in the scanning process, including on-premises systems, cloud environments, and remote endpoints.

### Regular and consistent scanning

One-off scans provide limited utility. Continuous vulnerability monitoring through recurring scans is essential to reflect the ever-changing threat landscape. Cybersecurity architects need to champion regular scan schedules balancing frequency with business constraints.

Monthly or weekly cadence provides predictable cycles keeping visibility current amid rapid change. Critical assets may warrant more frequent checks. New cloud workloads require prompt on-boarding into routines.

Consistent cycles establish a rhythm prioritizing scanning amid business-as-usual. Irregular, sporadic scans undermine sustainable risk reduction. Cybersecurity architects need to architect reliable scan automation that minimally impacts operations.

With continuous assessment, new issues are caught early before exploits emerge. Regular scanning combined with streamlined remediation instills resilience through perpetual vigilance, essential in chaotic threat environments.

By championing regular, consistent scanning, architects provide ongoing illumination of risks that security teams and leadership rely upon for rapid, data-driven response. Disciplined routines reinforce protection.

Therefore, cybersecurity architects must conduct scans at regular intervals to catch new vulnerabilities as they emerge. Consider the frequency of scans based on the organization's threat landscape and asset criticality.

### Credentialed scanning

While uncredentialed scans provide value, authenticated scans that use supplied credentials enable far more exhaustive and accurate analysis. Cybersecurity architects should advocate credentialed scanning where feasible for heightened insights.

By allowing scanners full administrator access, they can probe more system functions, registry settings, directory services, and files. Privileged scans eliminate huge swaths of false positives through deeper inspection.

However, credentialed scans introduce risks of misuse and service disruption. Cybersecurity architects need to institute strict access controls and monitoring to secure credentials. Scheduling scans during maintenance windows or on cloning target systems also mitigates risk.

The enhanced visibility and precision of credentialed scanning warrants the additional complexity for critical assets. Cybersecurity architects need to implement scans that grant just enough, but not excessive, privilege to maximize value while minimizing necessary exposure.

Therefore, cybersecurity architects must perform credentialed scans for a deeper analysis. Credentialed scans can log in to systems to evaluate them more thoroughly, providing a more accurate assessment of the vulnerabilities.

## Prioritize findings

The volume of scanning results necessitates prioritizing remediation based on potential business impact. Cybersecurity architects need to drive consistent risk ratings aligned with CVSS and other frameworks to focus on limited resources.

By objectively ranking findings using severity scores, teams can address the most urgent high-risk flaws first, such as remote code execution, while deferring lower risks to maintenance windows.

However, cybersecurity architects still need to ensure lower-ranked vulnerabilities get addressed over time rather than being ignored indefinitely. Curating reports filtering by severity facilitates judicious workflow.

Tagging scan results with risk levels provides crucial context alongside raw findings. Paired with asset criticality and threat intelligence, calibrated prioritization enables optimal risk reduction with finite resources.

With structured triage enabled by frameworks, architects empower security teams to make informed decisions aligning remediation urgency with potential harms.

Therefore, cybersecurity architects must classify vulnerabilities based on risk levels to prioritize remediation efforts. Utilize CVSS scores as a guide.

## Integration with patch management

For vulnerabilities identified through scanning to be mitigated, architects need to tightly integrate assessments with patching processes into cohesive workflows.

Scan results tagged with associated CVEs should automatically trigger scripts that verify the patched status or initiate updates through patch management platforms. Ticketing systems connect scanning and IT teams.

Integrations also enable tracking of vulnerabilities from discovery through remediation completion. Dashboards present consolidated data on the vulnerability life cycle.

Joining scanning and patching breaks down silos enabling rapid, comprehensive responses based on objective risks. Architects need to connect these disjointed domains into coordinated vulnerability management.

With continuous findings triggering actionable patching, architects institute consistent processes, thus reducing weaknesses at scale. Integrations bridge the gap from flaws to fixes.

Therefore, cybersecurity architects must link vulnerability scanning results with patch management processes to ensure that detected vulnerabilities are addressed promptly.

### Perform scanning after significant changes

While regular scans establish routine assessment, major events warrant immediate off-cycle scanning to validate protections. Cybersecurity architects need to mandate vulnerability checks while following significant environmental changes.

Major code releases, infrastructure migrations, new external connectivity, and platform upgrades all introduce potential weaknesses that standard routines could miss.

Scanning provides prompt confirmation that alterations did not unintentionally regress security posture. New cloud workloads in particular necessitate quick validation after deployment.

As per policy, architects need to require vulnerability assessments while following predefined events such as new software roll-outs or architecture shifts. Change triggers the actuation of scans outside cadences.

With scanning rigorously integrated after disruptive events, architects can enable teams to decisively confirm defenses remain intact despite transformation. Change scanning sustains confidence.

Therefore, cybersecurity architects must perform additional scans after any significant network or system changes to identify any new vulnerabilities that have been introduced.

### Diverse toolset

Relying on any single scanning tool risks blindspots from signature gaps or implementation flaws. Cybersecurity architects need to advocate layered scanning from diverse solutions for enhanced perspective.

Different scanning tools leverage unique signature libraries, inspection approaches, and detection algorithms. Orchestrating tools such as Nessus, OpenVAS, and Qualys generate more comprehensive findings.

Static scanning from one perspective only provides part of the picture. Dynamic scanning through actual penetration techniques surfaces different exposures. Both capabilities add value.

With a toolchain of scanners leveraging varied techniques, architects can minimize the chances of any one product missing a subtle but critical flaw. A diversity of assessment capabilities provides security breadth.

However, managing multiple tools adds overhead. Cybersecurity architects balance depth with efficiency to maximize vulnerability insight while maintaining operational manageability.

Therefore, cybersecurity architects must use a combination of scanning tools to get different perspectives on the security posture, as no single tool is guaranteed to catch every issue.

## Validate the results

Due to scan engine imperfections, cybersecurity architects should implement processes to manually verify critical or ambiguous findings to confirm their validity and business impact.

While scanners automate much of the heavy lifting, the nuance of complex weaknesses warrants human validation to avoid false positives derailing operations. Scans occasionally misinterpret behaviors.

For highly sensitive systems, cybersecurity architects need to enact a policy that requires manual verification by technical SMEs before remediation. Even high-fidelity scans contain assumptions.

Validating the results also quantifies the actual business impact of vulnerabilities based on compensating controls such as firewall rules. Manual analysis augments automated assessments with human insight.

By combining automation with human oversight, architects can enable the scalable identification of vulnerabilities while minimizing disruptive false positives. Selective validations provide quality assurance and refine scan engines.

Therefore, cybersecurity architects must manually validate critical or complex vulnerabilities to confirm findings and reduce false positives.

## Compliance with regulations

For heavily regulated industries, aligning vulnerability management to frameworks such as PCI DSS, HIPAA, and SOX demonstrates rigorous security hygiene to auditors.

Compliance mandates often dictate specific scanning requirements such as quarterly external scans or reasonably up-to-date internal scanning. Cybersecurity architects need to design controls and reports to satisfy essential standards.

However, auditors ultimately seek evidence of a comprehensive program holistically reducing risks through scanning, remediation, and oversight. Compliance provides a baseline, not an end state.

With practices mapped to essential regulations, cybersecurity architects can validate program maturity through compliance ratings. But regulations alone cannot guarantee effective, tailored security.

By balancing compliance checkboxes with business risk reduction, architects can demonstrate that security drives value, not simply adherence. Vulnerability management demands outpace what regulations stipulate.

Therefore, cybersecurity architects must adhere to industry standards and compliance requirements, tailoring scanning practices to meet specific regulatory mandates.

## Lab

This lab demonstrates the vulnerability scanning process when using an open source tool to scan a system for known vulnerabilities.

The prerequisites are as follows:

- A controlled lab environment with a target system for scanning, such as a virtual machine with known vulnerabilities

- A vulnerability scanning tool, such as OpenVAS or Nessus (in this example, we'll use OpenVAS; if you have the virtual machine that we used in other labs, you can reuse it)

Let's look at the steps:

1. Preparation and setup:

    A. Install and configure OpenVAS on a scanning machine or use the Kali virtual machine from the previous chapter's lab.

    B. Update the vulnerability definitions database (Greenbone Security Feed for OpenVAS).

2. Configure the target system:

    A. Set up a target system, ensuring it has a mix of services running to simulate a realistic environment.

    B. Prepare a virtual machine as the target system with a vulnerable operating system installed, such as Metasploitable or Windows XP.

    C. Ensure the virtual machine's network setting is in a bridged or host-only configuration to allow scans from your OpenVAS host.

    D. Document the known state of the system, including any intentionally configured vulnerabilities.

3. Security considerations:

    A. Verify that the target system is isolated from the production environment to prevent scan interference.

    B. If necessary, obtain permission to scan the target systems since vulnerability scanning can cause systems to become unresponsive.

4. Scan execution:

    A. Initiate a scan against the target system from OpenVAS, choosing a predefined scan configuration suitable for the target:

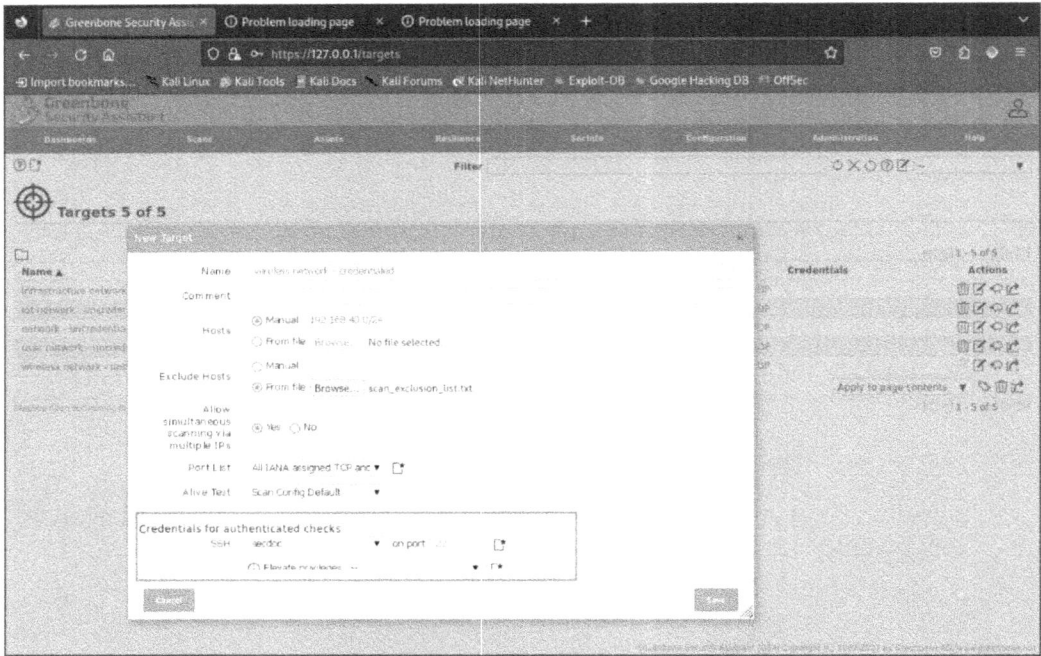

Figure 11.1 – Choosing a predefined scan configuration

B.    Monitor the progress of the scan, noting any issues or interruptions:

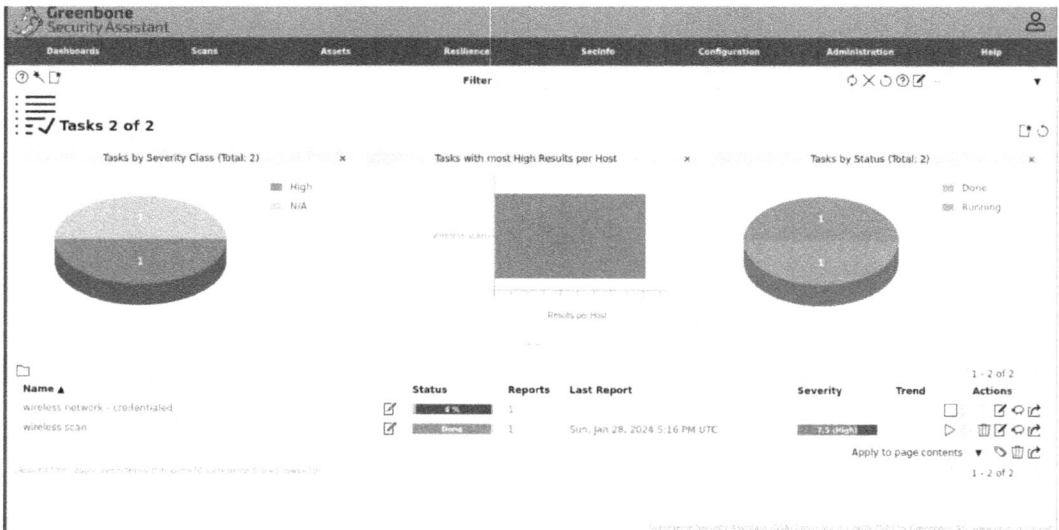

Figure 11.2 – Progress of the scan

5.  Results analysis:

A.  Review the scan results, filtering out false positives and noting down true positives:

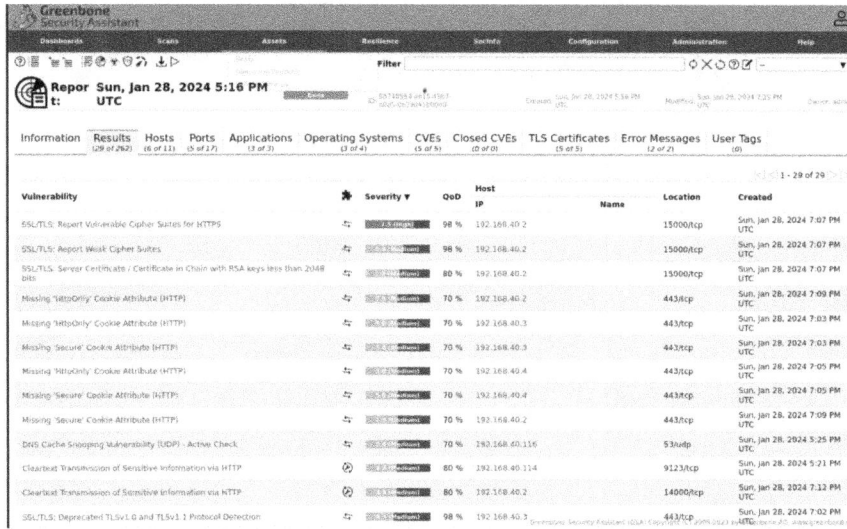

Figure 11.3 – Reviewing the scan results

B.  Prioritize the vulnerabilities based on risk levels:

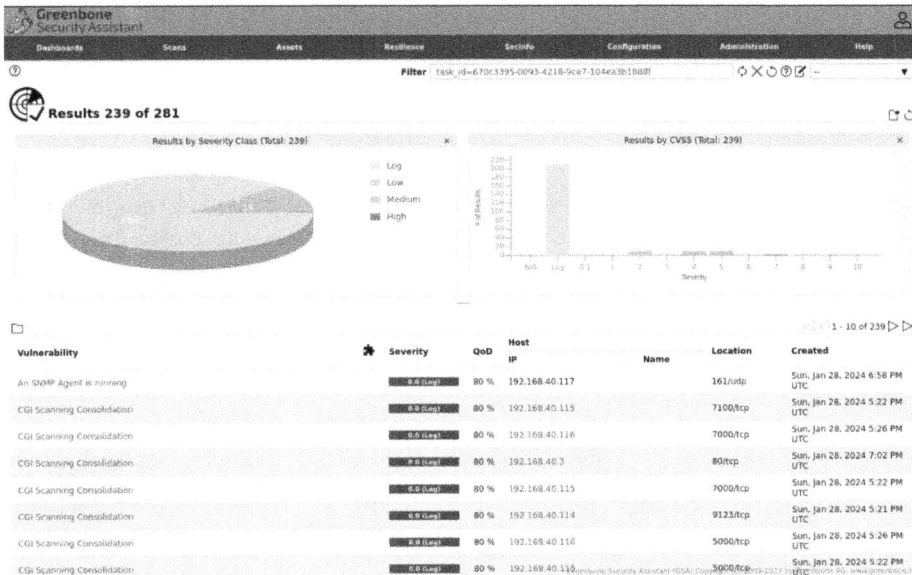

Figure 11.4 – Prioritizing vulnerabilities

6.  Reporting and action:

    A.  Generate a report detailing the findings, including severity, vulnerability description, and remediation recommendations:

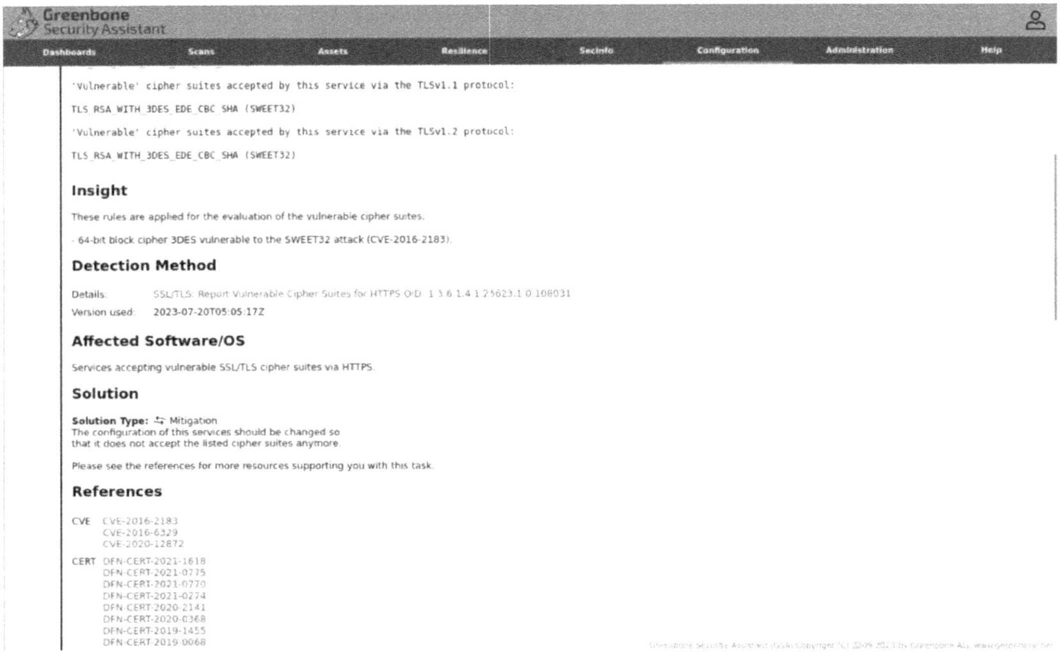

Figure 11.5 – Generated report

    B.  Develop a remediation plan to address the detected vulnerabilities.

By following these detailed steps, you can effectively utilize OpenVAS to identify and prioritize vulnerabilities within your organization's systems. This process should be a key component of your regular security maintenance activities, providing critical insights into your environment's security posture and where improvements can be made.

## Example scenarios

Let's look at example 1 – a routine network scan:

- **Situation**: An organization schedules a monthly vulnerability scan for its internal network

- **Action**: The security team uses OpenVAS to conduct the scan, review the results, and initiate remediation actions

- **Outcome**: Regular scans help the organization keep on top of emerging vulnerabilities, maintaining a robust security posture

Now, let's look at example 2 – post-deployment scan:

- **Situation**: A new application is deployed in the production environment

- **Action**: A post-deployment scan is conducted to ensure that no new vulnerabilities have been introduced

- **Outcome**: The scan reveals several configuration errors, which are promptly fixed, preventing potential security breaches

Incorporating these best practices into the vulnerability scanning process ensures a thorough and efficient approach to identifying and mitigating potential security threats. Regularly scheduled scans, combined with strategic remediation, form a strong foundation for an organization's proactive defense strategy.

# Summary

In this chapter, we explored a wide range of cybersecurity best practices that are essential for organizations to implement to strengthen their security posture. From foundational practices such as least privilege and patch management to critical measures such as MFA, vulnerability scanning, and security training, implementing these guidelines enables organizations to build robust defenses aligned with their business needs.

However, these best practices are most impactful when woven together into a cohesive cybersecurity program, not applied in a piecemeal fashion. Cybersecurity architects hold a crucial responsibility to holistically govern the adoption of complementary best practices that provide defense in depth across people, processes, and technology. By thoughtfully combining standards with business objectives, architects can curate tailored best practice toolkits scaling to their unique environment.

Cybersecurity best practices are not merely recommended actions but are the synthesis of expert experience, regulatory requirements, and lessons learned from past security incidents. They represent the collective wisdom of the cybersecurity community and serve as a critical foundation for building robust security postures.

In practice, interplay exists between established standards and technology-specific best practices. While standards offer a baseline for compliance and a universal language across diverse operational landscapes, technology-specific best practices offer granular, actionable steps tailored to specific tools, platforms, and environments.

Just as Sun Tzu emphasized adaptively applying strategy to circumstances, cybersecurity architects must remain flexible in applying best practices to avoid rigid dogma. While frameworks provide a strong foundation, prudent customization to address specific organizational terrain often determines victory or defeat. Through comprehensive implementation guided by business goals, cybersecurity architects can construct resilient architectures where security empowers operations, not impedes them.

In certain scenarios, the need may arise to prioritize one set of practices over another. This necessity is often dictated by the unique requirements of the operational environment, emerging threats, or regulatory changes. For instance, an emergent zero-day exploit in a widely used software component may necessitate an immediate deviation from the regular patch management cycle, prioritizing rapid mitigation over the established process.

Similarly, a regulatory body might release new compliance requirements that supersede existing protocols, compelling organizations to realign their cybersecurity strategies. In technology-specific contexts, a new best practice may emerge from the evolution of the technology itself, demanding swift adaptation.

The dynamic nature of cybersecurity insists that while adherence to best practices is vital, so is the agility to adapt to new circumstances. This book has armed you with the knowledge to discern when to adhere strictly to standards, when to employ technology-specific best practices, and, crucially, when to navigate the gray area in between.

Ultimately, the greatest strength lies in an organization's ability to combine the rigor of best practices with the flexibility to adapt them as necessary. By embracing this dual approach, cybersecurity practitioners can ensure they provide the highest levels of protection in a rapidly shifting digital world.

Equipped with both strategic and practical knowledge, you can now spearhead the adoption of essential best practices, transforming disjointed controls into a robust human firewall and integrated defense system. With vigilant governance, continuous adaptation, and a nuanced application of standards, organizations can implement cybersecurity excellence at scale.

As a cybersecurity architect, how you design, analyze, and approach architecting solutions and technology is critical to the job function. In the next chapter, we will discuss and build upon previous chapters regarding the need for cybersecurity architects to be adaptable to the business and other organizational goals while still providing the best solution or mitigating the overall risk.

# 12
# Being Adaptable as a Cybersecurity Architect

*"What the ancients called a clever fighter is one who not only wins, but excels in winning with ease."*

*– Sun Tzu*

*"If quick, I survive. If not quick, I am lost. This is death."*

*– Sun Tzu*

*"To secure ourselves against defeat lies in our own hands, but the opportunity of defeating the enemy is provided by the enemy himself."*

*– Sun Tzu*

*"Plan for what it is difficult while it is easy, do what is great while it is small."*

*– Sun Tzu*

*"Ponder and deliberate before you make a move."*

*– Sun Tzu*

In the previous chapter, we explored the implementation of essential cybersecurity best practices that strengthen an organization's security posture when applied comprehensively. Adoption must be governed holistically with business objectives in mind to construct resilient architectures. Just as Sun Tzu emphasized adaptability in strategy, cybersecurity architects should remain flexible in applying best practices to avoid rigid dogma. With thoughtful customization and continuous adaptation guided by business goals, organizations can implement tailored best practice toolkits enabling operations securely. Equipped with strategic and practical knowledge, architects can now spearhead the adoption of complementary best practices, transforming them into robust integrated defenses.

The art of cybersecurity demands adaptability and nuance, as the great general Sun Tzu emphasized. *"What the ancients called a clever fighter is one who not only wins, but excels in winning with ease,"* he noted. Victory relies on implementing security judiciously, not dogmatically.

This aligns with the essence of this chapter – the need for cybersecurity architects to remain agile in applying controls. As Tzu stated, *"If quick, I survive. If not quick, I am lost. This is death."* Architects must react swiftly to evolving threats and business needs through rapid mitigation and balanced security.

Tzu also noted that opportunities for victory come from adversaries themselves: *"To secure ourselves against defeat lies in our own hands, but the opportunity of defeating the enemy is provided by the enemy himself."* By understanding the motives and methods of attackers, architects can implement pointed defenses while enabling business.

Additionally, Tzu emphasized preparedness and forethought: *"Plan for what is difficult while it is easy, do what is great while it is small."* Architects must architect with the future in mind, scaling defenses before threats escalate. As Tzu said, *"Ponder and deliberate before you make a move."*

These concepts and the need to be adaptable tie back to the concept of the **OODA loop** discussed in *Chapter 7*. The ability to be adaptable allows the cybersecurity architect to be able to flow through the OODA loop quickly to pivot and deal with the cybersecurity threats being posed to the enterprise.

This chapter will equip architects to align security as a strategic enabler, not an impediment. By mastering balanced implementation, swift mitigation, and adaptable controls, architects can secure organizations with efficiency and precision, as Tzu described.

The chapter covers the following topics:

- What is adaptability?
- Be a reed in the wind
- Mitigation of risk
- Finding balance

# What is adaptability?

Adaptability in cybersecurity refers to the ability to adjust strategies, tactics, and responses effectively in the face of changing circumstances, threats, and technologies.

## The imperative of adaptability in cybersecurity

Adaptability is not just beneficial, but fundamentally necessary for cybersecurity professionals to secure organizations in a rapidly evolving digital landscape. This agility provides resilience against several dynamics.

## Evolving threat landscape

New attack techniques such as ransomware as a service, supply chain compromise, and deepfakes continuously emerge. Adaptive defense incorporating deception tools, vendor assessments, and authentication enhancements is required to counter novel threats.

## Technological advancements

Innovations such as 5G networks, edge computing, and cryptocurrencies create new risks. Architects must continuously assess and integrate new solutions such as microsegmentation, encrypted overlays, and hardware security modules to secure emerging tech. Take, for example, the rise of cloud computing, **Internet of Things (IoT)** devices, and AI creates new challenges and opportunities in cybersecurity. An adaptable approach allows for the integration of new technologies and methodologies into cybersecurity practices.

## Regulatory and compliance changes

As regulations such as the **California Consumer Privacy Act (CCPA)** and **New York State Department of Financial Services (NYDFS)** cybersecurity mandates arise, architects need to nimbly adjust controls and governance processes to satisfy new rules efficiently through policy automation and reporting dashboards.

## Targeted attack strategies

Threat actors personalize tactics to exploit organizational weak spots through reconnaissance. Adaptability requires using threat intel to rapidly shift defenses by strengthening detected vulnerabilities before targeted exploits occur.

## Resource and capability constraints

With finite budgets, architects need to prioritize adaptively, focusing controls on assets with the highest business impact and redirecting investments as conditions evolve. Cloud-based security measures scale cost-efficiently.

## Complex IT environments

Adapting centralized visibility, management, and AI-based analytics is key to securing heterogeneous on-prem, cloud, and hybrid ecosystems. Unified logging, role-based access, and next-gen network monitoring provide resilience.

## Insider threats and human factors

Combating unpredictable insider risks requires tailored behavioral analytics, access controls, and training attuned to workforce risk profiles. Adaptive, integrated technical and human-centric controls mitigate this threat.

In essence, agility in continuously assessing and evolving defenses provides the resilience needed to counter increasingly sophisticated and dynamic threats across ever-more complex environments. Adaptability is the cornerstone of effective modern cybersecurity architectures.

## Cultivating adaptability in application security architecture

As modern software continuously evolves with new frameworks, languages, and paradigms, cybersecurity architects need to champion adaptable application security architectures scaling to meet these dynamic shifts. Rigid and static defenses expose even the most robust applications to emerging threats.

We will explore strategies and technologies that allow architects to implement application security controls that automatically adjust to changes. We will also examine techniques to rapidly reconfigure protections in response to new vulnerabilities.

With an emphasis on DevOps integration, threat modeling, and policy flexibility, architects can construct application security ecosystems that reliably realign defenses amid technology and threat transformations. By fostering adaptability, they sustain protections through turmoil.

### Challenges in application security

Several dynamics make adaptability critical to application security:

- **Accelerated release cycles**: Frequent iterations and continuous delivery necessitate controls keeping pace

- **Microservices and APIs**: Granular components and increased interconnection expand the threat landscape

- **Dynamic languages and frameworks**: The complex risks presented by innovations such as serverless and reactive programming

- **Open source dependencies**: The need to rapidly address vulnerabilities in incorporated third-party libraries and components

- **Insider threats**: The unpredictable risks stemming from compromised or malicious insiders

These realities underscore why rigid application security architectures fail – threats are too fluid and environments too complex. Only adaptable systems sustain defenses amid unrelenting change.

### Enabling adaptive policy and compliance

Mandating universal static standards is untenable in dynamic applications. Architects need to drive policies and compliance processes built for continuous, automated adaptation:

- **Parametrized compliance checking**: Dynamically assess controls and configurations against policies as code and infrastructure evolve

- **Conditional access**: Restrict permissions contextually based on variables such as user role, device posture, and data classification

- **Pipeline integration**: Incorporate compliance gates natively into CI/CD, releasing only compliant builds

- **Dashboard reporting**: Centralize enterprise-wide visibility into compliance gaps to enable responsive enhancement

By policy design, compliance evolves from periodic audits to continuous self-assessment that is woven into development pipelines. Automation provides the adaptation speed imperative to managing risk amid change.

## Threat modeling and segmentation

Applications must be modeled from an attacker perspective, identifying high-risk components and segregating them through adaptive segmentation controls:

- **Asset inventory**: Maintain a real-time catalog of components, dependencies, data flows, and trust boundaries

- **Attack surface analysis**: Model potential threats and highest risk elements, such as sensitive data processing

- **Live segmentation**: Separate components into least-privilege environments using embedded rules that are enforced dynamically

- **Microperimeters**: Secure higher-risk functions into isolated containers or trust zones with tightly restricted access

Cybersecurity architects can enable adaptive segmentation of critical systems and data by integrating threat intelligence. By continually monitoring threat feeds, vulnerability scanners, and other external risk data, organizations can automatically detect infrastructure changes and vulnerabilities. Segmentation policies then reactively isolate high-value components to limit lateral movement threats.

For example, when new sensitive servers come online, the segmentation platform automatically places them in a secure zone with restricted access. Similarly, any systems found to contain critically vulnerable services can trigger policy changes to cordon off the compromised components.

Unlike legacy network architectures with static, coarse-grained perimeters, this intelligent segmentation takes an identity and role-aware approach. Granular microperimeters enforce least privilege around components based on vulnerability profiles, data classifications, workflows, and risk scores.

By leveraging automation informed by continuous risk intelligence, architects can contain modern infrastructure sprawl more surgically. The segmentation remains fluid and focused on separating critical assets from constantly evolving attack paths.

## *Empowering rapid response*

Adaptable infrastructure sustains the rapid recalibration of access, data flows, and interconnection to address emerging threats:

- **Dynamic authorization**: Allow on-demand modification of user access and permissions through policy engines

- **Selective rollback**: Roll back or disconnect risky functionality immediately while keeping applications operational

- **Compartmentalization**: Construct compartmentalized architectures to prevent lateral threat movement

- **Fail-safe defaults**: Implement fail-safe access defaults that can be elevated contextually for least privilege

With built-in agility, security tools, configurations, and system architectures can adapt quickly based on threat intelligence to counter imminent attacks. Response speed compounds the benefits of composable application platforms.

## *Fostering a culture of adaptability*

Beyond technical controls, cultivating an adaptable culture minimizes human-driven risk:

- **Communications emphasizing change management**: Convey the dynamism of the environment and the importance of adapting to teams

- **Training on secure coding**: Incorporate extensive hands-on labs using leading practices against contemporary threats

- **Actionable monitoring**: Implement context-aware alerting, tightly coupled with issue ticketing to drive rapid mitigation

- **Incentives for vulnerability discovery**: Encourage reporting of new threats through recognition, gamification, and rewards

Cybersecurity architects should focus on building adaptive capacity across both technical and human controls. By avoiding rigid policies and static designs, architects can enable security programs that evolve with the threat landscape.

Through ongoing education and communication, architects can condition teams to embrace shared responsibility and proactive learning. By conveying the growing complexity of emerging attack types, architects can spur motivation to participate in continuous enhancement of defenses.

Architects should aim for organizational agility, that is, the ability to recalibrate protections in step with technology innovations and newly observed attack patterns. Whether iterating perimeter designs, adding new monitoring capabilities, or improving incident response plans, adaptability is key.

Fostering this culture of organizational resilience provides the foundation for sustained progression. Alongside adaptive technology and analytics, proactive and educated teams create **human firewalls** that scale and strengthen protections over time. Architects should champion comprehensive readiness initiatives that resist complacency and bridge silos in the name of continuous security advancement.

With adaptive architectures, compliance becomes near real time rather than periodic. Segmentation evolves intelligently, containing emerging risks. Controls self-adjust via automation to counter focused attacks. By implementing application security that bends but does not break, architects master the art of malleable protection.

# Be a reed in the wind

In the multifaceted domain of cybersecurity, the role of the cybersecurity architect is analogous to that of an architect in the physical world, requiring a careful balance between aesthetic design and structural integrity. Like a reed that bends with the wind to avoid breaking, a cybersecurity architect must exhibit flexibility, adapting to changing business landscapes, emerging threats, and evolving technologies without compromising on the overarching goal of risk mitigation. This section elucidates the necessity for adaptability in cybersecurity architecture and strategies for achieving this while aligning with organizational objectives.

## The principle of adaptive security architecture

In order to understand the concept of the cybersecurity architect as a *reed in the wind*, it is essential to grasp the principle of adaptive security architecture. This paradigm emphasizes the ability to quickly adjust and respond to new threats, integrating predictive, preventive, detective, and responsive capabilities.

To align with the concept of the cybersecurity architect as a *reed in the wind*, cybersecurity architects must embrace the principle of adaptive security architecture. This approach emphasizes constructing defenses with agile capabilities to swiftly detect threats and adjust protections.

Adaptive security is founded on four interconnected capabilities:

- **Predictive capabilities**: By leveraging threat intelligence and data analytics, architects can continuously model the evolving threat landscape to anticipate new attacker tradecraft, vulnerabilities, and targets. By preparing for possible future scenarios, organizations can get ahead of threats before incidents occur.

- **Preventive measures**: Architects need to ensure foundational security controls are already in place to cover known threats based on best practices. Preventive hygiene such as patching, access management, and perimeter defense preempt basic attack vectors.

- **Detective controls**: To identify novel threats, adaptive security relies on advanced monitoring systems such as SIEMs, deception tools, and machine learning analytics to quickly detect anomalies and potential incidents for rapid response.

- **Responsive strategies**: Architects must develop and regularly test robust incident response plans to deploy containment and mitigation countermeasures immediately upon threat detection. Speedy intervention limits damage.

By fusing predictive, preventive, detective, and responsive elements into a cohesive architecture, security teams gain the agility to counteract threats early and adjust defenses on the fly. Adaptive security enables architects to become the metaphorical reed bending in the wind.

## Architectural flexibility in alignment with business goals

The crux of architectural flexibility lies in the cybersecurity architect's ability to tailor security strategies that both protect the organization and facilitate its business objectives. This balance necessitates a deep understanding of the business, including its operational workflows, strategic direction, and risk appetite.

A key imperative for cybersecurity architects is the ability to implement flexible security architectures tailored to an organization's unique business needs and strategic direction. This necessitates aligning security decisions with business priorities and risk tolerances:

- **Business-driven security**: Rather than taking a one-size-fits-all approach, architects must drive security strategies according to specific business goals and workflows. Controls should empower business operations instead of hampering productivity with rigid prohibitions. Close collaboration with business stakeholders is essential.

- **Risk assessment aligned with business impact**: Not all assets and threats warrant the same level of protection. Architects need to conduct risk assessments evaluating potential business impacts, and then allocate security resources proportional to actual risks. Critical business processes and sensitive data require extra safeguards, while excessive controls for lower-risk activities divert resources inefficiently.

By maintaining architectural flexibility rooted in business objectives, cybersecurity can evolve from a perceived inhibitor to a strategic enabler. Architects must balance security with productivity through pragmatic controls tailored to organizational needs. Adaptive alignment of cyber defenses with operational realities and risk tolerances enables architects to become the metaphorical reed bending in the wind.

## Adaptation to organizational changes

Organizations are living entities that continually evolve, driven by market trends, regulatory changes, and internal strategies. A cybersecurity architect must be adept at anticipating and responding to these changes.

Organizations undergo continual transformation in today's dynamic business landscape. As strategists securing the enterprise, cybersecurity architects must adeptly realign security programs to evolving environments. Adaptability hinges on several capabilities:

- **Regulatory adaptability**: Architects need to maintain current knowledge of changing regulations and compliance mandates. Security controls and governance processes must be adjusted to satisfy new requirements efficiently without overextending resources.

- **Technological evolution**: As new technologies emerge, architects must comprehend their implications for the threat landscape and attack surface. Controls need re-evaluation and potential redesign to account for different risk profiles introduced.

- **Business continuity and resilience**: While securing the organization, architects must ensure controls do not hamper essential business operations, especially during disruptions. Controls such as BCDR planning and cloud-based redundancy provide continuity amid crises and evolve with the business.

By proactively realigning security architecture in step with organizational transformations, architects position cybersecurity as an enabler of change, not a barrier. Keeping controls aligned through regulatory, technological, and operational shifts allows architects to nimbly bend like the metaphorical *reed in the wind*.

## Case studies – architectural adaptability in action

This subsection contains case studies illustrating how cybersecurity architects have successfully adapted security solutions in the face of specific business transformations, such as mergers and acquisitions, digital transformation initiatives, and shifts in regulatory landscapes:

- **Merger and acquisition**: When Company A acquired Company B, its cybersecurity architects had to rapidly assess and integrate each organization's disparate security tools and policies into a unified architecture. By leveraging central management platforms and carefully coordinating transition timelines, it achieved a streamlined architecture that met expanded needs.

- **Cloud migration**: As the business aimed to migrate critical systems to the cloud, architects conducted in-depth assessments determining essential security capabilities for the new environment. They worked closely with cloud providers to architect cloud-native controls, allowing massive upscaling while complying with regulations.

- **New privacy regulations**: Emerging data privacy regulations forced a redesign of identity and access controls governing sensitive information. Architects implemented contextual access management and data masking to balance security with usability under the new rules. Detailed audits ensured controls were fine-tuned for compliance.

- **Digital transformation**: A push toward modernized digital operations required architects to holistically re-evaluate security platforms to match new tech stack capabilities and user needs. By emphasizing automation, API integration, and DevSecOps, they delivered robust security enabling rapid iteration.

These case studies demonstrate how architects must remain agile regarding security transformations required by business evolutions. Adaptability sustains resilient protection amid continual change.

## Embracing adaptability as a cybersecurity virtue

To be successful, a cybersecurity architect must personify the notion of being a *reed in the wind* – possessing the strength to protect yet the flexibility to adapt. By embracing this philosophy, cybersecurity architects can craft security solutions that not only withstand the forces of change but also leverage them to enhance the organization's security posture. As the cybersecurity landscape continues to shift, the reed that bends in the wind will stand tall, maintaining its integrity and purpose amid the tempests of digital transformation.

## The OODA loop revisited

As has been noted several times through this book, there is an intricate tapestry of cybersecurity, and the path from an entry-level position to the role of a cybersecurity architect is both complex and nuanced. This chapter discusses the impacts and perspectives through the lens of adaptability, drawing inspiration from Sun Tzu's wisdom, the OODA loop, and other concepts discussed previously. As Sun Tzu metaphorically speaks of the limitless combinations of primary elements to create myriad expressions, so too does the field of cybersecurity offer endless permutations of its core principles.

### The OODA loop – a framework for adaptability in cybersecurity

The OODA loop, conceptualized by military strategist John Boyd, is a decision-making process comprising four stages: *Observe*, *Orient*, *Decide*, and *Act*. The OODA loop offers a structured yet flexible framework for cybersecurity professionals to accelerate their career development through each stage.

### Entry-level to mid-level

In entry-level roles, professionals should continuously absorb emerging information on threats, tools, and techniques (*Observe*). They need to connect new knowledge to practical job applications and hands-on projects (*Orient*). Key decisions involve pursuing foundational certifications and training to build core capabilities (*Decide*). Actions revolve around hands-on labs, cyber ranges, and experimentation (*Act*).

### Mid-level to advanced

At the mid-level, cybersecurity experts need to closely monitor threat intelligence, innovations, and specializations (*Observe*). They should analyze how their skills align strategically with organizational security needs (*Orient*). Decisions include diving deeper into a niche or expanding breadth (*Decide*). Actions mean leading projects, earning certifications, and presenting insights to executives (*Act*).

## Advanced to architect

At the higher tiers, observations focus on business objectives, risk environments, and architecture best practices. Orientation requires planning integrated, holistic security programs. Decisions involve embracing leadership roles and emerging tech. Actions revolve around briefing executives, guiding teams, and architecting resilient cybersecurity ecosystems.

By cycling through observation, orientation, decisions, and actions tailored to their career stages, cybersecurity professionals can accelerate their trajectory to attain strategic leadership roles. Just as fighter pilots leveraged OODA loops to outmaneuver rivals, cyber experts can gain career advantages through this agile framework.

### *Real-life application of the OODA loop*

In my own journey, the OODA loop played a pivotal role. From an early career in technology roles, I constantly observed the evolving cybersecurity field. I oriented myself by understanding how my skills fit into the larger security landscape. Decisions were made to pursue specific certifications and roles that enhanced my expertise. Finally, acting on these decisions involved transitioning into roles that progressively built my experience toward the role of a cybersecurity architect.

In my mid-career, I observed the growing importance of compliance and risk management. This orientation led me to decide to acquire a master's degree in information assurance, thus acting to formalize my cybersecurity knowledge. This decision was pivotal in transitioning from technical roles to strategic cybersecurity leadership.

Just as Sun Tzu highlighted the infinite combinations of basic elements to create diverse outcomes, the journey in cybersecurity is about creatively combining core principles and adapting them to unique career paths. The OODA loop serves as a critical framework for this adaptability, guiding cybersecurity professionals through the constant flux of the industry. By mastering this cycle of observation, orientation, decision, and action, a cybersecurity architect can not only anticipate changes but also lead them, crafting a resilient and dynamic career in the ever-evolving world of cybersecurity.

### *Deep dive into the OODA loop for cybersecurity professionals*

The OODA loop offers a powerful framework for cybersecurity professionals to enhance adaptability in career development and strengthen organizational resilience against evolving threats.

## Observe

Continuously monitoring emerging technologies, attack trends, and industry best practices is crucial to identifying new career growth opportunities and areas needing updated security skills. Regularly assessing internal systems and processes provides visibility into potential vulnerabilities requiring mitigation.

## Orient

Strategically analyzing observations enables contextual understanding. Professionals can recognize how new developments such as AI or quantum computing may impact their roles and skills. Security teams gain insights into how observed threats such as supply chain attacks could exploit organizational weaknesses.

## Decide

Informed orientation sets the stage for career moves aligning with industry shifts, such as pursuing certifications in cryptographic security. Observation-guided threat assessments facilitate proactive decisions to bolster defenses, such as implementing vendor risk assessments.

## Act

Finally, prompt execution of decisions is key. Enrolling in advanced training, rotating into emerging tech units, or changing roles accelerates career growth. Rapidly deploying updated data retention policies, access controls, or network monitoring enables resilient security.

By cycling through this loop, security professionals gain career adaptability and strengthen organizational risk management. The OODA loop philosophy facilitates continuous alignment of skills, systems, and protections with the ever-changing cyber landscape.

### *Case study – application of OODA loop in a cybersecurity career*

The OODA loop provides a highly effective framework for making critical career and security decisions amid dynamic conditions. This case study will demonstrate applications of the OODA loop philosophy in two scenarios: navigating a cybersecurity career evolution and responding to a ransomware attack.

It will highlight how the loop's stages of observing, orienting, deciding, and acting can be leveraged to rapidly adapt to changes and threats. The first scenario will showcase using OODA for a strategic career transition into the high-demand field of cloud security. The second scenario will illustrate the loop's utility in enabling agile incident response by quickly assessing a ransomware attack, deciding on mitigation plans, and containing damages through decisive action.

Scenario 1 involves transitioning to a role in cloud security:

- **Observe**: Noticing the rising demand for cloud security experts and the increasing adoption of cloud services by businesses
- **Orient**: Evaluating how your current skills align with cloud security and identifying gaps
- **Decide**: Deciding to pursue specialized training or certifications in cloud security
- **Act**: Enrolling in a course, obtaining the certification, and seeking roles or projects related to cloud security

Scenario 2 involves responding to a ransomware attack:

- **Observe**: Quickly gathering information about the nature of the attack and its impact

- **Orient**: Assessing the threat in the context of the organization's existing security measures and vulnerabilities

- **Decide**: Choosing the best course of action, such as isolating affected systems, initiating backups, or contacting law enforcement

- **Act**: Executing the response plan and mitigating the attack's impact

Together, these scenarios will provide tangible examples of how cybersecurity professionals can apply the OODA loop mindset to enhance their career maneuverability while strengthening organizational resilience against ever-evolving threats. By internalizing OODA across both individual career growth and risk management, security practitioners gain a profound advantage in today's intensely dynamic threat landscape.

Being a *reed in the wind* means exploring applications of the OODA loop framework to cultivate personal and organizational adaptability in cybersecurity. It revisits how the OODA cycle of *Observe, Orient, Decide,* and *Act* can accelerate career development through tailored actions aligned to experience levels. Examples demonstrate using OODA dynamism when responding to ransomware by rapidly gathering context, planning mitigations, and containing impacts. A case study showcases leveraging OODA for strategic transitions into growing fields such as cloud security by closing skill gaps. Together, these scenarios underscore the immense value of embedding OODA-aligned agility to match cybersecurity capabilities with ever-changing threats. By internalizing constant orientation, assessment, and adaptation, both individuals and organizations gain profound abilities to maneuver through turbulent conditions.

# Mitigation of risk

In the realm of cybersecurity, the role of a cybersecurity architect transcends the mere selection of security tools and technologies; it encompasses the holistic design, analysis, and strategic integration of solutions that align with and support the business's objectives. A paramount aspect of this role is the consistent focus on mitigating risk. This chapter builds on previous discussions of adaptability and delves into how a cybersecurity architect orchestrates risk mitigation strategies effectively while aligning with organizational goals.

## Foundations of risk mitigation in cybersecurity architecture

At its core, the role of a cybersecurity architect is to enable the mitigation of organizational risks through architectural strategies. Effective risk mitigation relies on several key foundations:

- **Risk assessment frameworks**: Architects need to leverage comprehensive risk analysis frameworks such as NIST or ISO to systematically identify assets, threats, and vulnerabilities. Risks must be accurately assessed and prioritized based on potential business impacts. This grounds strategies in data-driven decisions.

- **Threat modeling**: Threat modeling methodologies such as STRIDE and PASTA allow architects to proactively design targeted defensive measures into architectures that specifically address relevant threats. By gaming out attacker perspectives, impactful controls emerge.

- **Layered defense mechanisms**: No single control offers impenetrable security. Architects must advocate for defense in depth with overlapping preventive, detective, and responsive controls. With this multi-layered model, multiple defenses must fail for an attack to succeed, greatly reducing risk surface.

By combining continuous risk assessments, adversarial threat modeling, and layered defenses into cybersecurity architecture, architects can enable robust and tailored risk mitigation that is customized to an organization's terrain. This empowers the confident pursuit of business goals.

## Strategic risk mitigation aligning with business objectives

To enable business success, cybersecurity architects must cultivate risk mitigation strategies aligned with organizational goals and risk tolerance. This involves balancing security with operational realities:

- **Enabling secure business practices**: Rather than erecting rigid barriers, architects should collaborate with business leaders to embed frictionless controls directly into workflows. Solutions such as single sign-on and contextual access strengthen protection while maintaining productivity.

- **Cost-effective solutions**: Not all risks warrant expensive controls such as failover clusters or threat monitoring platforms. Architects need to advocate for pragmatic solutions proportional to potential losses. Striking the right balance sustains security investments.

- **Risk appetite and tolerance**: Every organization exhibits unique risk preferences based on strategic priorities. Architects must tailor mitigation to stay within the boundaries set by executive risk appetite. For example, a fintech firm may accept higher risks for rapid innovation.

Through adaptive risk mitigation calibrated to operational objectives and executive risk preferences, security architects position cybersecurity as an enabler, not an impediment. This empowers organizations to confidently pursue business goals.

## Integrating risk mitigation across the organization

To enable pervasive risk reduction, cybersecurity architects must champion the integration of mitigation efforts throughout all organizational facets. This requires both technological and cultural rigor:

- **Collaborative risk management**: Architects need to spearhead collaborative bodies, such as risk management committees, involving leadership across units. Together, they can develop consistent taxonomies assessing threats based on unified metrics. Central platforms provide enterprise-wide visibility enabling coordinated responses.

- **Cultural integration**: Through training tailored to each role, architects reinforce secure mindsets into daily tasks such as vetting emails and validating login prompts. Employee-focused messaging conveys how individuals contribute to collective risk management. Reinforcing positive behaviors builds accountability.

On the technical side, pervasive logging, asset management, and identity governance sustain visibility and access controls across on-premises, cloud, and endpoints. Automation platforms such as SOAR scale policy enforcement.

With the comprehensive integration of risk management into technology, processes, and culture, mitigation becomes a collective responsibility woven into the organizational fabric. Architects enable risk management to persist as a competitive advantage.

## Evolving mitigation strategies in a dynamic threat landscape

With attackers continuously adapting tactics, cybersecurity architects must champion agile risk mitigation capabilities that keep pace with the threat landscape:

- **Adaptive security controls**: Static rule-based controls have limited utility against sophisticated threats. Architects need to implement machine learning systems such as Darktrace that model normal network patterns, automatically detecting and responding to anomalous threats. Such AI-based controls adapt to unique environments.

- **Continuous monitoring and improvement**: Point-in-time assessments provide limited visibility. Architects must architect continuous monitoring via SIEM aggregation, endpoint detection and response, and log analytics. This enables identifying and containing emerging threats in real time.

Furthermore, integrating threat feeds and regularly revisiting risk registers and mitigation strategies maintains an updated understanding of exposures. Simulations and purple teaming validate controls against rising threats.

Through adaptive controls and ongoing enhancement, architects enable risk mitigation to persist as a competitive advantage, even against relentlessly evolving threats. Continuous, data-driven strategies sustain resilient security postures that are aligned with business success.

## Case studies – dynamic risk mitigation in practice

Let us look at a few case studies:

- **Entering new markets**: Expanding internationally required adjusting data governance strategies to address unique regional privacy regulations. Architects implemented automated data masking and access controls to enable localized compliance.

- **Adopting new tech**: Transitioning communications to **Voice over Internet Protocol (VoIP)** opened new attack vectors. Architects counteracted with improved network segmentation, expanded monitoring for anomalous voice traffic, and encryption to effectuate secure adoption.

- **Responding to incidents**: A successful ransomware attack necessitated reviewing disaster recovery postures. Architects instituted immutable backups with isolated recovery environments and tested restored systems against contemporary threats.

- **Mergers and acquisitions**: A corporate acquisition required rapidly integrating disparate security tools into a unified architecture with centralized visibility, role-based access controls, and consolidated alerting to improve incident response.

These examples showcase how architects must continuously recalibrate risk mitigation implementations in response to evolving business environments and threats. Adaptability is key to sustaining optimal risk postures.

## The harmonization of risk mitigation and business strategy

The role of a cybersecurity architect in risk mitigation is integral to sustaining business integrity and success. By adopting a strategy that is deeply rooted in the organization's mission and operational needs, a cybersecurity architect ensures that risk mitigation is not an afterthought but a driving force for secure innovation and growth. Through the adept application of frameworks, collaboration, and continuous adaptation, the architect provides not only a shield against threats but also a catalyst for resilient and secure business practices.

This section emphasizes the pivotal role architects play in holistically integrating risk management into organizational culture, technology, and processes. It advocates fostering collaborative bodies overseeing threat landscapes enterprise-wide alongside grassroots reinforcement of secure behaviors. Continuously evolving mitigation postures are underscored to match rising threats, with highlights on leveraging automation, AI-based adaptive controls, simulations, and threat intelligence. Case studies demonstrate recalibrating defenses, whether entering new markets, adopting emerging tech, or responding to incidents. Together, they showcase the creativity and adaptability required of architects to harmonize optimal risk reduction with business success through tailored, evergreen mitigation capabilities that are woven into the organizational fabric.

# Finding balance

In the high-stakes realm of cybersecurity, architects hold a crucial responsibility to bridge the gap between robust technical defenses and ever-evolving organizational needs. This requires mastering the delicate art of balance, adaptively striking the right equilibrium between security and operational realities.

Much like the strategic flexibility emphasized in Sun Tzu's teachings, cybersecurity architects must remain agile in applying controls to match unique threat environments and business priorities. A rigid, one-size-fits-all approach often hampers productivity or leaves gaps while strict prohibitions invite workaround risks.

To overcome these pitfalls, architects must become strategic advisors who fully comprehend organizational aims, risk appetites, and changing technologies. With this integrated advantage, they can craft tailored solutions aligning security as an enabler, not an impediment.

For example, by implementing single sign-on or step-up authentication, access controls are strengthened without disrupting workflows. Prioritizing patching for mission-critical applications balances risk reduction with operational needs.

Through continuous collaboration, meticulous fine-tuning, and situational compromises, cybersecurity architects master the art of balance. They fulfill the paradoxical mandate of maximizing security while supporting productivity, adaptively aligning protections with ever-evolving threats and business landscapes.

## The art of balancing security and business objectives

Mastering cybersecurity necessitates adaptability in calibrating protections to avoid impeding business innovation and agility, as Sun Tzu underscored. As he emphasized, adaptability and strategic flexibility are essential to victory. This wisdom profoundly aligns with the cybersecurity architect's paradoxical mandate – maximizing security while enabling business success. Architects must implement security judiciously, rapidly mitigate risks, transform threats into opportunities, and proactively plan as key strategies to strike that balance.

### Strategic implementation of security

Rather than implementing blanket prohibitions, architects need to collaborate with business leaders to embed controls seamlessly into workflows. For example, integrating multi-factor authentication into single sign-on platforms strengthens access management while maintaining productivity.

### Rapid mitigation and response

Architects must champion automation and orchestration to enable swift response to events such as vulnerabilities and data exposures. Employing SIEM dashboards and SOAR playbooks allows for one-click incident containment. Prompt intervention is crucial to continuity.

### *Leveraging threats as opportunities*

By gaming out attacker perspectives, impactful controls emerge. For instance, an uptick in supply chain cyber attacks led an architect to design a vendor risk assessment program, improving resilience while optimizing partnerships.

### *Preparedness and proactive planning*

Architects need to continuously evaluate controls against emerging attack trends, using threat intelligence and red teaming. As technologies evolve, architects must assess and redesign architectures accordingly, rather than waiting for disruption. This sustains future-ready defenses.

By mastering balance through strategic implementation, agile response, opportunity creation, and proactive planning, cybersecurity architects can fulfill the paradoxical mandate of maximizing security while enabling business success.

## Adaptive security architecture

To cultivate strategic adaptability, cybersecurity architects must champion adaptive security architectures that dynamically recalibrate defenses. By embracing the *reed-in-the-wind* mentality of strength with flexibility, architects can balance robust protection with business agility. This entails implementing capabilities spanning predictive threat modeling, preventive access controls, advanced behavioral anomaly detection, and rapid incident response playbooks. Adaptive architectures fuse these elements to constitute agile defenses that shift in response to changing conditions. Whether from new regulatory obligations, the adoption of emerging technologies, or the evolution of attack tactics, adaptive architectures provide the foundation to reshape security seamlessly without obstructing productivity. With comprehensive visibility, governance, and user experience considerations woven throughout, adaptive designs enable fearless advancement secured by architected defenses that are nimble enough to match relentless change.

### *Predictive capabilities*

Continuous threat modeling and intelligence analysis allow architects to foresee potential attack vectors and vulnerabilities. They can proactively address risks before incidents unfold.

### *Preventive measures*

Foundational controls such as patching and access management provide wide protection against known tactics. However, they must be implemented flexibly to enable rapid strengthening as new threats emerge.

### *Detective controls*

Implementing advanced monitoring, such as machine learning-driven anomaly detection, provides visibility into novel attacks that bypass preventive controls. Emerging threats are quickly identified.

## Responsive strategies

With robust incident response playbooks and containment protocols, damage from detected threats is swiftly mitigated. Architects architect for seamless investigation, remediation, and recovery capabilities.

An adaptive posture sustains a balance between security and operations. Defenses dynamically bend to match evolving threats without impeding productivity or innovation. Architects must champion adaptive architectures to implement protection judiciously, not dogmatically.

## Architectural flexibility in alignment with business goals

To evolve security into a strategic enabler that sustains innovation, cybersecurity architects must foster deep alignment with business objectives and risk preferences through tailored architectures. The following key strategies enable this synergy.

### Business-driven security

Architects need to become trusted advisors, not antagonists, collaborating closely with business leaders to embed controls seamlessly into workflows for frictionless protection. Solutions must empower and secure operations simultaneously.

### Risk assessment aligned with business impact

Not all assets warrant identical controls. Architects need to conduct contextual risk analysis based on potential business impacts and executive risk appetite to implement calibrated controls. Critical processes receive additional safeguards.

### Cultural alignment

Through tailored education that reinforces secure operational habits, architects can nurture employee accountability, promoting security as an enabler. Positive messaging builds partnerships, not adversarial relationships. Promoting security successes creates confidence.

With architectural flexibility rooted in business intimacy and data-driven assessments, rather than one-size-fits-all diktats, cybersecurity evolves into a strategic asset that secures the organization while fueling – not limiting – progress and innovation.

Balance arises from this unison, weaving cyber resilience intrinsically into the business fabric. Architectural agility sustains robust security postures while enabling operational success.

## Adaptation to organizational changes

As business environments continually evolve, cybersecurity architects must champion adaptive strategies that realign security implementations to changing conditions while avoiding disruption. This requires technological and procedural dexterity across several domains.

### Regulatory compliance

Architects need to maintain a current understanding of evolving regulations, adjusting controls and governance workflows to satisfy new mandates efficiently through automation and policy updates.

### Technology integration

Emerging technologies such as IoT and quantum computing introduce new risk considerations. Architects must proactively security-assess new solutions and redesign controls ahead of integration to preempt incidents.

### Business continuity

During contingencies such as outages, architects must architect redundancy and automated failover mechanisms that secure alternate operations. Disaster recovery controls balance resilience with cost through tiered data retention and recovery time objectives.

By continuously recalibrating security architectures and control frameworks to address shifting regulations, technologies, and business demands, cybersecurity sustains alignment despite volatility. Adaptability enables architects to overcome impediments.

## Achieving work-life balance as a cybersecurity architect

We cannot discuss balance for a cybersecurity architect without discussing the balance that needs to be made between work and home life. In cybersecurity, the responsibilities of an architect can be all-consuming with ever-present threats demanding vigilance. However, just as cybersecurity architects must strike a balance between security and business objectives, they also need a balance between professional and personal realms to avoid burnout.

While the need to strike a balance between work and home is not unique to cybersecurity architects, it can sometimes be absent from the discussion. Mastering work-life integration sustains excellence and mental acuity, enabling clearer judgment amid chaos. Like a resilient reed bending before fierce winds, architects thrive by practicing flexibility across life's domains.

### Understanding work-life imbalance risks

The always-on nature of cybersecurity imposes heavy demands on time, energy, and mental focus. Without proper work-life boundaries, architects risk the following:

- Burnout from unrelenting strain
- Impaired decision-making due to fatigue
- Reduced job satisfaction and creativity
- Prioritizing work over health and relationships
- Increased anxiety from unstructured schedules

Prolonged imbalance takes a toll both professionally and personally (mentally, physically, and emotionally). However, with intentional practices, architects can harmonize responsibilities across spheres.

### Enabling integration through remote work

Remote and hybrid arrangements allow architects to maintain productivity with reduced commutes and location-shifting. This enables both professional and personal priorities through flexibility:

- Schedule focused deep work during peak energy hours for efficiency

- Adjust hours as needed for obligations, such as school or family functions

- Reduce relocation barriers to pursue opportunities anywhere

- Create home offices that optimize comfort and ergonomics

With intentional planning, remote work provides the autonomy to thrive professionally and personally. Architects can shape environments that enable both realms.

### Cultivating integration through wellness

Neglecting self-care corrodes work-life balance. Architects must champion routine wellness habits that fortify physical, mental, and emotional health:

- Take regular time off for comprehensive rejuvenation. Disconnect from work.

- Practice mindfulness techniques such as meditation to reduce stress.

- Maintain healthy sleep routines, leaving ample time for restoration.

- Pursue enriching hobbies and relationships that are unrelated to work.

Making well-being a non-negotiable priority lays the foundation for sustaining professional excellence with integrity and joy. Health enables architects to show up fully in all life domains.

### Achieving harmony through time optimization

With careful time stewardship, architects can maximize the sharing of duties at home and effectiveness at work by doing the following:

- Systemize common tasks for efficiency using guides and checklists

- Schedule focused blocks for energy-intensive deep work

- Limit low-value meetings through clear agendas and regular cadences

- Automate repetitive processes using scripts and macros

With more time yielded through optimization, architects gain the flexibility to devote care where it matters most, at work and at home.

Like mastering the delicate equilibrium between security and operations, finding balance across life's facets is an ongoing journey. But with intentionality, architects can achieve integration, enabling both professional impact and personal fulfillment. Just as the flexible reed withstands storms, work-life harmony sustains architects through turbulence.

## Exercise examples

This subsection delves into detailed scenarios and exercise examples, demonstrating how a cybersecurity architect can be adaptable and apply the OODA loop framework for enhanced adaptability in mitigating and preventing risks and threats.

### Scenario 1 – rapid response to emerging ransomware threat

This exercise aims to instill adaptability in cybersecurity architects through the application of the OODA loop in managing a simulated ransomware attack on an organization's critical infrastructure. The emphasis is on developing a flexible and rapid response to evolving cyber threats.

### Exercise setup

The setup is a virtual environment that is set up for dynamic adaptability:

- **Virtual environment**: Set up a controlled, isolated virtual network mimicking the organization's critical systems and do the following:

  - Aim to emulate your organization's critical network environment in detail

  - Incorporate diverse systems (Windows and Linux servers, workstations, and network devices) to simulate a real-world heterogeneous network

- **Ransomware simulation**: Deploy a benign ransomware simulation tool within this virtual environment and do the following:

  - Choose a ransomware simulation tool such as RanSim or Infection Monkey

  - Ensure the tool is designed for safe, ethical testing without causing actual harm

  - Deploy the simulation tool in your virtual environment

  - Install and configure the tool to mimic real ransomware behavior, understanding that adapting to unexpected scenarios is key to cyber defense

- **Monitoring tools**: Implement network monitoring and endpoint protection tools to detect abnormal activities, such as the following:

  - **Network monitoring**: Deploy tools such as Wireshark or SolarWinds for real-time network surveillance

  - **Endpoint protection**: Implement solutions such as Microsoft Defender or Norton on all virtual machines

- **Configure alerts**: Set alerts for activities that are indicative of ransomware (e.g., unusual network traffic, file changes, unauthorized encryption attempts), highlighting the importance of early detection in a dynamic threat landscape

**Steps for conducting the exercise**

Let us look at the steps:

- **Initiate ransomware simulation**: Launch the simulation tool to mimic an attack, ensuring a comprehensive impact on the network to mirror a real-world breach.
- **Observe**: Vigilantly monitor for alerts and signs of the ransomware simulation, recognizing the need for prompt and flexible responses to evolving cyber threats.
- **Orient**: Analyze the nature of alerts to confirm ransomware activity. Evaluate the affected systems and potential impacts, adapting your understanding to the unfolding scenario.
- **Decide**: Formulate immediate and adaptable containment actions. Establish a flexible communication protocol for incident reporting.
- **Act**: Implement containment strategies effectively, demonstrating adaptability in crisis management. Document the incident comprehensively, noting the adaptive measures taken.

**Post-exercise analysis**

The analysis consists of the following steps:

- **Review the simulation**: Analyze the simulated attack's spread and defenses, focusing on adaptability to unexpected challenges and threat evolution
- **Evaluate response actions**: Assess the agility and effectiveness of your response strategy, identifying gaps and areas for improvement
- **Document lessons learned**: Summarize insights and potential enhancements to strategies, emphasizing the importance of continuous learning and adaptation in cybersecurity

**Exercise summary**

This exercise underscores the criticality of adaptability in cybersecurity architecture. Through simulating and responding to a ransomware attack, professionals enhance their skills in dynamic threat assessment and response, preparing them for real-world cybersecurity challenges.

## Scenario 2 – adapting to cloud migration

The objective of this exercise is to cultivate adaptability in cybersecurity architects during the transition of an organization's IT infrastructure to a cloud-based environment. The focus is on maintaining security integrity and operational efficiency throughout the migration process.

## Exercise setup

First, we establish a cloud simulation environment:

- Utilize platforms such as AWS, Azure, or GCP to create a cloud environment that mirrors real-world scenarios
- This setup will serve as a testing ground for migration strategies and security implementations in a cloud context

Next, we emulate legacy systems within the cloud environment:

- Replicate existing on-premises infrastructure within the cloud simulation
- This step is crucial to understanding the challenges and nuances of transitioning from a traditional to a cloud-based infrastructure

## Steps for conducting the exercise

Let us look at the steps:

- **Observe**: Monitor the cloud environment, focusing on the performance, security posture, and compatibility of legacy applications. Stay informed about evolving cloud security best practices and tools, emphasizing the need for continuous learning in an ever-changing cybersecurity landscape.
- **Orient**: Assess the practicality, risks, and advantages of migrating specific applications and datasets to the cloud. Recognize and plan for necessary modifications in security policies and controls that are pertinent to cloud environments, highlighting the adaptability required in policy formulation and implementation.
- **Decide**: Develop a migration strategy, prioritizing applications and data based on business importance and security implications. Choose suitable cloud security tools and configurations that demonstrate the ability to adapt security measures to different cloud environments and requirements.
- **Act**: Execute the migration plan, beginning with non-critical applications to minimize potential disruptions. Continuously oversee the cloud environment for security concerns and performance issues, adapting strategies as needed to ensure a secure and efficient migration process.

## Post-exercise analysis

The cybersecurity architect successfully directs the migration process, adeptly integrating security into the new cloud infrastructure while ensuring operational efficiency. This exercise highlights the importance of adaptability in managing the dynamic and complex process of cloud migration in cybersecurity.

**Exercise summary**

This exercise demonstrates the critical role of adaptability in cybersecurity, especially in the context of cloud migration. The ability to effectively transition IT infrastructure while maintaining robust security measures and operational efficiency is an essential skill for contemporary cybersecurity architects.

The previous section underscores work-life balance as an essential capability for cybersecurity architects to sustain excellence amid relentless demands. It examines risks of imbalance such as burnout, impaired judgment, and anxiety while providing tactics to cultivate integration. Enabling flexibility through remote work, prioritizing wellness routines, and optimizing time management can empower architects to thrive professionally and personally. Detailed scenarios demonstrate applying adaptability principles when responding to ransomware attacks with agile containment strategies. Additional cases showcase adapting security postures effectively during complex initiatives such as cloud migrations. Together, these examples showcase the multifaceted nature of the adaptability required of architects spanning technological and personal realms to secure organizations while achieving fulfillment. Just as resilient reeds bend without breaking, intentional balance across facets sustains impact amid turbulence.

## Summary

This chapter emphasizes the critical importance of adaptability in cybersecurity, drawing parallels with Sun Tzu's principles of strategic flexibility. Adaptability remains imperative for cybersecurity architects to secure organizations amid relentless change. Just as Sun Tzu emphasized strategic flexibility, architects must implement protections judiciously, not dogmatically. Rigid adherence risks leaving gaps while inflexible prohibitions hamper operations.

Technologically, architects need to architect adaptive security ecosystems that fuse predictive capabilities such as threat modeling, preventive fundamentals such as access controls, detective measures such as AI-powered anomaly detection, and responsive incident playbooks. This layered, agile architecture dynamically recalibrates defenses against shifting conditions.

Architecturally, solutions must align with business workflows, risk appetites, and compliance needs. By collaborating with stakeholders, architects embed frictionless controls natively into processes through secure-by-design paradigms. They eschew one-size-fits-all mandates in favor of pragmatic, right-sized implementations that are tailored to environments.

Strategically, architects must think long-term, scaling defenses before threats escalate through continuous simulation, red teaming, and skills development. But they also need to act decisively, containing incidents rapidly and turning threats into opportunities. OODA loop proficiency accelerates this cycle of observation, orientation, decision, and action.

Finally, personal adaptability enables professional excellence. Like the supple reed weathering storms, integrating wellness and life balance fosters resilience against workplace turbulence. Architects thrive by practicing flexibility across all facets.

This chapter underscores that cybersecurity architects must maintain a state of comprehensive vigilance and preparedness, exhibit nimble responsiveness to threats, customize solutions collaboratively, and center themselves personally to excel in their roles. This holistic approach to adaptability is presented as essential to mastering the art of protection in the context of unrelenting change in the cybersecurity landscape.

Since the next chapter is the final chapter before the conclusion and summary, it is a culmination of what has been presented within the book. The chapter asks you to apply the lessons learned within the context of strategies that have been successfully used to design, develop, and architect solutions within organizations. The chapter will also help you understand the implications of these considerations and strategies as they relate to business or other goals and how to best mitigate risk.

# 13
# Architecture Considerations – Design, Development, and Other Security Strategies – Part 1

*"Strategy without tactics is the slowest route to victory. Tactics without strategy is the noise before defeat."*

*– Sun Tzu*

*"If you know the enemy and know yourself, you need not fear the result of a hundred battles. If you know yourself but not the enemy, for every victory gained you will also suffer a defeat. If you know neither the enemy nor yourself, you will succumb in every battle."*

*– Sun Tzu*

*"Victorious warriors win first and then go to war, while defeated warriors go to war first and then seek to win."*

*– Sun Tzu*

*"To conquer the enemy without resorting to war is the most desirable. The highest form of generalship is to conquer the enemy by strategy."*

*– Sun Tzu*

*"Weak leadership can wreck the soundest strategy."*

*– Sun Tzu*

The previous chapter emphasized the criticality of adaptability for cybersecurity architects to secure organizations amid continuous change. The key focus areas include designing adaptive security architectures, aligning protections with business needs, thinking strategically yet acting decisively, and cultivating personal flexibility.

Architects need to integrate predictive, preventive, detective, and responsive capabilities into adaptable ecosystems recalibrating defenses dynamically. Controls must align seamlessly with organizational workflows, risk tolerance, and compliance obligations.

Strategic planning and rapid response are both imperative and are enabled by OODA loop proficiency. Finally, resilience stems from personal adaptability balancing professional demands with wellness. Holistic vigilance, nimble responsiveness, tailored customization, and personal centering enable architects to master adaptable protection.

This chapter serves as *Part 1* of a comprehensive culmination of the principles, methodologies, and strategies that have been discussed throughout this book while focusing on their application in designing, developing, and architecting robust solutions within organizations. *Part 1* will help you understand and integrate these considerations into your strategy, specifically around design and life cycle. Master strategist Sun Tzu noted *"Strategy without tactics is the slowest route to victory. Tactics without strategy is the noise before defeat."* This underscores the essence of this chapter – the need for cybersecurity architects to harmonize strategic vision with tactical implementation.

Architects must intimately *"know the enemy and know [themselves],"* comprehending threats along with organizational terrain, workflows, and risk tolerance. With this context, they can conquer adversaries through strategies, securing the enterprise before battle rather than reactively going to war and then seeking victory.

By leveraging time-tested tactics such as layered controls, least privilege access, and threat modeling, architects can enact sound strategies tailored to environments. As Tzu warned, *"Weak leadership can wreck the soundest strategy."* Architects must exemplify versatile leadership by fusing security expertise with business acumen.

Just as Tzu emphasized strategy and adaptability as being essential to victorious campaigns, cybersecurity architects must master these to secure enterprises amid relentless threats through architectures that are resilient yet tailored. By internalizing this wisdom, architects can conquer adversaries through flexible strategy and proficient execution.

This chapter covers the following topics:

- Technical design
- Life cycle

# Technical design

In the context of security architecture and solution development, **technical design** is a critical phase where theoretical concepts meet practical implementation. It involves translating requirements into a detailed plan that guides the creation of a system or solution. This section delves into the intricacies of technical design, highlighting its importance in aligning with organizational goals and security requirements.

## Fundamentals of technical design

Technical design forms the crucial bridge between strategic cybersecurity plans and their tangible implementation as resilient architectures and solutions. Core focus areas include system architecture, data architecture, interface design, overarching security architecture, and future-ready adaptability.

Robust technical design requires synthesizing business workflows, data classifications, user needs, compliance obligations, and security priorities into comprehensive diagrams and specifications. Proven frameworks provide guidance while examples demonstrate real-world implementations.

By delving into key technical design fundamentals, architects gain the foundational knowledge to transform high-level security objectives into concrete architectures and solutions fulfilling complex organizational requirements while sustaining robust protection aligned to business goals.

Developing robust technical designs requires architects to synthesize business objectives, user needs, compliance obligations, and security priorities into comprehensive schematics encompassing core elements:

- **System architecture**: Architects define modular infrastructure diagrams delineating server roles, network zones, trust boundaries, and data flows based on business processes and security protocols. Designs balance performance, access, and resilience:

  - **References: The Open Group Architecture Framework (TOGAF), Department of Defense Architecture Framework (DoDAF)**, and AWS Well-Architected Framework

  - **Example**: Architecting on-premises and cloud infrastructure into tiered zones governing data access based on classification levels

- **Data architecture**: Data schemas, structures, and storage designs adhere to proper classification, encryption, retention policies, access controls, and continuity requirements per data type. Rights are scoped to least privilege:

  - **References**: OWASP Data Security Cheatsheet and CIS Controls v8 Data Protection

  - **Example**: Designing encrypted data stores with role-based access and immutable backup protocols per data type such as PII or intellectual property

- **Interface architecture**: External and internal application, system, and user interfaces enforce secure authentication, session management, and access levels aligned to roles. API designs limit integration risks:

  - **References**: OWASP Authentication and Session Management Cheatsheets

  - **Example**: Enforcing MFA and single sign-on for web/mobile apps, along with API request scoping

- **Security architecture**: Multi-layered controls such as firewalls, proxies, sandboxing, IAM, VPNs, and logging are woven together to align with zero-trust models and compliance obligations:

  - **References**: Zero Trust Model, CIS Controls, NIST 800-53, and ISO 27001

  - **Example**: Layering next-generation firewalls, WAFs, sandboxing, VPNs, and analytics tools to secure hybrid environments

- **Future-ready design**: Built-in infrastructure automation and composable architectures allow for the rapid adaptation of configurations and controls when threats evolve or new systems integrate:

  - **References**: NIST Framework for Cyber-Physical Systems and MITRE ATT&CK

  - **Example**: Architecting **Infrastructure as Code** (**IaC**), microservices, and API-driven interconnection for rapidly adapting to new threats

By synthesizing business goals with security capabilities in technical designs, architects can bridge the gap from concepts to solutions, fulfilling organizational needs while ensuring resilient protection against dynamic threats.

## *Aligning with organizational goals*

For cybersecurity technical designs to enable business success, architects must align decisions with overarching organizational objectives. Case study analysis reveals how design choices impact outcomes, highlighting the need to balance technical feasibility with strategic priorities. Close stakeholder engagement ensures designs address business needs.

By examining examples where technical designs proved either too rigid causing business disruption or too lax compromising security, architects gain perspective into crafting flexible solutions. Collaboration with business leaders, IT stakeholders, and compliance teams helps transform security from impediment to enabler through designs securing operations intrinsically. With alignment, technical designs unlock organizational potential rather than restrict it.

## The role of design in business success

Cybersecurity technical design plays a pivotal yet subtle role in enabling organizations to fulfill strategic goals securely. Realizing this potential requires comprehending the nuanced interplay between security and operations:

- **Cybersecurity as a business enabler**: The most impactful architects position security not as a restrictive barrier but as an enabler that seamlessly powers business objectives. By embedding controls intrinsically into workflows and infrastructure through secure-by-design methodologies, security empowers progress.

- **Technical and business goals alignment**: Effective design bridges the gap between security capabilities and business aims. For example, implementing single sign-on improves productivity by reducing redundant authentication yet strengthening access management. Loose coupling and APIs enable legacy modernization without compromising security. With deep business alignment, security elevates potential.

Technical excellence alone creates brittle security. However, combined with intimate business knowledge and secure-by-design principles, architects can implement adaptive solutions unlocking organizational possibilities. This underscores the art of mastering design to fuel success.

## Case study analysis – the impact of design choices

By examining case studies where technical design decisions enabled either strong security posture or business success, as well as cases where misalignment proved detrimental, architects gain perspective into crafting balanced solutions:

- **Examples of rigid designs**: Target implemented overly stringent supply chain network monitoring that alerted continuously on benign anomalies, disrupting operations. Lack of collaboration with business teams and threat modeling resulted in ineffective controls.

- **Examples of lax designs**: Equifax focused on rapid feature delivery without sufficient security testing or architectural controls. Malicious code injected in updates resulted in a major breach halting innovation progress entirely until root causes were addressed.

- **Examples of aligned designs**: Citigroup struggled with productivity declines from disjointed legacy systems. Architects facilitated secure legacy modernization through APIs and single sign-on, improving efficiency.

By analyzing examples where designs were too restrictive, too permissive, or aligned, architects can calibrate solutions for their unique environment, maximizing security while enabling business success. This case-based wisdom grounds effective designs.

## Balancing technical feasibility with strategic priorities

Cybersecurity architects must adeptly balance technical feasibility with desired business outcomes. This involves reconciling constraints with aspirations through pragmatic solutions and stakeholder collaboration:

- **Evaluating technical realities**: Architects need to conduct exhaustive assessments to determine realistic options given legacy systems, budgets, skill gaps, and compliance obligations. This grounds designs in what is technically achievable.

- **Understanding business vision**: By engaging diverse stakeholders, architects gain insights into long-term strategic goals, risk appetite, and pain points. This knowledge anchors designs to desired business outcomes.

- **Reconciling constraints and aspirations**: With technical and strategic vantage, architects can implement elegant solutions aligning infrastructure capabilities with business vision. For instance, API gateways securely connect valuable legacy data to modern apps for improved insights without disruption.

Through meticulous constraint-possibility mapping and sustained stakeholder involvement, cybersecurity architects can craft technical designs that fulfill grand visions while respecting real-world limitations. This balancing act enables security to power progress.

## Engaging with stakeholders for effective design

Cybersecurity technical designs reach their full potential through sustained, collaborative engagement with business leaders, IT teams, and compliance stakeholders. This transforms security from an impediment to an enabler:

- **Engaging business leadership**: Architects should participate in leadership team meetings, summarizing cyber risk landscapes, upcoming regulatory changes, and security program maturity. This enables designing for business strategy.

- **IT and compliance collaboration**: Cross-functional workshops involving IT, security, and compliance teams facilitate knowledge sharing on technical possibilities, control gaps, and regulatory obligations. Collaborative designs emerge.

- **Security as a business advantage**: Well-designed identity management improves workforce productivity through reduced login friction. Builds that integrate security testing accelerate release velocity long-term by preventing downstream defects.

- **Integrating security into business processes**: Architects can collaborate with owners to embed controls into processes naturally, such as access request workflows or automated policy enforcement in CI/CD pipelines. This bakes in security by design.

Sustained stakeholder involvement, not siloed design, allows architects to craft solutions that transform security into an asset enhancing operations, performance, and compliance. This unlocks the full potential of cybersecurity technical design.

## Crafting flexible and adaptive solutions

To sustain protections amid business transformations, cybersecurity architects must champion flexibility and adaptability within technical designs. This equips organizations to securely evolve as threats and technologies rapidly change:

- **Architecting for adaptability**: Solutions should integrate IaC, APIs, microservices, and modular designs that allow controls and systems to be reconfigured, replaced, or extended on demand without disrupting operations.

- **Embracing new capabilities**: Designs need built-in capacity to natively incorporate emerging capabilities such as zero-trust network access, cloud-based threat analytics, or deception tooling to counter novel threats.

- **Case study examples**:

  - A healthcare system's modular network architecture could dynamically rearrange network segments, access tiers, and monitoring levels to safely accommodate IoT devices, cloud workloads, and remote users at scale while maintaining compliance

  - A financial firm could rapidly onboard new fintech acquisitions into their centralized identity and access framework through API integrations, quickly enforcing consistent access policies at scale

Architecting for continual change sustains security resilience. With forward-looking flexible designs, organizations can securely adapt and evolve amid dynamic threats and innovations.

## Cybersecurity design as an enabler of progress

When properly strategized, cybersecurity technical design synergizes security capabilities with business objectives, unlocking organizational potential constrained by threats or outdated technology. The future-leaning architect plays a pivotal role in orchestrating this transformation:

- **Strategic design synergies**: Unified identity frameworks provide improved workforce mobility and collaboration at new levels by enabling secure access consistently across legacy and modern tools. Comprehensive data protections foster innovation using sensitive datasets that would otherwise be inaccessible.

- **Future-oriented mindset**: Architects must think beyond immediate capabilities, architecting scalable foundations that extend protections to emerging tech such as quantum computing or ambient environments. Designs that embrace modern paradigms such as IaC and everything-as-API position organizations to securely innovate fearlessly.

With visionary design blending security seamlessly into the business fabric, cybersecurity elevates from a restrictive necessity to a strategic multiplier that enhances nearly all aspects of operations. Architects hold the power to unleash potential by converging security with progress through foresight and synergy.

In summary, aligning cybersecurity technical designs with organizational goals is essential for ensuring that security measures aid rather than hinder business success. Through careful consideration, stakeholder collaboration, and a balance between technical feasibility and strategic priorities, architects can create cybersecurity frameworks that not only protect but also empower organizations.

## Security by design

Incorporating security into the initial stages of technical design is not just a best practice but a necessity in the modern digital landscape. This section explores the principles of *security by design* and the importance of designing for compliance, ensuring that systems are not only secure but also adhere to regulatory standards.

### Security-by-design principles

To effectively secure complex environments, cybersecurity architects must champion security-by-design principles while holistically embedding protections early across system layers, not bolting on after deployment:

- **Integrating security early**: Rather than deferring security to late phases, architects need to analyze risks and requirements during initial solution envisioning. Subject matter experts then implement layered controls intrinsically across software, infrastructure, network, and data layers in tandem with feature development.

- **Benefits of early integration**: Deeply embedding security controls into architecture and workflows from inception results in drastically reduced vulnerabilities and attack surfaces. Products designed securely from the start significantly reduce the need for expensive redesigns or disruptive patches later.

- **Comprehensive security across layers**: Holistic security permeates all aspects of designs. Data is encrypted, tokenized, and anonymized with minimal access. Code undergoes static analysis, libraries are vetted, and infrastructure is hardened. Network micro-segmentation and monitoring provide further resilience.

By infusing security expertise through collaboration from the earliest phases, architects shift security left into the DNA of solutions, resulting in inherently secure systems aligned with business needs. This exemplifies the art of security by design.

### Integrating compliance into technical design

To satisfy evolving regulations in healthcare, finance, energy, and other sectors, cybersecurity technical designs must deeply embed compliance considerations across areas such as access controls, logging, availability, and encryption:

- **Understanding regulatory requirements**: Architects must maintain current knowledge of laws such as GDPR, HIPAA, PCI DSS, and CCPA. They need to integrate legal teams early to guide compliant designs proactively, avoiding rework.

- **Focus areas for compliant design**: Key considerations include stringent access management to ensure least privilege, detailed activity logging to prove controls, and uptime SLAs. Encrypting data end-to-end ensures protection in transit and at rest, as per standards.

- **Compliance implementation strategies**: Architects weave compliance into designs through encryption schemes to secure sensitive data types, role-based access scopes to limit data exposure, and expansive logging to enable audits of critical operations.

- **Leveraging proven frameworks**: Mapped implementations demonstrably satisfying ISO 27001, NIST 800-53, CIS Controls, and other codified standards expedite audit processes and provide compliance guardrails.

- **Robust patch cycle**: While patching and system updates may seem innocuous compared to headline-grabbing threats, diligent update cycles constitute the backbone of cyber resilience. By implementing structured life cycles that constantly test and roll out remediations in sync with update availability, architects can shift patching from a rote chore to a key enabler in securing operations. Even so, beyond the spotlight occupied by sophisticated cyber attacks, rigorous update cycles addressing essential hygiene issues provide the silent guardians for sustaining organizational security postures amid turbulence. Just as strong foundations enable towering achievements in the physical world, timely patching secures innovation by strengthening digital groundwork, protecting enterprises from threats both common and extraordinary.

With compliance intrinsic to technical design, organizations gain the freedom to innovate securely and serve expanded markets, turning regulatory obligations into strategic advantages.

## Bringing security design principles to life

Hands-on labs and workshops enable cybersecurity architects to tangibly experience integrating security and compliance into technical designs, cementing knowledge through practice.

## Lab exercise – designing user data management for GDPR compliance

This lab exercise aims to design a **General Data Protection Regulation (GDPR)**-compliant user data management system. You will apply the principles of data protection, minimization, encryption, and access control to align with GDPR requirements.

*Scenario overview*: You have been tasked with designing a system for managing user data in a way that complies with the GDPR. The system must handle personal data securely, ensuring privacy and control for users.

Here are the steps:

1. Understand the GDPR requirements:

    I.      Research and discussion:

    - **Task**: Research GDPR requirements relevant to user data management
    - **Focus points**: Consent, data minimization, right to access, right to be forgotten, data portability, and data breach notifications
    - **Outcome**: Compile a list of GDPR requirements that the design must adhere to

    II.     Analyze the implications:

    - **Activity**: Discuss how each GDPR requirement affects system design decisions
    - **Considerations**: The impact on data storage, processing methods, and user interface design

2. Design for data minimization and consent:

    I.      Data minimization:

    - **Task**: Develop a data model that only collects essential user data
    - **Activity**: Create entity-relationship diagrams illustrating the minimal data requirements
    - **Outcome**: A data model that aligns with the principle of data minimization

    II.     Create a consent mechanism:

    - **Task**: Design a user consent mechanism for data collection and processing
    - **Activity**: Sketch interface mockups showing how users provide, revoke, or modify consent
    - **Outcome**: Interface designs demonstrating a clear and user-friendly consent process

3. Implement encryption and access controls:

    I.      Data encryption:

    - **Task**: Plan for the encryption of user data both at rest and in transit
    - **Activity**: Outline the encryption methods and protocols to be used
    - **Outcome**: A documented encryption strategy ensuring data security

    II.    Access control:

- **Task**: Design access control mechanisms to secure user data
- **Activity**: Develop access control policies and roles
- **Outcome**: A detailed access control plan detailing user roles and data access permissions

4. Ensure user rights and breach notification:

    I.    User rights:

- **Task**: Implement features for user rights, such as the right to access, the right to be forgotten, and data portability
- **Activity**: Design system functionalities that allow users to request their data, request deletion, and export their data
- **Outcome**: Functional specifications for user rights implementation

    II.    Breach notification:

- **Task**: Establish a protocol for data breach notification
- **Activity**: Create a breach notification plan detailing the response steps and communication channels
- **Outcome**: A comprehensive breach notification procedure in compliance with GDPR

5. Review and testing:

    I.    Peer review:

- **Task**: Conduct peer reviews of the designs and plans
- **Activity**: Exchange designs with peers for feedback while focusing on GDPR compliance and practicality
- **Outcome**: Refined designs that incorporate peer feedback

    II.    Compliance testing:

- **Task**: Test the design for GDPR compliance
- **Activity**: Simulate scenarios to test the effectiveness of data protection, consent, user rights, and breach notification mechanisms
- **Outcome**: A report on compliance testing results and any identified gaps or improvements

This lab exercise provided hands-on experience in designing a GDPR-compliant user data management system. By applying GDPR principles in a practical scenario, you can gain a deeper understanding of privacy-focused design, preparing you for challenges in real-world implementations.

## Workshop – holistic security design

*Objective*: This workshop aims to guide you through the process of designing a system where security is integrated at every level, fostering an understanding of security as an integral component of system design.

*Overview*: Participants must work in teams to design a system (for example, a web application, a network infrastructure, or a cloud-based service) with a focus on embedding security throughout every aspect of the system.

Follow these steps:

1. Introduction and team formation:

    I.    Workshop introduction:

    - **Activity**: Brief the participants on the objectives and structure of the workshop

    - **Topics**: The importance of holistic security, an overview of security principles, and the role of security in system design

    II.   Form teams:

    - **Activity**: Divide participants into teams

    - **Consideration**: Ensure a mix of skills and experience in each team

2. System conceptualization:

    I.    Define the system:

    - **Task**: Each team selects a type of system to design (for example, web application, network infrastructure, and so on)

    - **Activity**: Brainstorm and outline the basic functionality and components of the system

    II.   Identify security needs:

    - **Task**: Identify potential security threats and requirements for the chosen system

    - **Activity**: Create a list of security considerations specific to the system

3. Design security at each level:

   I.   System architecture:

   - **Task**: Design the system architecture with security as a foundational element
   - **Activity**: Develop architectural diagrams that incorporate security components such as firewalls, intrusion detection systems, and secure communication protocols

   II.  Data security:

   - **Task**: Plan for data security measures such as encryption, access controls, and data integrity checks
   - **Activity**: Design data flow diagrams that include these security measures

   III. Application security:

   - **Task**: Integrate security into the application layer
   - **Activity**: Define secure coding practices, input validation, and session management strategies

   IV.  Network security:

   - **Task**: Design network security measures
   - **Activity**: Include network segmentation, VPNs, and secure wireless protocols in the network design

   V.   Identify security needs

   - **Task**: Identify potential security threats and requirements for the chosen system
   - **Activity**: Create a list of security considerations specific to the system

4. Address compliance and standards:

   I.   Compliance requirements:

   - **Task**: Identify relevant compliance standards (for example, GDPR, HIPAA, and PCI DSS) that apply to the system
   - **Activity**: Integrate compliance requirements into the system design

   II.  Best practice standards:

   - **Task**: Refer to industry best practices and standards (for example, NIST or ISO/IEC 27001)
   - **Activity**: Ensure the design aligns with these standards

5.  Presentation and review:

   I.  Team presentations:

   - **Task**: Each team presents their system design

   - **Activity**: Highlight how security is integrated at each level of the system

   II.  Group review:

   - **Task**: Conduct a collaborative review of each design

   - **Activity**: Provide feedback focusing on the effectiveness and integration of security measures

6.  Reflection and discussion:

   I.  Group discussion:

   - **Task**: Discuss the challenges and lessons learned from the exercise

   - **Topics**: The importance of holistic security, trade-offs in design, and innovative security solutions

Through this workshop, participants will gain hands-on experience in designing systems with security as a core element. This exercise underscores the importance of considering security at every stage of system design, encouraging a mindset shift toward viewing security as an integral, not peripheral, part of system architecture and development.

## Technical design process

The technical design process provides a structured approach to translating abstract requirements into concrete and compliant architectures. Key phases include exhaustive requirements analysis to capture business needs across dimensions such as capabilities, workflows, and compliance obligations. These organic needs are then systematically codified into technical specifications forming the foundation for architecture development. Hands-on labs provide tangible experience with requirement elicitation and translation. By following a rigorous technical design process, architects can develop solutions precisely tailored to fulfill organizational requirements. The articulation of clear technical specifications grounded in business realities is crucial for effective cybersecurity design efforts.

### *Requirement analysis*

Developing cybersecurity technical designs requires methodically progressing through key phases to translate requirements into robust and compliant architectures.

Through workshops with business teams, architects elicit comprehensive requirements across dimensions such as system capabilities, workflows, user roles, and compliance needs. Structured capture prevents gaps.

## Lab exercise – requirements gathering and translation for system design

*Objective*: This lab exercise focuses on practicing the skills of requirements gathering and translating them into technical specifications for a fictional system. Participants will engage in creating questionnaires, conducting mock interviews, and codifying requirements into structured technical specifications.

*Overview*: Participants will assume the role of architects tasked with developing a fictional system. The exercise involves interacting with "stakeholders" to gather requirements and then translating these requirements into technical specifications.

Follow these steps:

1.  Understand the fictional system:

    I.   Introduction to the scenario:

         • **Task**: Brief participants on the fictional system scenario (for example, an e-commerce platform, an internal employee portal, or a customer relationship management system)

         • **Outcome**: Ensure all participants have a clear understanding of the system's purpose and scope

2.  Prepare for requirements gathering:

    I.   Create questionnaires:

         • **Task**: Develop questionnaires aimed at uncovering stakeholder needs and expectations

         • **Guidance**: Include questions on system functionality, performance, security, user experience, and integration with other systems

         • **Outcome**: A comprehensive set of questionnaires tailored to different stakeholder groups

    II.  Role assignments:

         • **Task**: Assign roles to participants, dividing them into architects and stakeholders

         • **Outcome**: Prepare teams for the stakeholder interview process

3.  Conduct stakeholder interviews:

    I.   Mock interviews:

         • **Task**: Architects conduct interviews with stakeholders using the prepared questionnaires

         • **Activity**: Role-play interviews to simulate real-world interactions

         • **Outcome**: Gathered information on system requirements from various perspectives

4.  Translate requirements into technical specifications:

    I.   Codify the requirements:

      • **Task**: Translate the gathered requirements from natural language into structured technical specifications

      • **Activity**: Identify and document key functionalities, system constraints, performance criteria, and security requirements

      • **Outcome**: A set of clear and structured technical specifications for the fictional system

    II.  Identify dependencies and risks:

      • **Task**: Analyze the requirements to identify any dependencies and potential risks

      • **Activity**: Create a matrix or diagram showing interdependencies and risk profiles

      • **Outcome**: A comprehensive understanding of the system's dependencies and risk landscape

    III. Establish acceptance criteria:

      • **Task**: Define criteria for the successful implementation of each requirement

      • **Activity**: Develop measurable and testable criteria for requirement validation

      • **Outcome**: A set of acceptance criteria that align with business needs and technical feasibility

5.  Presentation and feedback:

    I.   Present the specifications:

      • **Task**: Each team presents their translated technical specifications

      • **Activity**: Share and discuss the rationale behind the decisions made during the translation process

      • **Outcome**: Peer feedback and insights into different approaches to requirement translation

    II.  Reflective discussion:

      • **Task**: Engage in a group discussion on the challenges and lessons learned from the exercise

      • **Topics**: Discuss the importance of aligning technical requirements with business needs and the complexities involved in requirement translation

      • **Outcome**: Enhanced understanding and appreciation of the intricacies of requirements gathering and translation in system design

This lab exercise provides architects with practical, hands-on experience in the crucial process of requirements gathering and translation. By engaging in this simulated scenario, participants will develop skills that are essential for creating tailored solutions that address organizational realities and successfully meet stakeholder needs.

## Architectural design

Architectural design represents the creative stage of the technical design process, where requirements are translated into comprehensive infrastructure diagrams, specifications, and components. This involves synthesizing business needs with technical possibilities to create innovative, tailored solutions. Key focus areas include developing holistic system architectures that map out modules, connections, and data flows based on use cases. Rigorous security architecture is woven throughout encompassing controls, protocols, and integrations needed to manage risk. Interface design clearly defines how users and systems will interact with the solution being developed. Compliance requirements are embedded intrinsically across all architecture layers. With a meticulous yet creative architectural design process, cybersecurity professionals can develop tailored technical solutions that securely fulfill complex organizational needs. Architectural design transforms requirements into tangible and compliant technical realities.

### Crafting cybersecurity architectural designs

During architectural design, architects creatively synthesize requirements into comprehensive technical solutions that balance security, compliance, performance, and usability:

- **Develop a holistic system architecture**: Architects diagrammatically map out infrastructure schematics delineating how servers, data stores, networks, endpoints, and modules interconnect based on workflows and data flows. The architecture provides the blueprint.

- **Design the security architecture**: Security architecture is woven throughout the design, incorporating controls such as firewalls, IAM, encryption, monitoring, and integrations needed to manage risk and meet compliance obligations based on frameworks.

- **Create interface designs**: The specifics of how users and external systems will interface with the solution are defined considering access methods, privilege levels, MFA, and security protocols to enable secure interactions.

- **Embed compliance requirements**: Compliance considerations around access management, availability, confidentiality, and audit logging are embedded natively into architecture components to fulfill legal and industry mandates intrinsically.

With comprehensive architecture designs, cybersecurity practitioners bring requirements to life as innovations for securing organizations. Meticulous designs manifest technical visions.

## Lab exercise – creating an architectural blueprint for a cloud-based service

*Objective*: This lab exercise is designed to guide participants through the process of creating an architectural blueprint for a cloud-based service. The exercise focuses on understanding cloud architecture components, designing a scalable and secure cloud infrastructure, and documenting the architecture effectively.

*Overview*: Participants will design a cloud-based service architecture while considering key aspects such as scalability, security, availability, and cost-effectiveness. The chosen cloud platform can be AWS, Azure, or GCP.

> **Note**
>
> While architectural diagrams often focus on internally managed systems, solutions increasingly integrate third-party **Software-as-a-Service (SaaS)** offerings spanning domains such as call centers, CRM, HRIS, and more. Adopting managed cloud services requires adjusting architecture perspectives – rather than building from scratch, the priorities become securely interfacing with external providers, wrapping inherited controls with compensating policies, and safeguarding organizationally managed components.
>
> Hybrid diagrams must map trust boundaries at intersection points between proprietary infrastructure and external vendor environments. Data flow models guide protection efforts around sensitive information traversing networks. Customizing isolation, monitoring, and encryption for interfaces and integrations helps secure critical connections that link services.
>
> By specializing in architecture designs and diagrams for hybrid ecosystems that blend owned and rented systems, architects can accelerate innovation through cloud services while ensuring security policies stretch across blurring perimeters. Just as mapping old and new cities was key for growth in ancient times, updated architecture renderings guide the modern secure adoption of third-party systems.

Follow these steps:

1. Introduce the cloud services and architectures:

    I. Cloud service models:

    - **Task**: Learn about different cloud service models (IaaS, PaaS, and SaaS)

    - **Activity**: Discuss the characteristics and use cases of each model

    II. Cloud architecture fundamentals:

    - **Task**: Understand the basics of cloud architectures, including key components and design principles

    - **Activity**: Review examples of cloud architecture diagrams

2.  Define the requirements and scope:

    I.    Determine the service scope:

    - **Task**: Define the scope of the cloud-based service (for example, a web application, a data processing service, and so on)

    - **Activity**: Outline the core functionalities and objectives of the service

    II.   Identify technical requirements:

    - **Task**: Identify key technical requirements such as performance, scalability, security, and data management

    - **Activity**: Create a requirements document that will guide the architectural design

3.  Design the cloud architecture:

    I.    Select cloud components:

    - **Task**: Choose appropriate cloud components (for example, compute instances, storage options, and database services)

    - **Activity**: Map the components to the requirements defined earlier

    II.   Design for scalability and availability:

    - **Task**: Incorporate scalability and high availability into the design

    - **Activity**: Plan for auto-scaling, load balancing, and multi-zone/multi-region deployment

    III.  Ensure security and compliance:

    - **Task**: Design the architecture with security and compliance in mind

    - **Activity**: Include security controls such as firewalls, IAM policies, encryption, and monitoring services

4.  Document the architecture:

    I.    Create the architectural blueprint:

    - **Task**: Create a detailed architectural blueprint of the cloud-based service

    - **Activity**: Use diagramming tools to visually represent the architecture, including all components and their interactions

    II.    Write an architecture overview:

- **Task**: Write an overview document explaining the architecture
- **Activity**: Describe the purpose of each component and how the architecture meets the service requirements

5.    Review and iteration:

    I.    Peer review:

- **Task**: Conduct peer reviews of the architectural blueprints
- **Activity**: Exchange blueprints with peers for feedback on design choices, scalability, security, and compliance

    II.    Iterative improvement:

- **Task**: Refine the architectural blueprint based on feedback
- **Activity**: Make adjustments to the design to address any shortcomings or inefficiencies identified during the review

6.    Presentation:

    I.    Final presentation:

- **Task**: Present the final architectural blueprint to the group
- **Activity**: Explain the rationale behind design decisions and how the architecture fulfills the defined requirements

This lab exercise equips participants with practical experience in designing a cloud-based service architecture. By going through the process of defining requirements, selecting appropriate cloud components, and creating a detailed blueprint, participants gain a deeper understanding of cloud architectures and their critical role in modern cloud services.

## Data and interface design

In addition to high-level system architecture, cybersecurity technical design requires detailed data and interface plans to realize security and compliance requirements. Data design involves modeling logical schemas and physical flows based on principles such as encryption, partitioning, and access controls tailored to data types. Interface design creates wireframes or interactive prototypes depicting user and system touchpoints. These validate usability, integrate necessary security workflows, and fulfill compliance rules. With meticulous data and interface models, technical designs transform from conceptual visions into tangible and compliant application blueprints ready for engineering implementation. Data and interface design makes architecture real.

## Realizing data and interface architectures

In addition to overall system architecture, cybersecurity technical design requires detailed data and interface plans that fulfill security and compliance needs:

- **Data modeling and mapping**: Architects design logical and physical data models depicting structured databases and flows between data stores. Schema design considers security principles such as encryption, partitioning, access controls, and retention policies tailored to data types such as PII.

- **Interface mockups and prototyping**: To validate usability and security workflow integration, simple wireframes or interactive prototypes depicting key user interfaces must be created. User stories validate designs against required tasks and compliance rules. Software-assisted prototyping accelerates iteration.

The following figure shows a couple of examples of wireframes:

Figure 13.1 – Sample interface markups

With comprehensive data and interface models, technical designs evolve beyond conceptual to tangible application blueprints anchored to security requirements and ready for engineering implementation. Data and interfaces make designs real.

## *Security integration*

**Security integration** represents a pivotal phase of technical design where protective controls and protocols are woven throughout all layers of the architecture. This security architecture delineates technologies such as firewalls, web proxies, SIEM systems, and access management tools and how they work in tandem to reduce risks. Architects assign controls to address vulnerabilities identified during threat modeling. Security workflows such as authentication, access management, and encryption are tightly integrated with system functions and data flows. Compliance requirements are embedded natively into designs through access restrictions, audit logs, availability enhancements, and encryption schemes. With comprehensive security intrinsic across infrastructure, applications, data, and interfaces, organizations can innovate securely, transforming cybersecurity from inhibitor to enabler.

## Intrinsic security integration

Woven throughout technical design, security architecture entails controls, systems, and protocols integrated to reduce risks and uphold compliance:

- **Allocating controls**: Guided by threat modeling identifying system vulnerabilities, architects assign controls such as WAFs, sandboxes, SIEMs, and deception tech to mitigate exposures across infrastructure, apps, data, and networks
- **Embedding security workflows**: Technical designs tightly integrate authentication, access management, encryption, availability enhancement, and other security workflows into system functions and data pipelines, rather than bolting them on after

With security intrinsically woven into technical fabrics, organizations can innovate fearlessly. Being able to bolt on security gives way to embedded protection by design.

## Lab exercise – implementing security controls in design and conducting threat modeling

*Objective*: This lab exercise is designed to provide hands-on experience in integrating security controls, specifically encryption and access controls, into a system design. Additionally, it will involve conducting threat modeling to identify potential security vulnerabilities.

*Overview*: Participants will work on a hypothetical system design, implementing encryption and access controls as core components. They will also engage in threat modeling to assess and mitigate potential security risks associated with the system.

Let's get started:

1. Introduction to security controls and threat modeling:

    I.   Security controls overview:

    - **Task**: Learn about various security controls, focusing on encryption and access controls
    - **Activity**: Discuss how these controls safeguard information integrity, confidentiality, and availability

    II.  The basics of threat modeling:

    - **Task**: Understand the principles of threat modeling:

- **Activity**: Review methodologies such as **Spoofing, Tampering, Repudiation, Information Disclosure, Denial of Service, and Elevation of Privilege (STRIDE)** to identify potential threats

2.  System design and security requirement analysis:

    I.    Design a hypothetical system:

    - **Task**: Sketch a basic design for a hypothetical system (for example, a web application or a network system)

    - **Activity**: Outline key components, such as user interfaces, databases, and network connections

    II.   Identify security requirements:

    - **Task**: Determine the security requirements for the system

    - **Activity**: List the requirements for data protection, user authentication, authorization, and secure communication

3.  Implement encryption:

    I.    Plan data encryption:

    - **Task**: Plan how to implement encryption in the system

    - **Activity**: Choose encryption algorithms for data at rest and in transit (for example, AES, TLS, and so on)

    - **Outcome**: A documented encryption strategy for the system

    II.   Integrate encryption into the design:

    - **Task**: Incorporate the encryption mechanisms into the system design

    - **Activity**: Update the system design so that it includes encryption modules and processes

    - **Outcome**: An updated system design that includes detailed encryption implementation

4.  Implement access controls:

    I.    Design access control mechanisms:

    - **Task**: Develop an access control plan for the system

    - **Activity**: Define roles, permissions, and access policies

    - **Outcome**: A comprehensive access control strategy for the system

    II.    Integrate access controls into system design:

- **Task**: Embed the access control mechanisms into the system architecture
- **Activity**: Update the system design to reflect user authentication and authorization processes
- **Outcome**: A system design with integrated access control mechanisms

5.   Conduct threat modeling:

    I.    Perform threat analysis:

- **Task**: Conduct a threat modeling exercise using the STRIDE methodology
- **Activity**: Identify potential threats for each component of the system
- **Outcome**: A list of identified threats and their potential impact

    II.    Mitigate identified threats:

- **Task**: Develop mitigation strategies for the identified threats
- **Activity**: Propose solutions or design changes to address each identified threat
- **Outcome**: A threat mitigation plan outlining how to address each potential security issue

6.   Review and refinement:

    I.    Peer review:

- **Task**: Review each other's designs and threat models
- **Activity**: Provide feedback on the integration of security controls and the comprehensiveness of the threat model
- **Outcome**: Refined system designs and threat models based on peer feedback

    II.    Group discussion:

- **Task**: Discuss the challenges and lessons learned from the exercise
- **Topics**: The importance of encryption and access controls and insights from threat modeling
- **Outcome**: A deeper understanding of implementing security controls in system design and conducting effective threat modeling

This lab exercise offers a practical approach to understanding and implementing critical security controls in system design. Through encryption and access control integration, along with thorough threat modeling, participants gain valuable skills in creating secure system architectures that can withstand potential security threats.

# Implementing technical designs

The realization of cybersecurity technical designs requires methodical and collaborative implementation across diverse teams. Architects need to lead engagements with engineering groups that can translate architectures into functioning solutions. They must also coordinate with stakeholders. ensuring consistent delivery when matching designs. Standardized development procedures, open communication channels, active issue resolution, and gate-based progress tracking enable smooth implementation. With architects guiding the materialization of designs by empowered and aligned teams, organizations can bring impactful security innovations tailored to needs to fruition efficiently. Diligent collaboration transforms visions into realities.

## *Development strategies*

Cybersecurity architects need to lead engagements with engineering teams using methodologies that optimize the successful implementation of technical designs. Strategic approaches include Agile principles for complex initiatives, which emphasize continuous requirements validation and incremental delivery of modular solutions. However, the waterfall methodology provides the necessary rigorous gates and documentation for compliance. Hybrid models aim to balance Agile velocity and waterfall rigor. Hands-on labs provide tangible experience in executing designs using these approaches. Architects also leverage libraries of standardized policies, controls, and subsystems to accelerate development while customizing implementations to address unique risks. By guiding teams through strategic development approaches, architects can transform security designs into functioning solutions rapidly yet reliably, realizing the full innovation potential.

### Strategic development approaches

Cybersecurity architects must lead engagements with implementation teams using methodologies that optimize the successful realization of technical designs:

- **Agile versus waterfall comparison**: For complex initiatives, Agile principles such as continuous requirements validation and incrementally delivering modular solutions excel. However, waterfall's rigid gates and documentation ensure compliance. Hybrid provides balance.

- **Standardization and customization**: Architects leverage libraries of repeatable policies, controls, and subsystems for acceleration but customize implementations addressing unique risks. Reference models such as the AWS or Azure architecture blueprints provide baselines.

With strategic, collaborative development approaches, architects can transform cybersecurity designs into functioning solutions, thereby fulfilling their full security and innovation potential rapidly and reliably.

### Lab exercise – implementing a technical design using Agile methodology

*Objective*: This lab exercise is designed to guide participants through the process of implementing a technical design using Agile methodology. The focus is on understanding and applying Agile principles to translate a technical design into a working solution.

*Overview*: Participants will be divided into teams, each tasked with implementing a specific technical design (for example, a software application, a network solution, and so on). The exercise will involve iterating through Agile cycles, including planning, development, testing, and review.

Let's get started:

1.  Introduction to Agile methodology:

    I.    Agile principles and practices:

       • **Task**: Learn about the key principles and practices of Agile methodology, including iterative development, self-organization, and continuous improvement

       • **Activity**: Discuss the Agile Manifesto and its implications for software development

    II.   Overview of Agile frameworks:

       • **Task**: Introduce different Agile frameworks, such as Scrum and Kanban

       • **Activity**: Discuss the specifics of Scrum, including roles (Scrum master, product owner, and team members), artifacts (product backlog, sprint backlog, and increment), and events (sprint planning, daily stand-up, sprint review, and sprint retrospective)

2.  Team formation and role assignment:

    I.    Form Agile teams:

       • **Task**: Divide participants into small Agile teams

       • **Outcome**: Each team should have roles assigned, including a Scrum master and product owner

    II.   Define the project:

       • **Task**: Assign each team a specific technical design project

       • **Outcome**: A clear understanding of the project goals and deliverables

3.  Agile project planning:

    I.    Product backlog creation:

       • **Task**: Develop a product backlog that lists all features, functions, requirements, and enhancements

       • **Activity**: Break down the technical design into manageable pieces of work (user stories)

       • **Outcome**: A prioritized product backlog

II.   Sprint planning:

- **Task**: Conduct a sprint planning session to plan the work for the first sprint
- **Activity**: Select items from the product backlog to include in the sprint backlog
- **Outcome**: A defined sprint goal and a sprint backlog

4.  Sprint execution:

I.   Daily Scrum:

- **Task**: Hold daily Scrum meetings to track progress and address any impediments
- **Activity**: Each team member reports what they did yesterday, will do today, and any obstacles
- **Outcome**: An updated sprint backlog with all impediments identified

II.   Development work:

- **Task**: Implement the selected backlog items
- **Activity**: Code, design, test, and integrate features
- **Outcome**: Potentially shippable product increment

5.  Sprint review and retrospective:

I.   Sprint review:

- **Task**: Conduct a sprint review at the end of the sprint
- **Activity**: Demonstrate completed work to stakeholders and collect feedback
- **Outcome**: Feedback incorporated into the next sprint plan

II.   Sprint retrospective:

- **Task**: Hold a sprint retrospective to reflect on the past sprint
- **Activity**: Identify what went well and what could be improved, and plan for improvements in the next sprint
- **Outcome**: Actionable insights for continuous improvement

6.  Iteration and incremental development:

I.   Iterative development:

- **Task**: Repeat the sprint cycle, implementing feedback and new items from the product backlog
- **Outcome**: Continuous development and improvement of the project

II.    Final presentation:

* **Task**: Present the final product and share experiences of using Agile methodology

* **Outcome**: A functional product and insights into the Agile development process

This lab exercise provides hands-on experience in implementing technical designs using Agile methodology. Participants learn to work collaboratively and iteratively, adapting to changes and continuously improving their product and process, embodying the core principles of Agile development.

## Testing and validation

Rigorous testing and validation ensure cybersecurity technical designs are realized as intended with built-in resilience across edge cases. Architects need to mandate layered test suites that assess functionality, interoperability, scalability, security, and compliance. Test cases validate expected flows and behaviors along with resilience against abnormal conditions such as invalid inputs or overloaded systems. Success criteria should encompass usability, performance benchmarks, and standards adherence. Controlled staging environments mirror production, allowing for holistic assessments. By championing comprehensive testing, architects ensure designs materialize into solutions that deliver robust protection aligned with business needs across scenarios. Stringent validation provides the final safeguard, crystallizing designs into secure technical realities.

### Comprehensive testing and validation

To fulfill design potential, rigorous multidimensional testing processes assessing functionality, resilience, and compliance must validate the final implementations:

* **Automated testing frameworks**: Architects design automated test suites deployed in CI/CD pipelines to validate that application functionality, interoperability, and scaling needs are fulfilled as expected with each build.

* **Security and compliance testing**: Static analysis, penetration testing, attack simulations, and compliance audits conducted in staging environments identical to production verify that security and policy requirements are intrinsically met.

* **Validation criteria**: Test cases cover happy-path scenarios along with boundary conditions such as invalid inputs, overloaded systems, or simulated component failures to validate resilience. User acceptance criteria ensure usability and utility.

By championing layered testing regimes throughout development and staging, architects ensure designs securely materialize into solutions that deliver robust protection aligned with business needs across scenarios. Validations crystallize visions into reality.

## *Deployment and monitoring*

The final phase of implementing cybersecurity technical designs is controlled deployment into production with extensive monitoring validating performance and protection. Architects oversee the process of staging solutions using immutable infrastructure principles and blue/green deployment models, minimizing downtime risks. Observability platforms enable continuous verification that solutions function as intended after deployment while meeting benchmarks. Issues trigger automated rollbacks. With meticulous deployment strategies and monitoring upholding system health post-launch, organizations can confidently transition innovative security designs into live production to power business processes with resilience. Deployment and monitoring finalize the manifestation of secure technical realities.

### Controlled deployment with continuous monitoring

The production deployment of technical designs requires meticulous orchestration and live monitoring to uphold performance, security, and compliance:

- **CI/CD integration**: Technical designs are integrated into CI/CD pipelines, enabling automated testing, staged rolling deployment, and rollback mechanisms to minimize disruption risks

- **Blue/green and canary models**: New environments are staged in parallel to production before switching over or gradually shifting a percentage of traffic to test for issues

- **Observability platforms**: Holistic monitoring provides real-time visibility into health, security, and compliance across all components, enabling rapid response to deviations from expected baselines

By championing mature deployment strategies and continuous monitoring capabilities, architects can ensure solutions perform as designed securely after launch. Rigorous implementation practices finalize designs into resilient realities.

## Case studies and real-world applications

Industry-specific case studies provide invaluable perspectives into tailoring cybersecurity technical designs to address sector-specific challenges. Architects gain insights into customizing healthcare data protections, securing highly sensitive financial systems, and more by analyzing implementations across industries. Additionally, case studies of updating existing architectures demonstrate strategies for pragmatically evolving designs in step with emerging priorities. Major overhauls may shift entire platforms while targeted component swaps streamline upgrades. Regardless of the starting point, the continuous incremental adaptation of existing technical architectures allows organizations to innovate fearlessly while upholding security. By studying tailored designs and evolution strategies across diverse real-world environments, architects can skillfully craft customized solutions fulfilling unique organizational needs today while sustaining flexibility to adapt securely for tomorrow.

### *Industry-specific designs*

Technical designs manifest differently across sectors due to unique workflows, compliance demands, and risk environments. Architects must tailor designs to industry intricacies:

- **Healthcare security design**: Architecting around **electronic health records** (**EHR**) systems necessitates fine-grained access controls, detailed activity logs, and immutable backups to balance privacy with availability

- **Financial services design**: Securing high-value transactions mandates stringent controls such as microservices for isolation, hardware-backed cryptography, and real-time fraud analysis integrated across legacy mainframes, the web, and mobile channels

By studying examples across industries, architects can gain some perspective into customizing designs addressing sector-specific challenges such as data sensitivity, uptime requirements, and regulation obligations.

### *Adapting designs to evolving landscapes*

Existing designs face pressures to evolve in tandem with new threats, technologies, and business priorities. Architects must pilot updates strategically:

- **Major design overhauls**: Significant changes necessitate meticulous version deprecation schedules coordinated across teams to minimize disruption – for example, gradually shifting legacy monoliths to microservices architectures.

- **Component replacements**: Targeted subsystem swaps such as next-generation firewalls streamline enhancements while maintaining system integrity. Canary deployments provide safeguards.

Design reviews balance innovation with pragmatism. By continuously adapting existing architectures, organizations can innovate fearlessly while upholding security.

Technical design is a multi-faceted discipline that requires a deep understanding of organizational goals, security imperatives, and technological capabilities. Through the practical labs and examples provided, you are now equipped with the skills to conceptualize and implement robust technical designs that are not only functionally effective but also secure and aligned with business objectives. This chapter serves as a foundation for building advanced skills in architectural and solution development, ensuring that you are prepared to meet the challenges of the ever-evolving tech landscape.

# Life cycle

The architecture life cycle represents a structured framework that guides the design, implementation, and evolution of technology solutions. It provides a vital sequence of stages progressing from initial conceptualization to final deployment and beyond. By methodically following this life cycle, architects can develop tailored solutions that fulfill complex requirements while upholding security. The initial phases focus on gathering comprehensive requirements, drafting high-level models, and creating detailed technical designs. The middle stages encompass solution development, rigorous testing, and validation. The final phases involve secure deployment and ongoing monitoring, maintenance, and enhancement. Each stage necessitates integrating appropriate security controls and compliance measures. Through step-by-step progression across the architecture life cycle, practitioners can craft innovative yet resilient technical realities from conception to implementation. A structured life cycle sustains secure solutions.

## Conceptualization phase

The conceptualization phase represents the starting point of the architecture life cycle and is where high-level functional and security requirements are gathered before detailed technical design commences. This involves extensive stakeholder workshops to compile needs across dimensions such as system capabilities, workflows, user types, and compliance obligations. Structured methodologies elicit comprehensive requirements minimizing late-stage issues. Collaborative tools facilitate remote gathering and centralized access. This stage establishes the foundation guiding subsequent design efforts. With exhaustive upfront conceptualization, architects gain a stable understanding of organizational problems to be solved along with mandatory protective measures. Capturing core requirements and constraints provides direction before you embark on technical specifics.

### Understanding conceptualization

Conceptualization represents the first phase of the architecture life cycle and is where an abstract vision of the solution takes shape through exploratory research and stakeholder engagement:

- **Phase objectives**: The conceptualization phase focuses on gathering high-level functional and technical requirements that will guide subsequent design efforts. Additionally, mandatory security and compliance needs are compiled.

- **Key activities**: Extensive stakeholder workshops elicit needs across dimensions such as capabilities, workflows, user types, and regulations. Market research and expert consultations further inform requirements. Structured methodologies prevent critical gaps at this foundational stage.

By thoroughly investigating the problem space before technical design commences, architects can make fully informed decisions. Conceptualization assembles the puzzle pieces for coherent solution shaping.

### Initial security considerations

To deeply embed comprehensive protection, architects must spearhead initial security activities during conceptualization to frame subsequent design efforts:

- **Integrating security early**: Rather than deferring security, architects need to facilitate preliminary threat modeling sessions identifying risks, adversaries, and vulnerabilities requiring controls. Early focus avoids costly late-stage issues.

- **Threat modeling and risk analysis**: Structured methodologies such as STRIDE, DREAD, and attack trees analyze risks in the context of requirements, such as planned data flows, technology stacks, trust boundaries, and compliance needs. These are all identified during conceptualization.

By seeding security considerations early before technical specifics, architects can deeply ingrain protection intrinsically across layers in the design phase. The conceptualization stage frames the security vision.

## Design phase

The design phase represents the creative stage where requirements gathered during conceptualization are translated into comprehensive technical architectures and specifications. Architects develop holistic infrastructure diagrams delineating components and connections based on intended workflows. Detailed security architecture is woven throughout encompassing controls and protocols needed to manage risks. Interface design clearly defines system touchpoints and their security mechanisms. Compliance requirements are embedded across layers intrinsically. With exhaustive upfront designs, complex organizational challenges can be solved through tailored security innovations. Diligent design crafts a clear technical vision that fulfills stakeholder needs securely.

### *Transitioning to design*

Following conceptualization, architects enter an exciting design phase, where they can bring stakeholder visions to life through impactful architectures. Transforming high-level needs into technical specifics, architects now detail intricate infrastructure diagrams, component interfaces, robust data models, and end-to-end workflows.

Leveraging design tools such as **Unified Modeling Language** (**UML**) and wireframing, ambiguity gives way to alignment and shared understanding. Creative possibilities take clearer shape, coalescing the team around a unified technical direction.

Through diligent yet adaptable planning, architects manifest specific innovations that balance functionality, security, and future growth. This design phase bridges the gap between abstract ideas and tangible systems that securely empower organizations.

With user needs thoroughly captured, architects now focus their expertise on crafting tailored technical specifications. They lead stakeholders from promising concepts to deliberate design blueprints engineered for current and emerging business objectives.

The design phase represents an opportunity to cement innovative foundations through structured methodologies. With their sights on the technical horizon, architects' passions and perseverance can unlock new potential.

## Bridging conceptualization to technical design

Following conceptualization, architects enter the design phase, translating abstract requirements into concrete technical architectures, interfaces, and specifications:

- **Creating detailed design plans**: High-level needs are codified into comprehensive diagrams and documentation depicting complete technical implementations spanning infrastructure, workflows, controls, and data flows

- **Employing design tools**: Methods such as UML, service diagrams, flowcharts, wireframes, and architectural blueprints standardize communications reducing ambiguity

The following is an example of a UML diagram:

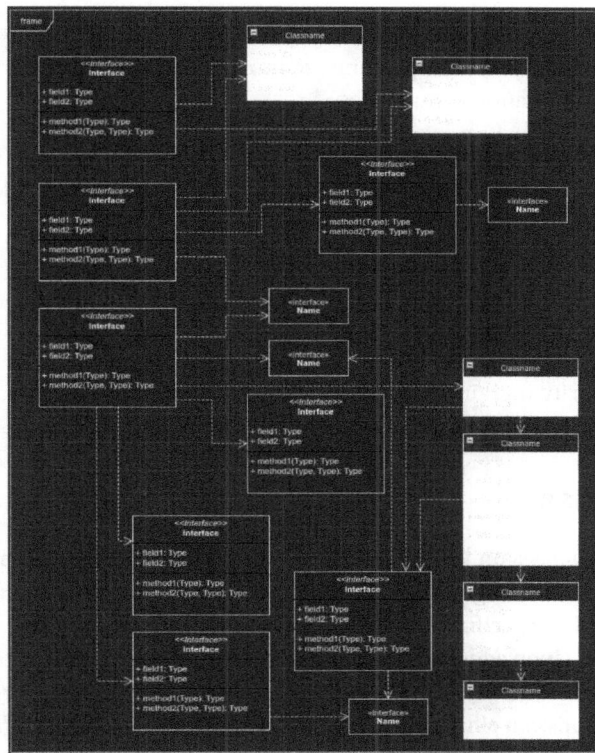

Figure 13.2 – Example UML diagram

The design phase bridges the gap between exploratory conceptualization and tangible technical realities. Structured design practices manifest nebulous requirements into specific architectures optimized for organizational risk and workflow.

## *Security in design*

The design phase represents the critical stage where security architecture is woven intrinsically throughout technical infrastructure and component specifications. Guided by threat modeling and risk analysis, controls are meticulously allocated to mitigate exposures across apps, data, systems, and networks. Authentication, access management, encryption, availability enhancement, and monitoring workflows are tightly integrated with core functionality rather than bolted on afterward. Compliance requirements are natively manifested across designs through access restrictions, audit logs, and encryption schemes tailored to data types. With security fundamentals firmly ingrained within designs instead of deferred, organizations can innovate fearlessly knowing solutions are resilient by design. Diligent security integration at the design stage enables innovation without peril.

### Ingraining security in design

The design phase represents the pivotal opportunity to intrinsically integrate security across technical architectures before implementation commences:

- **Security-by-design principles**: Rather than being bolted on later, controls such as encryption schemes, role-based access, secure coding practices, and micro-segmentation are woven into specifications embedding protection by design.

- **Case study examples**: A healthcare system had PHI leakage issues after deployment. By designing cryptographic schemes and granular access models into the revised data architecture during design, compliance was sustainably upheld.

With comprehensive security ingrained in technical DNA rather than deferred, organizations can confidently pursue innovations, knowing resilience is built-in by design from the start.

## Development phase

The development phase focuses on bringing technical designs to life through meticulous coding, configuration, and integration by engineers. Architects lead this collaborative stage using Agile sprints or waterfall gating to translate specifications into functioning solutions. Standard libraries accelerate development while customization addresses unique risks. Automated testing validates builds frequently against requirements. With architects closely guiding implementation teams through best practice methodologies, organizations can efficiently yet securely transform rigorous designs into operational realities fulfilling stakeholder needs. Diligent development realizes technical vision.

### *Implementing the design*

The development phase involves collaborative implementation by cross-functional engineering teams, thus bringing technical designs into functioning realities. Architects lead engagements using Agile sprints or waterfall models to systematically translate specifications into code, configurations, integrations, and data builds constituting complete solutions. Automated testing frequently validates interim outputs against requirements. While standardized libraries, frameworks, and policies accelerate development, customization addresses unique organizational risks. With architects guiding disciplined, phased implementation approaches, organizations efficiently yet securely manifest robust technical designs into operational systems fulfilling stakeholder needs. The development phase bridges designs with functioning solutions.

## Transforming designs into functioning realities

During development, architects lead meticulous engineering efforts, converting specifications into operational solutions using optimal approaches:

- **Tailored development strategies**: Based on factors such as team experience, compliance requirements, and solution complexity, architects choose Agile sprints, waterfall gating, or hybrid models balancing velocity and rigor
- **Security implementation oversight**: Architects continuously verify that controls and hardening measures specified in the design are integrated correctly during coding sprints and infrastructure configuration

With focused leadership through structured methodologies, architects shepherd seamless progression from robust on-paper designs into functioning systems that fulfill requirements securely.

### Continuous testing

Throughout development, continuous testing validates that solutions deliver the required functionality, performance, and security. Architects architect frameworks by conducting automated assessments with each build. Unit tests validate components as modular blocks. Regression testing uncovers architectural breaks or violations of requirements. Static analysis and dynamic penetration testing assess hardening and exploit resistance. These layered, automated test suites provide ongoing validation that implementations match designs securely. Continuous testing enables flaws to be addressed decisively long before solutions reach users. Disciplined testing sustains design integrity.

## Automating testing for continuous validation

During development, architects weave rigorous continuous testing frameworks into pipelines ensuring implementations match secure designs:

- **Automated testing security**: With each build, unit tests validate software functions. Static analysis and dynamic penetration testing unearth vulnerabilities such as injection points or configuration weaknesses, enabling early mitigation.
- **Testing tools and integration**: Frameworks such as Selenium and JUnit automate functionality and performance validation. Security orchestration tools integrate scanning and attack simulations into pipelines providing continuous verifications.

With automated, layered testing being intrinsic throughout development, organizations gain assurance that evolving implementations provably fulfill specifications and requirements without deviation or the introduction of weaknesses.

## Deployment phase

The deployment phase focuses on transitioning validated solutions from isolated development environments into live production securely and seamlessly. Architects oversee meticulous staging, canary releases, immutable infrastructure principles, and rollback mechanisms to minimize disruption risks. Extensive monitoring provides continuous verification that performance, protection, and compliance are upheld post-launch. Together with controlled deployment strategies, observability enables issues to be swiftly detected and contained. By guiding solutions through final robust validations into the live environment, architects fulfill the ultimate purpose of the architecture life cycle – securely manifesting designed innovations into business operations. Careful deployment realizes technical visions.

### Launching the system

The deployment phase signifies the transition from isolated development into live production operations. Architects oversee the meticulous execution of release plans using staging environments, canary launches, immutable infrastructure, and rollback protocols to minimize disruption risks. Extensive monitoring provides continuous validation that performance, security, and compliance are upheld after launch. Observability platforms give rapid issue detection and response. By guiding solutions through final failsafe validations into the live environment, architects fulfill the ultimate purpose of the architecture life cycle – securely bringing impactful innovations into business processes. Careful deployment transitions technical achievements into organizational capabilities.

### Deploying solutions securely into production

Guiding rigorously tested systems from isolated development into live environments requires meticulous deployment strategies that ensure smooth transitions:

- **Hybrid and cloud deployment**: Based on infrastructure needs, architects plan launches spanning on-premises, the cloud, containers, and serverless platforms. Gradual shifts enable testing.

- **Failsafe protocols**: Staged rollouts, canary releases, immutable infrastructure, automated rollbacks, and blue/green models all provide deployment safeguards minimizing disruption risks.

- **Security validation**: Pre-launch audits validate that production correctly implements controls specified in the design, such as encrypted connections, hardened configurations, and data governance schemes.

With incremental protocols and extensive validations, organizations can confidently transition innovations into live operations with full functionality, hardened security, and resilience by design.

## Maintenance phase

The maintenance phase focuses on continuously upholding solution health, security, and performance once live. Architects implement monitoring providing real-time visibility into all components spanning on-premises, cloud, and hybrid environments. Metrics validate SLAs while anomaly detection identifies emerging threats and issues. Ticketing integration enables swift response and remediation. Change management procedures govern updates, improvements, and issue resolution. Solutions evolve via CI/CD pipelines and rolling upgrades providing continuity. By championing comprehensive maintenance processes, architects can ensure innovations continue to securely fulfill business objectives over time with resilience. Diligent maintenance sustains achievements.

### Ongoing support and updates

The maintenance phase involves continuous upkeep and improvement of solutions once they're operational. Architects implement observability platforms providing real-time health visibility across on-premises, cloud, and hybrid components. Anomaly detection identifies performance deviations or emerging threats. Ticketing integration allows for swift responses and issues to be remediated. CI/CD pipelines enable rolling upgrades, improvements, and expansions, providing business continuity. Change management procedures govern modifications. With meticulous maintenance processes applied continuously post-launch, architects ensure solutions sustainably fulfill their purpose securely with resilience over time. Ongoing rigor preserves innovative achievements.

#### Sustaining innovation through continuous maintenance

Post-launch, rigorous processes uphold solutions to deliver capabilities and protections as intended over extended durations:

- **Holistic monitoring and upkeep**: Observability platforms provide real-time performance visibility across on-premises, cloud, and endpoint layers, enabling rapid detection and remediation of issues via integrated ticketing.

- **Evergreen security processes**: Regular patching, vulnerability scanning, penetration testing, and control upgrades respond to the evolving threat landscape. Access models and data governance adapt to changing regulations.

With meticulous ongoing maintenance intrinsically designed into solutions, organizations can future-proof innovations securely powering business objectives over long horizons.

### Case study – long-term maintenance

This case study exemplifies the extensive ongoing processes required to sustain large-scale solutions securely over decades across periods of immense technological and threat landscape change. It focuses on mainframe modernization via API integration with new channels and apps, enabling innovation atop stable legacy cores. Evergreen maintenance practices encompass tool upgrades, migration to new infrastructure types, security control revamps, and expertise renewals across generations using documentation. This long-term view highlights that through comprehensive maintenance planning, even aging solutions can be sustained as secure foundations that power innovation far into the future. Maintenance mastery enables prolonged achievement.

**Sustaining large-scale systems through proactive maintenance**

This case study of a decades-old financial mainframe modernized to support digital channels highlights extensive maintenance that integrates legacy with innovation:

- **Mainframe security and tooling refresh**: Rigorous version management, infrastructure upgrades, expertise renewals across generations using documentation, and control revamps such as adopting MFA, sustained core integrity over 40+ years

- **API integration of channels**: Exposing mainframe data and functions through APIs enabled modern web, mobile, and analytics solutions to be built while legacy data was leveraged securely via orchestration layers

This long maintenance view emphasizes architecting sustainably from inception to extend achievements through meticulous upkeep and integration of new paradigms atop stable cores.

This detailed exploration of the architecture life cycle provides a comprehensive understanding of each stage, underscoring the critical role of security at every step. By following this life cycle approach, architects and developers can ensure that their systems are not only functionally robust but also secure from potential threats.

# Summary

This chapter explored the core disciplines that enable cybersecurity architects to translate organizational needs into tailored technical realities that secure innovation. It emphasized aligning security intrinsically with business objectives early in conceptualization and design. Rigorous development and testing uphold initial visions. Measured deployment delivers functioning systems into production. Sustained maintenance and improvement preserve achievements.

Foundational concepts such as security by design, layered development testing, and maintenance as key enablers of adaptation were covered. Detailed analysis of architecture life cycle stages provided methodical guidance through each phase. Best practices, standards references, and practical labs reinforced techniques for eliciting comprehensive requirements and threats. Methods to systematically transform needs into robust layered designs, develop and validate systems, and then deploy and support solutions provided actionable upskilling.

The case studies provided showcased industry-specific implementations that meet unique data, risk, and regulatory challenges. Further examples demonstrated evolution strategies that balance innovation with legacy modernization. Together, this chapter provided both the strategic perspectives and tactical skills needed to shepherd secure innovations holistically from ideas to implementations. It aimed to help you lead complex projects with versatile skills that blend cybersecurity, software mastery, and business acuity tailored to dynamic environments.

While extensive architectural and technical expertise empowers cybersecurity practitioners to secure complex environments, the knowledge you've gained in this chapter reaches its full potential when paired with versatile project execution skills. Just as masterful blueprints enable towering achievements in engineering, the realized impact of robust security designs hinges on methodical scoping and management.

The next chapter provides that crucial but often overlooked layer – guidance and examples for architects to shepherd initiatives holistically. By codifying reusable artifacts, meticulously scoping projects balancing restraint and vision, and then adapting methodology to needs, success materializes. Outcomes manifest not just through technical mastery but through leadership orchestrating talents, objectives, and timelines seamlessly.

With diligent project oversight fused with architectural acumen, and pragmatism with inspiration, the smallest steps unfold into the greatest leaps. In the end, victories arise not solely from ingenious ideas but from synergy between vision and discipline applied to the complex at early junctures. Project leadership cements cybersecurity knowledge into organizational capabilities, thereby securing innovation from conception through completion.

# 14
# Architecture Considerations – Design, Development, and Other Security Strategies – Part 2

*"Victorious warriors win first and then go to war, while defeated warriors go to war first and then seek to win."*

*– Sun Tzu*

*"Who wishes to fight must first count the cost."*

*– Sun Tzu*

*"Plan for what it is difficult while it is easy, do what is great while it is small."*

*– Sun Tzu*

The prior chapter, *Part 1*, emphasized the focus on providing cybersecurity architects with integrated guidance on designing, developing, and managing solutions holistically from conception to production. Architects need to integrate predictive, preventive, detective, and responsive capabilities into adaptable ecosystems while recalibrating defenses dynamically. Controls must align seamlessly with organizational workflows, risk tolerance, and compliance obligations.

Strategic planning and rapid response are both imperative and enabled by OODA loop proficiency.

Strategist Sun Tzu noted that victory emerges from preparation in seeming peacetime, not reactionary scrambles once war has erupted. This principle applies directly to cybersecurity architects facing relentless threats. Meticulous blueprinting, scoping, and project approach orchestration in calm periods allow organizations to innovate fearlessly.

Part 2 of *Architecture Considerations – Design, Development, and Other Security Strategies* equips architects to adopt this mindset through comprehensive coverage spanning architectural reuse catalogs, controlled scoping procedures, and tailored project execution. By deliberately planning what is complex amid the simple, greatness emerges. With diligent scoping contrasting restraint and ambition, architects can secure innovative ecosystems that meet current and future needs.

Just as warriors win by calculating costs and then waging war while being prepared, prudent architects enable organizations to preempt threats by championing rigorous blueprints. With creativity melding reusable patterns to unique needs, scoped ambitions to realities, and structured methodology to talent, they manifest solutions that sustain productivity with resilience. In cybersecurity, the victorious innovate before threats even emerge by planning the complex through versatile architecture leadership.

This chapter covers the following topics:

- Blueprinting
- Scoping
- Project approach
- Next steps

# Blueprinting

In the context of software development and system architecture, **blueprinting** is the process of creating detailed plans or models for a solution or application. This section focuses on how to define and develop these blueprints, which act as standardized and repeatable guides for deployment. The goal is to ensure consistency, efficiency, and security compliance in the implementation process.

Blueprinting represents the practice of creating standardized architecture templates that codify proven security designs, patterns, and policy frameworks for consistent reuse across implementations. By intelligently leveraging blueprints as starting points, organizations can build and deploy solutions more efficiently with embedded resilience. Blueprints encapsulate accumulated wisdom so that each project doesn't need to be started from scratch. Elements cover cloud resource configuration, network topology, access management, encryption schemes, and more. Blueprints balance commonality with customization using modular libraries and policy as code. With adaptable solutions pre-architected in blueprint libraries, organizations gain velocity while upholding consistent security everywhere.

## Understanding blueprints

Blueprints encompass codified architecture templates, thereby enabling consistent reuse of proven designs across implementations, balancing standardization with customization. Modular libraries allow us to combine common elements, while policy as code allows us to tune it for our unique needs. Cloud blueprints facilitate **Infrastructure as Code** (**IaC**) to secure deployments. With blueprint libraries covering core patterns, organizations gain velocity while embedding consistent protection everywhere. Blueprints institutionalize scalable secure by design.

### Defining blueprints

In a technical context, blueprints are predefined, structured architecture templates that codify proven designs for consistent reuse across projects and environments. Let's take a closer look:

- **Objectives**: Blueprints embed configuration specifications, policies, and controls to provide a standardized starting point for infrastructure deployment and application development, enabling consistent security by design.

- **Key benefits**: By encapsulating accumulated architecture wisdom into adaptable templates, blueprints enable faster project launches without compromising resilience. They reduce design redundancies and promote the reuse of vetted patterns, minimizing risks.

At their core, blueprints foster consistent excellence and efficiency by transforming static specifications into dynamic modular libraries that are tenable across diverse implementations.

## Developing blueprints

Developing robust blueprints requires identifying common solutions for frequent scenarios, such as core network configurations, access management schemes, encryption implementations, and system hardening controls. These vetted patterns are codified into modular, adaptable templates using IaC and policy as code methodologies. Standardized designs undergo extensive review, ensuring they encapsulate proven security practices and configurations with adequate flexibility for customization. The resulting blueprint libraries enable the consistent reuse of validated architectures as starting points tailored efficiently for diverse projects. With diligent development centered around reusable elements, organizations can cultivate growing template collections while fostering excellence and compliance at scale.

### Blueprint design principles

Effective blueprints balance standardized content with customization capacity. Customization points allow unique extensions such as proprietary systems integration or specialized compliance controls through modular libraries, script parameters, and policy as code overlays. Iterative blueprint versions evolve via contributor feedback. Portability facilitates implementation across environments such as multi-cloud. Rigorously reviewed blueprints avoid bad patterns from being enforced across implementations. With adaptable blueprints encoding *secure by default* principles intrinsically, while providing flexibility for specialization, organizations gain velocity securely.

## Principles for blueprint development

Thoughtfully designed blueprints balance standardization with customization across key elements using proven methodologies:

- **Common components**: Blueprints encapsulate technical specifications such as infrastructure diagrams, access control models, encryption schemes, compliance controls, and security workflows constituting repeatable patterns.

- **Design considerations**: Modularity maximizes reusability across environments. Custom parameters and policy overlays provide adaptation while guardrails prevent over-engineering. Code-based abstractions enable portability across platforms.

With robust frameworks, structured templates, and modular libraries tuned for flexibility, blueprints institutionalize reusable *secure by default* building blocks that can be configured across diverse environments.

### *Security integration*

Integrating security intrinsically across multiple blueprint layers enables consistent protection by design. Network topology templates codify microsegmentation based on zero-trust principles. Server and system hardening blueprints automate least-privilege OS configuration. Testing blueprints validate functionality, security, and compliance. Modular libraries allow us to combine standardized controls such as SIEM integration with custom governance requirements. Hardened blueprints accelerate secure innovation at scale.

### Institutionalizing security in blueprints

To consistently uphold resilient protection, security is designed into the core of architecture blueprints across layers:

- **Security by design principles**: Rather than being overlaid afterward, controls such as encryption schemes, access management, and micro-segmentation are paramount elements within blueprints, holistically addressing risks from the start.

- **Configuration hardening**: Blueprints automate consistent provisioning of security postures across networks, servers, clouds, endpoints, and applications according to benchmarks such as CIS and STIG. Compliance checklists ensure controls align with regulations. Tools enable the integration of custom modules tailored to unique risks.

With *secure by default* principles thoroughly ingrained within standardized templates, organizations can inherently promote excellence across implementations.

# Blueprinting process

Creating impactful blueprints involves identifying common scenarios to standardize, designing modular template architectures hardened for security, and enabling flexibility for customization. A rigorous blueprinting process fosters consistent excellence and compliance.

## Initial drafting

The first blueprinting step involves identifying and standardizing common scenarios for reuse, such as core network configurations, identity management schemes, encrypted data stores, and system hardening stacks. Subject matter experts architect robust templated solutions to address these needs using IaC tools and policy as code frameworks. Standardization provides baseline uniformity while modularity and customization points foster flexibility. Compliance checklists guide the required control inclusion per data type and environment. The initial drafts form the blueprint's foundations and are later expanded into comprehensive libraries, thereby enabling consistent security and efficiency at scale.

### Creating blueprint foundation drafts

The first blueprinting step focuses on drafting templates that address common scenarios that require consistent implementations:

- **Requirements gathering**: Through stakeholder input and reviews of past projects, architects can identify frequently implemented needs such as core network and cloud configurations that are amenable to standardization

- **Initial design**: For each scenario that's been identified, subject matter experts can design structured draft blueprints for codifying configurations and controls to address essential requirements using IaC tools

These inaugural drafts establish the blueprinting foundations, setting the stage for expansion into comprehensive, secure modular libraries tailored for consistent reuse with customization.

## Refinement and detailing

After initial drafting, collaborative refinement adds layers of detail and flexibility to blueprints. Through meticulous detailing and refinement, streamlined template collections transform raw drafts into dynamic engines of excellence and efficiency.

### Expanding drafts through collaborative blueprint refinement

Initial blueprint drafts are expanded into detailed, production-ready templates through rigorous collaborative refinement:

- **Enriching standardization scope**: Analysis identifies new scenarios such as application onboarding and microservices deployment amenable to templatization for embedding consistent security and efficiency.

- **Adding modularity and flexibility**: Components become distinct modules for flexible reuse. Customization points allow adaptation through variables, script arguments, policy overlays, and modular options.

- **Layering in specifications**: Detailed definitions, data schemas, access protocols, encryption specifications, UX flows, and communications interfaces incorporate the required elements to form comprehensive technical implementations.

With sustained collaborative refinement, streamlined template libraries crystallize, enabling consistent security excellence across diverse implementations at scale.

While new organizations enjoy a blank canvas, most cybersecurity architects join enterprises with sprawling technology ecosystems already in motion. In these scenarios, success stems not from ground-up rearchitecting but from renovation – judiciously building upon existing foundations rather than demolishing them.

Meticulous reviews of current architectures, including diagrams, configurations, and team insights, reveal overlooked strengths to preserve alongside gaps that require reinforcement. Instead of rip-and-replace cycles, iteratively enhancing modular components accelerates improvement with minimized disruption.

By equipping architects with skills to analyze environments as-is, tailor reusable artifacts to unique needs, and then integrate advancements to minimize business impact, outcomes accelerate. With existing teams, knowledge and capabilities catalyze progress and security progresses not through revolution but through guided evolution – uplifting what works while methodically upskilling what does not. Just as restoration masters breathe new life into weathered artifacts, prudent architectural leadership sustains identity by honoring an organization's technical heritage amid change.

## Standardization and repeatability

To maximize consistency, blueprint standardization checklists validate that the required elements are embedded in each scenario. This includes approved network configurations, IAM integrations, and OS hardening packages. Reviews ensure designs enforce excellent security principles and avoid reinventing inferior solutions. Version control sustains repeatability as blueprints evolve. Testing frameworks continuously validate functionality and control integration. Detailed documentation and idempotent IaC implementations ease maintenance. Through rigorous standardization and enabling collaborative enhancement, blueprints are transformed from static specifications into dynamic, vetted solutions that boost secure velocity organization-wide.

### Creating standardized templates

To maximize consistent security and efficiency, blueprints must enforce standardization for common scenarios such as core network buildouts and identity management integration. Checklists validate the required elements, such as approved OS images, microsegmentation firewalls, role provisioning, and MFA. Extensive peer reviews prevent the repetition of suboptimal patterns across implementations. Version control sustains repeatability during ongoing blueprint evolution and enhancement. Testing frameworks continuously validate functionality, security, and compliance in templates. Detailed documentation eases maintenance. With comprehensive standardization and collaborative refinement, reusable blueprint libraries persist as engines dynamically drive resilient and efficient implementations across the enterprise.

## Maximizing consistency through blueprint standardization

To maximize the benefits, blueprints must provide standardized implementations to secure common scenarios consistently:

- **Constructing reusable libraries**: Modular templates are developed for needs such as public cloud deployment, microservices onboarding, and retail branch buildouts that cover core infrastructure, configurations, controls, and workflows.

- **Enforcing standardization**: Checklists validate that the required elements, such as multi-cloud procurement, SIEM integration, and encryption schemas, are included in each blueprint. Extensive peer reviews prevent inferior patterns from being repeated.

The result is a growing collection of streamlined, validated templates that codify proven security excellence for consistent reuse while allowing necessary customization.

### *Ensuring repeatability*

To maximize blueprint value over time, repeatability and adaptability are critical. This can be done through version control, modular libraries, and detailed documentation. With rigorous processes enabling enhancement and sustained consistency, blueprints persist as dynamic catalysts that scale excellence across the enterprise.

## Sustaining blueprint repeatability and evolution

For sustained value, repeatability and consistent evolution are pivotal. This can be through rigorous version control, documentation, testing, and modularity:

- **Collaborative refinement**: Review processes ensure changes reinforce rather than deviate from standards. Customization points such as script variables and modular libraries maximize flexible reuse across diverse projects.

- **Enabling maintenance**: Detailed documentation, IDE integrations, and IaC abstractions ease the process of updating blueprints over time. Testing frameworks continuously validate functionality and security.

With sustained processes safeguarding consistency amid ongoing enhancement, standardized blueprints persist as dynamic catalysts for excellence across the enterprise.

# Use cases and practical applications

Thoughtfully constructed cybersecurity blueprints embed proven artifacts, thereby accelerating secure innovation at scale. Forged from hard-won experience, prudent blueprints empower organizations to build the future fearlessly by standing on the shoulders of others who secured the past. With rigorous templates translating wisdom into action, architects can scale security, prevent regression, and propel progress beyond the bleeding edge.

## Case studies

By examining examples where meticulous blueprints enabled rapid compliant deployment or application modernization, architects can learn how to extract and template best practices for reuse.

Common insights include the value of peer reviews, the importance of modular flexibility, and the acceleration that can be achieved through standardizing infrastructure, policies, and integrations using code-based abstractions.

For instance, a financial firm created retail branch networking blueprints, reducing deployment time from weeks to days, while embedding consistent micro-segmentation controls.

Another example is an airline's cloud blueprint, which standardized hundreds of configuration checks, cutting cloud asset deployment time by 65% while enforcing hardened configurations uniformly.

## Hands-on exercise – developing a blueprint for a cloud-based application

*Objective*: This hands-on exercise is designed to guide participants through the process of developing a comprehensive blueprint for a cloud-based application. The aim is to create a detailed blueprint that can serve as a guide for implementation.

*Overview*: Participants must choose a scenario for a cloud-based application (for example, a **customer relationship management** (CRM) system, an e-commerce platform, and so on) and develop a complete blueprint that covers aspects such as architecture, data design, user interface, and security:

1. Scenario selection and requirements analysis:

    I. Choose a cloud-based application scenario:

    - **Task**: Select a specific type of cloud-based application to design
    - **Examples**: A CRM system, e-commerce platform, or cloud-based analytics tool
    - **Outcome**: A clear understanding of the application's purpose and target audience

    II. Conduct requirements analysis:

    - **Task**: Gather and document functional and non-functional requirements for the application
    - **Activity**: Use interviews, surveys, or existing case studies to gather requirements
    - **Outcome**: A comprehensive list of requirements for the application

2.  Architectural design:

    I.  Design the application architecture:

        • **Task**: Create an architectural diagram of the application

        • **Considerations**: Include components such as web servers, application servers, database servers, and cloud services

        • **Tool**: Use diagramming tools such as Lucidchart and Microsoft Visio

        • **Outcome**: An architectural diagram showing all components and their interactions

3.  Data design:

    I.  Develop a data model:

        • **Task**: Design the data schema for the application

        • **Activity**: Define data entities, relationships, data classification, and data flow diagrams

        • **Outcome**: A detailed data model for the application

4.  User interface design:

    I.  Sketch user interface mockups:

        • **Task**: Design the user interface for key screens of the application

        • **Activity**: Create mockups for interfaces such as the login page, dashboard, user settings, and main functional screens

        • **Tool**: Use tools such as Sketch and Adobe XD

        • **Outcome**: A set of user interface mockups

5.  Security planning:

    I.  Integrate security measures:

        • **Task**: Plan security measures for the application

        • **Considerations**: Include data encryption, user authentication, and authorization mechanisms

        • **Outcome**: A security plan detailing the security measures to be implemented

6.  Creating the blueprint document:

    I.  Compile the blueprint:

        • **Task**: Assemble all the components (architecture, data design, UI mockups, and security plan) into a comprehensive blueprint document

        • **Outcome**: A detailed blueprint document ready to be implemented

7.  Review and feedback:

    I.  Peer review:

- **Task**: Share the blueprint with peers for review

- **Activity**: Provide and receive feedback on the design, feasibility, and completeness of the blueprint

- **Outcome**: An improved blueprint that incorporates peer feedback

8.  Final presentation:

    I.  Present the blueprint:

- **Task**: Present the final blueprint to the group

- **Activity**: Explain the rationale behind design decisions and how the blueprint meets the requirements

- **Outcome**: A polished and comprehensive blueprint ready for the implementation phase

This exercise has provided you with hands-on experience in creating a detailed blueprint for a cloud-based application. It encompasses all critical aspects of system design, from architectural planning to security integration, ensuring a well-rounded and implementation-ready blueprint.

In conclusion, comprehensive blueprints encapsulate accumulated wisdom into dynamic libraries, thereby driving consistent excellence and efficiency. By codifying proven architectures using IaC and policy as code tools, blueprints embed resilient security intrinsically while providing adaptation mechanisms.

Meticulous blueprinting processes maximize standardization for common scenarios such as cloud deployments while allowing flexible customization. With mature blueprint libraries, organizations gain velocity securely through reusable *secure by default* building blocks that can be configured across diverse projects and environments.

# Scoping

**Scoping**, in the context of project and system design, refers to the process of defining and documenting the objectives, deliverables, tasks, costs, deadlines, and boundaries of a project. It is a critical phase in project management and system development that ensures clarity and alignment among stakeholders and helps in managing expectations and resources effectively.

## Understanding the importance of scoping

Scoping represents the critical process of aligning a project's vision and objectives with pragmatic realities such as timelines, budgets, resources, and capabilities. Clear scoping sets achievable goals, thus preventing overreach. It frames visions into actionable increments, delivering value. By scoping collaboratively, teams can clarify objectives, dependencies, roles, and measures of success. Structured scoping sustains focus, guiding effective planning and execution. With disciplined scoping, organizations can transform ambitions into defined roadmaps that are linked to tactical progress indicators and milestones.

## The role of scoping

Scoping entails framing visionary goals within pragmatic constraints to define achievable plans. It involves aligning business aims, timelines, budgets, resource availability, and execution feasibility into realistic roadmaps. Scoping elicits detailed requirements and success indicators from stakeholders and also helps with identifying risks and dependencies. Structured scoping sustains focus, transforming ambitions into defined development roadmaps. By providing a channel between vision and execution, disciplined scoping enables organizations to systematically progress from ideas to outcomes. Clearly scoped plans drive progress.

### Defining the crucial role of scoping

Scoping represents the structured process of aligning a project's visionary objectives with pragmatic realities to frame achievable plans and drive effective execution:

- **Scoping activities**: This involves eliciting detailed requirements, defining success indicators, mapping dependencies, analyzing feasibility constraints, establishing budgets and timelines, and delineating team roles and capabilities.

- **Significance of scoping**: Proper scoping provides guardrails that prevent common issues regarding vague goals, scope creep, premature optimization, inflated budgets, and timeline overruns. It sustains focus on core value delivery.

With pragmatic scoping framing vision, organizations can avoid pursuing ambiguous or impractical goals by systematically scoping vision into realizable increments, thereby delivering tangible outcomes.

## The process of scoping

The scoping process involves collaboratively aligning vision with reality via structured requirements gathering, solution modeling, defining success indicators, dependency mapping, budgeting, and scheduling. Cross-functional workshops elicit needs while avoiding overreach. Prototyping validates feasibility and user needs. Scope change processes sustain focus as dynamics shift. With methodical scoping, organizations can crystallize executable roadmaps, thereby driving progress.

## Initiation and requirements gathering

Scoping commences by gathering comprehensive requirements from stakeholders. Explicit requirements provide the raw material for framing pragmatic scope.

**Initiating scoping with comprehensive requirements elicitation**

The scoping journey begins by thoroughly gathering foundational requirements from all stakeholders through a structured initiative:

- **Identifying stakeholders**: Through stakeholder mapping, all affected groups are determined. This includes clients, user communities, subject matter experts, delivery teams, support personnel, and compliance partners.

- **Requirements gathering techniques**: Interviews, surveys, focus groups, and workshops elicit detailed needs. Personas and user stories frame problems from human perspectives. Structured analysis coalesces needs into capability and solution models that clarify must-haves versus nice-to-haves. Key needs are distilled into capability and solution models.

With broad inclusivity and rigorous elicitation, core requirements take shape, providing raw materials to pragmatically frame an achievable scope and thereby deliver stakeholder goals.

## *Defining scope*

Following requirements gathering, structured scoping workshops frame achievable plans. Core capabilities are carved into increments, thereby delivering value aligned with strategic goals. Success metrics define milestones, demonstrating progress. Budgets, timelines, resources, and feasibility guardrails establish pragmatic boundaries. Roadmaps compose defined increments, balancing vision with constraints. Clear scope enables focus and execution.

**Framing scope through structured definition**

Following requirements gathering, focused scoping workshops align vision with constraints to frame realistic plans:

- **Drafting scope statements**: Workshops facilitate collaborative drafting of scope documents, thus delineating business goals, core capabilities, success metrics, timelines, budgets, resources, constraints, and assumptions.

- **Establishing scope boundaries**: Discussions determine clear boundaries on what is explicitly included and excluded from scope based on must-haves versus nice-to-haves. This helps with avoiding scope creep.

The output provides carefully framed scope documents that codify the plan for transforming ideas into pragmatic reality through phased execution.

## Tools and techniques for effective scoping

Structured techniques enable goals to be translated into a well-framed scope. Requirements gathering leverages interviews, surveys, workshops, and user research to extract nuanced needs. Prototyping validates capabilities and constraints quickly. Estimation techniques size budget and timelines. Dependency mapping illuminates risks. With rigorous tools, scoped vision becomes achievable execution.

## Utilizing scope management tools

Specialized tools facilitate scope management from initial framing through requirements, planning, and documentation. Modeling tools visualize capabilities, dependencies, and timelines. Documentation editors streamline scope statement drafting. Requirements databases centralize stakeholder needs. Tracking tools monitor progress indicators. Whether manual or automated, appropriate tooling creates scope order out of ideas chaos.

### Leveraging tools to streamline scoping processes

Specialized tools help organize ambiguous ideas into structured scope by visualizing capabilities, requirements, timelines, and tasks:

- **Common scoping tools**: **Work breakdown structures (WBSs)** visually map scope elements. **Gantt charts** timeline tasks. Scope management software centralizes requirements and documents. Modeling tools depict capabilities.

- **Practical application**: As an example, using a WBS, architects can decompose a migration into phases such as planning, infrastructure, data transition, and testing. After, Gantt charts can be used to overlay timeframes, milestones, and team assignments to scope the schedule.

By leveraging target tools, abstract concepts can be transformed into concrete, measurable scope that includes visualized capabilities, scheduled tasks, and centralized requirements.

## Techniques for detailed scoping

Structured workshops facilitate detailed scoping by translating goals into defined capabilities, metrics, timelines, resources, constraints, and risks. Active requirements management identifies dependencies. Prototyping validates feasibility early. Estimation techniques size budget and schedule. Each capability increment undergoes meticulous scoping for clarity. With diligent detailing, organizations can progress from ideas to execution.

### Employing techniques for precise scoping

Following initial framing, detailed scoping workshops dive into specifics to galvanize execution planning:

- **Deconstructing capabilities**: Each capability increment is systematically decomposed into implementable tasks, measurable milestones, and value-adding deliverables

- **Prioritizing scope elements**: Milestones, functions, and tasks are then prioritized based on factors such as business impact, complexity, dependencies, and time sensitivity using methods such as **Must Have, Should Have, Could Have, Will Not Have (MoSCoW)**

- **Estimating budgets and timelines**: Techniques such as three-point estimation calculate the schedule and resources for scope elements, enabling tradeoff decisions

With meticulous detailing, ambiguity gives way to clarity, transforming ideas into scoped packages of prioritized capabilities for systematic execution.

## Managing scope changes

Despite best efforts, changes are inevitable, requiring processes to avoid uncontrolled scope creep. With rigorous change management, scoping sustains focus amid fluidity.

### Change control process

Rigorous change control processes avoid scope creep from uncontrolled expansions. Change requests undergo structured triage, analyzing impacts on schedule, resources, and value delivery. Changes that align with goals without overreach are integrated through deliberate scope adjustments. Scope fluidity is managed, not refused. With processes managing change, scoping sustains focus amid shifts.

### Implementing rigorous change control processes

Despite best efforts, changes are inevitable. Managing change requires structured processes that help avoid uncontrolled scope creep:

- **Assessing change impacts**: A structured change control process assesses requests for impacts on resources, budget, timeline, and value alignment. A change control board evaluates all change requests for effects on the timeline, budget, resources, benefits realization, and alignment with goals. Changes that reconcile new dynamics with minimal disruption are approved.

- **Scope documentation updates**: For approved changes, scope documents such as roadmaps and requirement artifacts undergo versioning to reflect additions. Stakeholders review changes, maintaining continuity. Scope documents are updated to reflect approved expansions. Open communications ensure stakeholder consensus on changes.

Through deliberate change control procedures, organizations can avoid scope chaos by integrating shifts systematically with oversight. Scope evolves; it doesn't balloon aimlessly.

### Communication and documentation

Open communication channels sustain alignment as scoping evolves. Stakeholders review scope changes, ensuring continuity of vision. Requirements documentation provides foundations, while scoping artifacts record current boundaries. Immutable audit trails demonstrate organized progress, not scope chaos. Consistent communications and documentation enable scoping agility.

### Consistent communication and documentation

Effective scoping requires sustained alignment through proactive communication and detailed documentation:

- **Open communication channels**: Regular touchpoints keep stakeholders apprised of scoping progress, changes, and roadmap direction. Feedback mechanisms foster collaboration and consensus.

- **Diligent documentation practices**: Requirements are documented immutably while scope statements and plans are versioned to reflect approved changes. Auditable documentation provides foundations.

With consistent inclusive communication and organized documentation, scope maintains integrity amid fluidity. Teams stay aligned to collaboratively drive progress.

## Practical exercise – scoping a sample project

This hands-on activity provides an interactive experience for applying scoping tools and techniques to frame execution plans for a hypothetical project. Participants assume the role of architects and utilize methods such as requirements workshops, prototyping, work breakdown structures, and Gantt charts to define comprehensive scope documents. By simulating collaborative scoping, participants gain firsthand experience in translating ambiguous ideas into structured, actionable roadmaps that guide systematic implementation. Through role-playing interactive scoping, professionals master proven strategies to align vision with achievable realities in their environments.

### Hands-on activity – scoping a hypothetical project

*Objective*: This hands-on activity is designed to give participants practical experience in scoping a hypothetical project. Participants will use various scoping tools and techniques to create a comprehensive scope document.

*Overview*: For this activity, participants will choose a hypothetical project, such as developing a new software application or launching a network upgrade. The task involves applying scoping principles to define the project's objectives, deliverables, tasks, and boundaries:

1. Choose a project scenario:

   I.   Selecting a project:

   - **Task**: Participants choose a hypothetical project (for example, software development, network upgrade, and so on).
   - **Outcome**: A clear understanding of the project's basic idea and objectives

2. Stakeholder identification:

   I.   Identify stakeholders:

   - **Task**: List all potential stakeholders involved in the project (for example, end users, developers, IT staff, and management)
   - **Outcome**: A comprehensive list of stakeholders

3.  Gather requirements:

    I.    Identify business and technical goals:

    •  **Task**: Collect project requirements via interviews, surveys, or research

    •  **Tools**: Utilize questionnaires and interviews to extract detailed project needs

    •  **Outcome**: A documented list of project requirements

4.  Developing the scope statement:

    I.    Drafting the scope statement:

    •  **Task**: Write a scope statement that includes project objectives, deliverables, tasks, exclusions, constraints, and assumptions

    •  **Outcome**: An initial draft of the project scope statement

5.  Creating a WBS:

    I.    Develop a WBS:

    •  **Task**: Break down the project into smaller, manageable parts (tasks and subtasks)

    •  **Tool**: Use a WBS template or piece of software

    •  **Outcome**: A detailed WBS that outlines all tasks required to complete the project

6.  Prioritizing tasks and setting boundaries:

    I.    Prioritize tasks:

    •  **Task**: Identify and prioritize tasks based on impact, resource availability, and deadlines

    •  **Outcome**: A prioritized list of project tasks

    II.   Define scope boundaries:

    •  **Task**: Clearly state what is included and excluded in the project

    •  **Outcome**: Defined project boundaries to prevent scope creep

7.  Documenting and reviewing the scope document:

    I.    Finalize the scope document:

    •  **Task**: Compile all the information into a comprehensive scope document

    •  **Outcome**: A complete and detailed project scope document

II.    Peer review:

- **Task**: Exchange scope documents with peers for review

- **Activity**: Provide and receive feedback on the clarity, completeness, and feasibility of the scope document

- **Outcome**: A refined scope document that incorporates peer feedback

8.   Reflection and discussion:

I.    Group discussion:

- **Task**: Discuss the challenges and insights that were experienced during the scoping process

- **Topics**: Discuss the importance of thorough scoping, stakeholder engagement, and handling changes in project scope

- **Outcome**: Enhanced understanding of the scoping process and its impact on project success

This hands-on activity provided you with valuable experience in creating a comprehensive scope document for a hypothetical project. Through this exercise, you've learned how to effectively define, document, and communicate the scope of a project, ensuring clarity and alignment among all stakeholders.

In this section, you gained an in-depth understanding of the scoping process, its importance, and the best practices for defining and managing project scope effectively. This knowledge is crucial in ensuring that projects are delivered on time, within budget, and according to specified requirements.

# Project approach

In the realm of project management, various methodologies can be employed, each offering distinct advantages that are suited to different types of projects. This section explores several project approaches, providing insights into how and why certain methodologies are more effective under specific circumstances. By examining real-world examples, you will learn how to discern and select the most appropriate approach for your projects while considering factors such as project size, complexity, team dynamics, and organizational needs.

## Overview of project methodologies

Myriad methodologies exist for executing projects, each with its unique strengths and weaknesses. The traditional waterfall methodology provides linear order, while Agile emphasizes adaptability. Emerging methods such as DevOps focus on speed and collaboration. Factors such as team experience, compliance needs, and solution complexity inform approach selection. Hybrid models blend rigor with agility. By matching execution style to the environment, organizations can sustain focus on delivering value while not following doctrine. Pragmatism drives methodology.

## *Traditional versus Agile methodologies*

Traditional methodologies such as waterfall follow a linear, sequential path with structured phases that do not overlap. Requirements are gathered fully upfront before design begins. Testing happens only after development is complete. Change during the project is difficult. Traditional methods emphasize detailed planning, documentation, and upfront design.

In contrast, Agile methodologies utilize short, iterative cycles called sprints to develop smaller increments of the product. Requirements can evolve through collaboration during the project rather than being locked down initially. Working software is delivered frequently, with testing integrated throughout the development life cycle. Agile values customer collaboration, responding to change, and software being delivered frequently.

Some pros of the traditional waterfall approach are that it provides a clear structure, milestones, and documentation. It works well when requirements are fixed and can't change. However, it lacks flexibility in terms of being adapted. Waterfall carries the risk that issues won't be found until the late testing stages.

Agile methods enable collaboration, frequent feedback, and adjusting quickly based on lessons learned. This allows for faster delivery of business value. However, Agile can be less predictable, requires more customer involvement, and needs a cultural shift for many organizations.

In summary, waterfall provides linear sequencing, while Agile focuses on iteration. The right approach depends on factors such as requirements uncertainty, the need for stakeholder collaboration, and willingness to change course. Agile fits smaller projects with shifting needs, while waterfall suits unchanging requirements. Hybrid models can also combine techniques from both methodologies.

## *Emerging methodologies*

In addition to foundational methodologies such as waterfall and Agile frameworks such as Scrum or Kanban, new approaches have emerged in recent years. These include lean, DevOps, and design thinking, among others. They aim to improve upon traditional techniques by focusing on waste reduction, collaboration, user-centric design, and continuous delivery:

- **Lean**: This methodology originated in manufacturing and focuses on maximizing value while minimizing waste. It emphasizes optimizing the whole process end-to-end. Key techniques include visualizing the workflow, limiting work in progress, and building quality in.

- **DevOps**: DevOps combines development and operations to enable continuous integration and delivery of software. It breaks down silos through collaboration and automation. Practices such as IaC, automated testing, and monitoring help detect issues faster.

- **Design thinking**: Design thinking centers around understanding users' needs and rapidly prototyping solutions. It applies design principles to problem-solving with phases that involve empathizing, defining, ideating, prototyping, and testing concepts.

**Application scenarios**

Lean principles work well for smoothing out inefficient processes with variability or waste. DevOps improves the speed and reliability of software delivery in complex, evolving environments. Finally, design thinking helps create innovative products that are tailored to customer needs through rapid experimentation and feedback.

These emerging techniques can complement traditional project management approaches. For example, lean or design thinking can be utilized for planning phases while Agile delivery sprints are implemented. Each has its strengths that are more suited to particular project challenges and environments.

# Deep dive into specific methodologies

Waterfall follows sequential stages without overlap: requirements, design, implementation, verification, and maintenance. Agile uses short iterations called sprints and incremental deliveries. Scrum is a framework under Agile that uses sprints and daily standups. Kanban focuses on visualizing the workflow and limiting work in progress.

## *Waterfall methodology*

The waterfall methodology is a linear, sequential approach to managing a project. It follows the rigid phases of requirements gathering, design, implementation, testing, and maintenance, none of which overlap. Waterfall emphasizes comprehensive upfront planning and documentation.

### Characteristics

The waterfall methodology follows a sequential, linear approach to managing a project. As mentioned previously, it consists of discrete phases that do not overlap: requirements, design, implementation, testing, and maintenance. Each phase must be completed fully before you move on to the next. Waterfall emphasizes detailed and rigorous planning upfront, along with comprehensive documentation.

This methodology provides a structured path with clear milestones and deliverables. However, it lacks the flexibility to change course later in the project life cycle. Once requirements are locked down in the initial stage, any modifications require the entire sequence of phases to be revisited.

### Best fit

Waterfall is well-suited for projects with fixed, well-defined requirements and low uncertainty. When the scope is clear and unlikely to change substantially, the predictability of waterfall makes it a lower-risk approach. It works best for large, complex projects where quality control is critical throughout each step.

Waterfall is less ideal for projects with ambiguous or rapidly evolving requirements. The lack of ability to adapt can lead to disconnects between delivered products and current customer needs after long development cycles.

## Case study

An example of the waterfall methodology's success is NASA's Mars Curiosity rover project. Given the scientific nature of the project, engineers could define requirements fully upfront before building the complex rover. There was minimal uncertainty, so waterfall provided the rigor needed for mission-critical development. The step-by-step process, which included extensive testing and verification, ensured a high-quality end product ready for space deployment.

In summary, waterfall provides structure through sequenced phases but lacks agility. When requirements are fixed, its linear nature can lead to predictable, successful outcomes. However, uncertainty limits its adaptability in dynamic environments.

### *Agile methodology*

Agile methodologies utilize short, iterative cycles called sprints to develop smaller increments of a product. Rather than a sequential path, Agile focuses on incorporating customer feedback and continuously delivering working software. Examples of agile frameworks include Scrum and Kanban.

Some key characteristics of Agile include valuing the following:

- Individuals and interactions over processes and tools
- Working software over comprehensive documentation
- Customer collaboration over contract negotiation
- Responding to change over following a plan

It emphasizes fluid requirements gathering, lightweight development cycles, continuous testing, and rapid adaptation. Popular frameworks that adopt Agile principles are Scrum, Kanban, and Extreme Programming.

### Best fit

Agile methods are best suited for projects with ambiguous or rapidly changing requirements. Its iterative approach allows for flexibility and course correction throughout development, aligning the end product with evolving customer needs.

Agile also enables early risk mitigation since issues can be identified and addressed during initial sprints rather than late in the project. Close customer collaboration and rapid feedback loops make Agile a good fit for user-centric products.

### Case study

One example of a successful Agile implementation was at Spotify. Self-organizing squads owned product features end-to-end via 2-week sprints, continuous integration, and monitoring. This structure enabled high speed and autonomy, even as the product grew rapidly.

Spotify exemplified key Agile tenets – delivering working software frequently, adapting to change, and maintaining technical excellence and simplicity. By embracing Agile, Spotify could innovate and scale quickly in a demanding environment.

In summary, Agile provides rapid iterations and flexibility, whereas waterfall is more sequential. Agile fits projects with uncertainty and evolving requirements. Both methodologies have their strengths in particular contexts.

## Selecting the right approach

Factors such as project size, requirements uncertainty, the need for stakeholder collaboration, and organizational culture influence the choice between traditional and agile approaches. No single method fits every situation.

### *Factors that influence methodology choice*

Factors that influence the choice between waterfall and Agile include the project's size, requirements uncertainty, the need for stakeholder collaboration, and the organization's culture. Agile tends to work better for smaller projects with shifting requirements.

### Project size and complexity

The size and complexity of a project play a significant role in determining the right methodology. For large, multi-year projects with strict quality and risk parameters, the structure of waterfall provides the necessary rigor. Its phased sequence allows many interdependent components to be coordinated.

For small-scale projects that require agility, Agile sprints better enable rapid adaptation. The overhead of extensive documentation and planning in waterfall slows time-to-market for simple projects.

### Team dynamics

Team structure and dynamics should align with the chosen approach. Waterfall requires defined roles, responsibilities, and hand-off between phases, while Agile needs cross-functional, collaborative teams who can work fluidly.

If a team is accustomed to waterfall processes, adopting Agile may require adjusting mindsets, building skills in areas such as continuous testing, and emphasizing accountability to the team over roles. Leadership style also influences the energy and empowerment levels suited for Agile.

### Organizational culture and needs

The existing culture and needs of an organization shape project methodology fit. Structured companies that value order may gravitate toward waterfall, while organizations that require agility and speed must be willing to embrace Agile principles.

If end users are involved, Agile's customer focus enables validating concepts and priority features early. For software companies, Agile delivery or DevOps may be critical for maintaining pace. Established industries such as manufacturing may adopt Agile with caution.

Considering organizational maturity, the growth stage, and business drivers helps determine appropriate project methodologies. Selecting approaches that align with culture and needs increases the likelihood of successful adoption.

In summary, factors such as team dynamics, organizational culture, business needs, and project risk profile all influence the ideal choice between different project management methodologies for an environment.

### *Decision-making framework*

A framework that can select an approach must be able to assess project attributes such as complexity, check organizational readiness, analyze team capacity, and determine customer needs.

When deciding between a traditional plan-driven methodology such as waterfall versus a more agile approach, there are several guidelines to consider:

- **Assess project attributes**: Is the project large, complex, or mission-critical? Are requirements clear or ambiguous? How much risk and uncertainty exists?

- **Check organizational readiness**: Is the culture accustomed to rigid structure or open to iteration? Are stakeholders willing to actively participate?

- **Analyze team capabilities**: Does the team have experience working cross-functionally? Can they adapt methods as needed?

- **Understand customer needs**: Are end users accessible for feedback? How quickly do they expect to see working functionality?

- **Consider external factors**: Are there regulatory constraints? How much change is anticipated in the market or technology?

The goal is to match the methodology with the unique characteristics and environment of the project. This requires understanding priorities such as predictability, speed-to-market, quality, and risk mitigation needs.

### Activity

Imagine that you are managing Project A to build a new software platform. The requirements are vague, funding is tight, and the market could change quickly. For fast results, Agile sprints make sense to deliver value iteratively.

Now, imagine Project B for a spaceship with stringent safety standards. Here, the waterfall methodology provides structure to methodically meet quality guidelines. The right methodology depends on the distinct project scenario.

This framework helps systematically assess and tailor project approaches, rather than prescriptively applying one standardized method. The best methodology combines aspects of predictability, agility, and discipline that are appropriate for the project's environment and objectives.

## Combining methodologies

Hybrid models take ideas from both traditional and agile methods. A project may use agile for development but waterfall for testing. Phased approaches plan with waterfall and then switch to short agile sprints.

### Hybrid approaches

Hybrid approaches combine elements of both the waterfall and Agile methodologies.

#### Concept

The goal of hybrid project management methodologies is to leverage the strengths of different techniques to best meet the needs of a specific project situation.

For example, a hybrid approach could utilize the waterfall methodology for planning and requirements gathering to leverage its structure, then switch to Agile sprints for development for more fluidity. Another hybrid model could use waterfall for backend system design and Agile for frontend customer-facing development.

Hybrid provides you with to flexibility to customize the process instead of rigidly following one methodology. It aims to balance predictability with agility, blending just enough of each based on the environment.

#### Implementation

To implement a hybrid approach, teams must deeply understand the project's goals, challenges, and priorities. This informs them of what combination of methodologies could work best. Teams also need buy-in and training on using the selected techniques appropriately.

It's critical to define how and when transitions between methodologies will occur upfront – for example, setting quality gates for moving from waterfall requirements to Agile sprints. Communication, coordination, and leadership commitment are essential for effective hybrid adoption.

#### Example

One example is an insurance firm that used waterfall for its claims system backend, which required stability and security. For the customer-facing website, they leveraged Agile with 2-week sprints and constant input from users.

This hybrid model allowed them to deliver an integrated product faster, meet regulatory standards, and continuously refine customer experience. The results combined predictable core systems with flexible frontends adapted to changing needs.

In summary, hybrid project management combines different techniques to meet project goals. Defining the transitions between methodologies is key to a successful implementation.

## Adapting to change

Regular feedback loops, retrospective meetings, and value-focused delivery allow teams to inspect and adapt their process over time as needs evolve.

### *Evolution of project management*

Project management has progressed from early-stage gate models to more flexible agile approaches. New hybrid techniques continue to emerge from lessons learned in the field.

### Trends and future directions

Project management has evolved substantially over the past decades. Early models such as waterfall were linear and sequential. As software development grew, lightweight Agile techniques emerged and focused on collaboration and iteration.

Here are some trends that are shaping the future of project management:

- Hybrid approaches that combine various aspects of predictive and adaptive models
- Increasing automation and the use of AI for routine project tasks
- More focus on value delivery over rigid processes
- Adopting holistic views that consider system dynamics and human factors
- Further blurring of development, operations, and experience roles
- Expanding agile mindsets beyond IT to broader business projects

As technology enables new ways of working and customer expectations rise, project management will continue to evolve. More flexible, data-driven approaches integrated with experiential design thinking will emerge.

### Adaptability

A key lesson from the evolution of project management is the importance of adaptability. Even Agile methods must continuously inspect and adapt to improve. No single process will fit all projects or remain optimal permanently.

Teams should regularly reflect on what works, what doesn't, and how their methodology could improve. Building skills in change management allows you to gracefully transition between approaches. A willingness to experiment, gather feedback, and adjust is critical.

The future of project management involves understanding no method is static. Continuous adaptation drives sustained success. With the right mindset and skills, teams can evolve their practices fluidly.

# Learning from real-world applications

Case studies of projects across industries provide insights into when specific methods work well or fall short. Teams learn from this experience to continually improve their project management practices.

## *Case study analysis*

Looking at in-depth case studies of projects using different methodologies provides valuable insights:

- **Agile case:** An online retail company needed to rapidly develop a new customer mobile app. Using 2-week sprints with continuous customer feedback throughout let them continuously refine and deliver features customers wanted most. The project took just 5 months and achieved high user satisfaction scores.
- **Waterfall case:** A team of engineers used the waterfall methodology to design a new aircraft engine. The sequential phases ensured rigorous testing and analysis at each stage before they moved to manufacturing. While the process took longer upfront, it resulted in an extremely reliable engine ready for deployment.

For each case, we can examine details such as requirements-gathering approaches, development life cycles, team structures, end products delivered, and measures of success. Comparing cases illuminates why different methods excel in some scenarios versus others.

### Lessons learned

Here are the key lessons that were learned from these cases:

- Customer involvement throughout Agile projects maximizes user value
- Waterfall provides discipline for complex systems engineering efforts
- Locking down requirements early in waterfall is high risk when end user needs aren't known
- Agile requires motivated teams with the right skills and mindsets
- Hybrid approaches can balance flexibility and rigor for specific projects

Thoughtfully evaluating real-world cases provides guidance on best-applying project management methodologies. Teams can continuously improve by learning from experience within their domains.

In summary, case study analysis yields important insights into effectively tailoring project management methods. Teams should learn from past examples to develop approaches optimized for their environments.

## Best practices

Best practices for project success emphasize having a supportive organizational culture, ensuring team buy-in, matching the approach to project needs, inspecting and adapting processes, and focusing on the continuous delivery of value, no matter the method.

Based on the strengths and weaknesses of the various methodologies discussed, here are some best practices:

- Match the methodology to unique project needs and constraints
- Involve end users early and continuously in the process
- Ensure the team is trained and buys into the selected approach
- Define objective, measurable targets and estimates for tracking progress
- Utilize prototyping and simulation to test concepts where possible
- Automate repetitive tasks through tools and streamlined workflows
- Inspect and adapt processes frequently, not just at the end of the project
- Combine predictive planning with flexibility to change course
- Focus on continuous value delivery throughout the project life cycle

Following these recommendations can optimize the outcomes for all methodologies – waterfall, Agile, or otherwise.

## Summarizing the key lessons learned

This overview aims to highlight that there is no one-size-fits-all methodology for project success. Waterfall provides structure but lacks agility, while Agile enables adaptation through iteration. Emerging methods such as lean and design thinking offer further techniques.

The right methodology depends heavily on the project's environment, requirements, timelines, and team capabilities. Often, hybrid approaches can blend aspects of various methods. Selecting and tailoring an approach is critical.

### Encouraging flexibility

Teams should remain flexible and open-minded so that they can adjust processes as projects evolve. Methodologies are not static. Inspecting and adapting to changing needs, while focusing on continuous value delivery, will lead to positive outcomes.

With an understanding of the available methods and best practices, project teams can make informed choices on approaches. Ultimately, this knowledge empowers better delivery of projects, from conception through completion.

This section provided an extensive analysis of traditional and emerging project management methodologies, including waterfall, Agile frameworks, DevOps, and design thinking. It contrasts linear, sequential waterfall with iterative Agile, detailing the strengths and weaknesses of each. We covered factors that influence methodology selection and emphasized aligning choices with team capabilities, organizational culture, and project goals. We also looked at hybrid models that fuse aspects of multiple approaches. We covered the best practices that apply broadly across methodologies. Additionally, we underscored the importance of continuous inspection, adaptation, and focusing on value delivery to sustain optimal outcomes amid changing needs. Combined, these concepts aim to help you deliberately tailor execution strategies, blending structured and adaptive techniques for success across diverse project scenarios. With thoughtful methodology shaping, organizations can confidently undertake technology initiatives to fulfill business objectives.

## Next steps

As the cybersecurity landscape continues to evolve with increasing complexity and sophistication, the role of the cybersecurity architect necessitates a continual advancement in knowledge and expertise. In this context, the importance of focus areas for ongoing learning is critical for cybersecurity architects looking to chart their next steps in this dynamic profession. A comprehensive roadmap must be established for those aiming to enhance their skills, stay abreast of the latest trends, and make significant contributions to the field.

The journey through the cybersecurity architectural profession, as outlined in this book, traverses a landscape rich in complexity and depth. From foundational cybersecurity principles to advanced architectural strategies, the profession demands a continuous pursuit of knowledge and skill enhancement. The next steps serve as a guide for cybersecurity architects at various stages in their careers, offering insights into potential next steps, specialization avenues, and strategies for sustained growth and mastery in the field:

- Strengthen existing skills:

    - **Core cybersecurity principles and domains**: Understanding the basics of cybersecurity, including its core principles and domains, is essential. This foundation is critical for all subsequent learning and professional development.

    - **Skills in compliance, vendor management, and risk assessment**: Mastery of compliance standards, effective vendor management, and comprehensive risk assessment is crucial. These skills ensure robust security measures and strategic decision-making.

    - **Certifications and education pathways**: Pursuing relevant certifications and educational qualifications is a lifelong endeavor. They validate expertise and open doors to advanced roles and responsibilities.

- Focus areas for ongoing learning:

  - **Emerging technologies:** Stay abreast of emerging technologies, such as AI, ML, and blockchain. Understanding their implications on security architecture is essential for future readiness.

  - **Industry-specific security**: Delve into industry-specific cybersecurity challenges and solutions, particularly in sectors such as finance, healthcare, or government, which have unique regulatory and security environments.

  - **Advanced threat modeling**: Developing advanced skills in threat modeling techniques can lead to specialization in areas such as predictive security analysis and targeted defense strategies.

  - **Cybersecurity research and development**: Engage in R&D to stay at the forefront of innovation in cybersecurity technologies and methodologies.

  - **Leadership and strategy development**: For those aspiring to be in leadership roles, focus on strategic planning, policy development, and organizational cybersecurity leadership skills.

  - **Communication**: Practice communication skills in diverse settings, including public speaking and cross-departmental collaboration.

  - **Mentoring and training others**: Participate in mentorship programs, either as a mentor or mentee, to refine these critical soft skills.

  - **Ethical and legal considerations**: Deepen your understanding of ethical hacking, cybersecurity law, and privacy regulations to navigate the complex legal landscape effectively.

  - **Personal growth and networking**: Invest in personal growth through mentorship, networking, and community involvement. Also, spend time outside the cybersecurity realm by gaining hobbies to help with work-life balance and personal growth.

In summary, excellence in cybersecurity architecture demands lifelong learning as threats and technologies continuously evolve. By strengthening their foundational knowledge, pursuing certifications, focusing on emerging domains, and engaging in research and development, architects can chart a course for impactful contributions. However, mastery also requires rounding soft skills through mentorship, communication practice, and networking. Personal growth matters too – activities beyond security provide balance against professional demands. With diligence across technical, soft, and self-improvement skills, cybersecurity architects can navigate a rewarding and fulfilling career, thereby securing our digital future. The journey traverses an expansive terrain, but each step builds the road.

# Summary

This two-part chapter served as a culminating synthesis that tied together various security architecture concepts that we looked at previously. It explored integrating predictive, preventive, detective, and responsive capabilities into adaptable ecosystems aligned with business needs and risk appetites. The core focus areas included tailoring technical designs and solutions to environments while upholding best practices using structured development life cycles.

The strategic importance of adaptability was underscored via examples that applied OODA loop principles for career development and incident response agility. Additional sections provided extensive analysis on strategically executing projects using methodologies such as waterfall, agile, or hybrid approaches based on unique needs. Guidance on the next steps you should take enabled you to chart growth strategically through skill-building, certifications, specializations, and leadership development.

This chapter crystallized the versatility that's required of architects to traverse technical excellence, be communication-savvy, have a strategic perspective, and excel in project execution. It aimed to help you customize architecture capabilities that match dynamic business objectives and ever-evolving threat landscapes. With insights bridging theory with practice across the solution life cycle, you now know how to integrate frameworks while securing innovation amid relentless change.

# Index

# <packt>

www.packtpub.com

Subscribe to our online digital library for full access to over 7,000 books and videos, as well as industry leading tools to help you plan your personal development and advance your career. For more information, please visit our website.

## Why subscribe?

- Spend less time learning and more time coding with practical eBooks and Videos from over 4,000 industry professionals

- Improve your learning with Skill Plans built especially for you

- Get a free eBook or video every month

- Fully searchable for easy access to vital information

- Copy and paste, print, and bookmark content

Did you know that Packt offers eBook versions of every book published, with PDF and ePub files available? You can upgrade to the eBook version at packtpub.com and as a print book customer, you are entitled to a discount on the eBook copy. Get in touch with us at customercare@packtpub.com for more details.

At www.packtpub.com, you can also read a collection of free technical articles, sign up for a range of free newsletters, and receive exclusive discounts and offers on Packt books and eBooks.

# Other Books You May Enjoy

If you enjoyed this book, you may be interested in these other books by Packt:

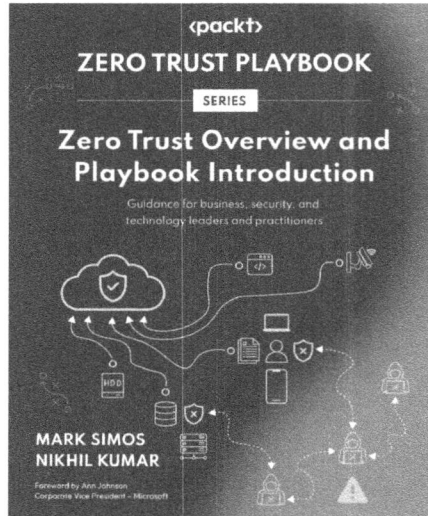

**Zero Trust Overview and Playbook Introduction**

Mark Simos, Nikhil Kumar

ISBN: 978-1-80056-866-2

- Find out what Zero Trust is and what it means to you
- Uncover how Zero Trust helps with ransomware, breaches, and other attacks
- Understand which business assets to secure first
- Use a standards-based approach for Zero Trust
- See how Zero Trust links business, security, risk, and technology
- Use the six-stage process to guide your Zero Trust journey
- Transform roles and secure operations with Zero Trust
- Discover how the playbook guides each role to success

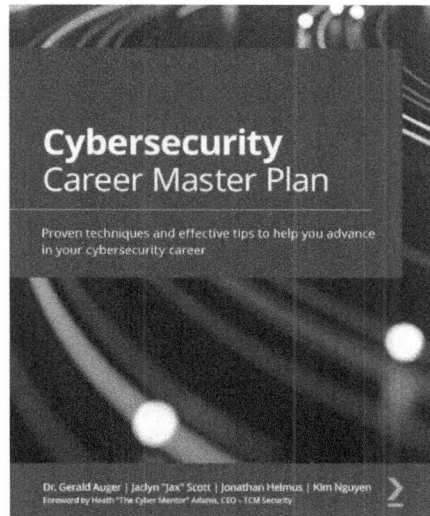

**Cybersecurity Career Master Plan**

Dr. Gerald Auger, Jaclyn "Jax" Scott, Jonathan Helmus, Kim Nguyen

ISBN: 978-1-80107-356-1

- Gain an understanding of cybersecurity essentials, including the different frameworks and laws, and specialties
- Find out how to land your first job in the cybersecurity industry
- Understand the difference between college education and certificate courses
- Build goals and timelines to encourage a work/life balance while delivering value in your job
- Understand the different types of cybersecurity jobs available and what it means to be entry-level
- Build affordable, practical labs to develop your technical skills
- Discover how to set goals and maintain momentum after landing your first cybersecurity job

## Packt is searching for authors like you

If you're interested in becoming an author for Packt, please visit `authors.packtpub.com` and apply today. We have worked with thousands of developers and tech professionals, just like you, to help them share their insight with the global tech community. You can make a general application, apply for a specific hot topic that we are recruiting an author for, or submit your own idea.

## Share Your Thoughts

Now you've finished *Cybersecurity Architect's Handbook*, we'd love to hear your thoughts! Scan the QR code below to go straight to the Amazon review page for this book and share your feedback or leave a review on the site that you purchased it from.

`https://packt.link/r/1803235845`

Your review is important to us and the tech community and will help us make sure we're delivering excellent quality content.

# Download a free PDF copy of this book

Thanks for purchasing this book!

Do you like to read on the go but are unable to carry your print books everywhere? d

Is your eBook purchase not compatible with the device of your choice?

Don't worry, now with every Packt book you get a DRM-free PDF version of that book at no cost.

Read anywhere, any place, on any device. Search, copy, and paste code from your favorite technical books directly into your application.

The perks don't stop there, you can get exclusive access to discounts, newsletters, and great free content in your inbox daily

Follow these simple steps to get the benefits:

1.  Scan the QR code or visit the link below

https://packt.link/free-ebook/9781803235844

2.  Submit your proof of purchase
3.  That's it! We'll send your free PDF and other benefits to your email directly

www.ingramcontent.com/pod-product-compliance
Lightning Source LLC
Chambersburg PA
CBHW081220220326
41598CB00037B/6848